Student Manual for Digital Signal Processing with MATLAB

John G. Proakis

Vinay K. Ingle

Department of Electrical and Computer Engineering
Northeastern University

Upper Saddle River, NJ 07458

Library of Congress Cataloging-in-Publication Data

Proakis, John G.
Student manual for Digital Signal Processing with MATLAB / John G. Proakis, Vinay K. Ingle
p. cm.
Includes index
ISBN 0-13-199108-6
1. Signal processing—Digital Techniques—Mathematics. 2. MATLAB. I. Ingle, Vinay K.
II. Title.
TK5102.9.1538 2006
621.382'2—dc22 2005056573

Vice President and Editorial Director, ECS: *Marcia Horton*
Associate Editor: *Alice Dworkin*
Executive Managing Editor: *Vince O'Brien*
Managing Editor: *David A. George*
Production Editor: *Wendy Kopf*
Director of Creative Services: *Paul Belfanti*
Art Director: *Jayne Conte*
Art Editor: *Greg Dulles*
Manufacturing Buyer: *Lisa McDowell*

Pearson Prentice Hall
Pearson Education, Inc.
Upper Saddle River, NJ 07458

Printed in the United States of America

10 9 8 7 6 5 4 3 2

ISBN 0-13-199108-6

Pearson Education, Inc., *Upper Saddle River, New Jersey*
Pearson Education Ltd., *London*
Pearson Education Australia Pty. Ltd., *Sydney*
Pearson Education Singapore, Pte. Ltd.
Pearson Education North Asia Ltd., *Hong Kong*
Pearson Education Canada, Inc., *Toronto*
Pearson Educación de Mexico, S.A. de C.V.
Pearson Education—Japan, *Tokyo*
Pearson Education—Malaysia, Pte. Ltd.

Contents

Preface

The Student Manual is intended for use as a companion for self-study to the textbook entitled *Digital Signal Processing, Principles, Algorithms, and Applications, Fourth Edition*, published by Prentice Hall. MATLAB is incorporated as the basic software tool for this manual. The Student Manual, along with the textbook, can be used by students, practicing engineers, and scientists who wish to obtain an introduction to the subject. It can also be used by people who have had a basic undergraduate course in DSP but who have not been exposed to DSP software tools, such as MATLAB, for analyzing and implementing DSP systems and algorithms.

For effective learning of the material and for understanding difficult concepts in DSP, it is important to integrate software tools with textual study. This integration makes it possible for students to simulate signals, systems (or filters), or algorithms and provides for "what-if" type of analyses to delve more deeply into the topics. This Student Manual, designed with this approach in mind, uses MATLAB as the tool. In addition to learning each topic, concept, or algorithm, the student should make every effort to use the provided MATLAB functions to develop a stronger intuition and a deeper understanding of the material.

The Student Manual treats traditional topics covered in an introductory DSP course. In each chapter of the manual, the reader is directed to review the topics covered in the corresponding chapter of the textbook. The basic topics and results treated in each chapter of the textbook are summarized in the manual, and the relevant MATLAB functions to be used in solving problems are introduced. At the end of each chapter of the Student Manual is a set of problems that the reader should solve, using MATLAB. Solutions to these problems are included in the manual.

The DSP textbook also contains numerous problems at the end of each chapter. Many of these problems can also be solved using MATLAB. If additional practice is desired, the reader is encouraged to solve some of these problems in MATLAB.

The various MATLAB functions cited in the Student Manual are available at:

`ftp://ftp.ece.neu.edu/pub/users/ingle/I&P_dsp_toolbox`

These functions require MATLAB version 5 or higher and the signal toolbox version 3 or higher. Further information about MATLAB and the related toolbox can be obtained from The MathWorks, Inc.

A self-contained more in-depth treatment of the use of MATLAB in the analysis and design of digital signal processing systems and algorithms is given in the book *Digital Signal Processing Using MATLAB*, written by Vinay K. Ingle and John G. Proakis, and published by Brooks/Cole–Thomson Learning.

John G. Proakis
Vinay K. Ingle
Boston, Massachucsetts

Part I

Student Manual

Chapter 1

Introduction

This Student Manual is intended to be used as a companion to the textbook entitled *Digital Signal Processing: Principles, Algorithms, and Applications*, 4th Edition, published by Prentice Hall. MATLAB is incorporated as the basic software tool for this manual. It is a very useful tool in the implementation of DSP systems and algorithms, in the analysis of discrete-time signals and systems, and in the design of digital filters. The manual, along with the book, can be used by students, practicing engineers, scientists who wish to obtain an introduction to the subject. It can also be used by people who have had a basic undergraduate course in DSP but who have not been exposed to DSP software tools, such as MATLAB, for analyzing and implementing DSP systems and algorithms.

A self-contained, more in-depth treatment of the use of MATLAB in the analysis and design of digital signal processing systems and algorithms is given in the book *Digital Signal Processing using MATLAB*, written by Vinay K. Ingle and John G. Proakis, and published by Brooks Cole, Thomson Learning.

1.1 Use of the Student Manual in Conjunction with the Book

The Student Manual is intended to direct the reader to the topics covered in each chapter and to provide some supplementary information to the treatment given in the book. In each chapter, the reader is directed to study the treatment of each topic in the book and to carry out the MATLAB examples. The reader could then return to the Student Manual and review the supplementary description of the corresponding topic. The various MATLAB functions cited in the manual are available at ftp://ftp.ece.neu.edu/pub/users/ingle/I&P_dsp_toolbox. These functions require MATLAB version 5 or higher and the signal toolbox version 3 or higher. Further information about MATLAB and the related toolbox can be obtained from The MathWorks, Inc.

The eight chapters of the book discuss traditional topics covered in an introductory DSP course. At the end of each of these chapters is a set of problems that the reader should solve. The Student Manual contains solutions to these problems.

1.2 Reading Assignment for Chapter 1

The reader should begin the course by reading the Preface, which describes the organization of the book, including the content of each chapter. Then, the reader should read Chapter 1 in the book, which provides an introduction to digital signal processing and introduces some basic concepts pertaining to sampling and quantization of analog signals.

1.3 About the Authors

JOHN G. PROAKIS

Dr. Proakis received the BSEE from the University of Cincinnati in 1959, the MSEE from MIT in 1961, and the Ph.D. from Harvard University in 1967. He is an adjunct professor at the University of California at San Diego and a professor emeritus at Northeastern University. He was a faculty member at Northeastern University from 1969 through 1998 and held the following academic positions: associate professor of electrical engineering, 1969–1976; professor of electrical engineering, 1976–1998; associate dean of the College of Engineering and director of the Graduate School of Engineering, 1982–1984; interim dean of the College of Engineering, 1992–1993; and chairman of the Department of Electrical and Computer Engineering, 1984–1997. Prior to joining Northeastern University, he worked at GTE Laboratories and the MIT Lincoln Laboratory.

His professional experience and interests are in the general areas of digital communications and digital signal processing and, more specifically, in adaptive filtering, adaptive communication systems and adaptive equalization techniques, communication through fading multipath channels, radar detection, signal-parameter estimation, communication-systems modeling and simulation, optimization techniques, and statistical analysis. He is active in research in the areas of digital communications and digital signal processing and has taught undergraduate and graduate courses in communications, circuit analysis, control systems, probability, stochastic processes, discrete systems, and digital signal processing. He is the author of the book *Digital Communications* (New York: McGraw-Hill, 2001, 4d ed.), and coauthor of the books *Digital Signal Processing* (Upper Saddle River, NJ: Prentice Hall, 2007, 4th ed.), *Algorithms for Statistical Signal Processing* (Upper Saddle River, NJ: Prentice Hall, 2002), *Discrete-Time Processing of Speech Signals* (New York: IEEE Press, 2000), and *Communication Systems Engineering* (Upper Saddle River, NJ: Prentice Hall, 2002, 2d ed.), *Digital Signal Processing Using MATLAB V.4* (Boston: Brooks/Cole–Thomson Learning, 2000), and *Contemporary Communication Systems Using MATLAB* (Boston: Brooks/Cole-Thomson Learning, 2000). Dr. Proakis is a fellow of the IEEE. He holds five patents and has published over 200 papers.

VINAY K. INGLE

Dr. Ingle received the B.Tech. from the Indian Institute of Technology, Bombay, India, in 1976, the MSEE from the Illinois Institute of Technology, Chicago, in 1977, and the Ph.D. from Rensselaer Polytechnic Institute in 1981. He joined the Department of Electrical and Computer Engineering at Northeastern University in 1981, where he is currently an associate professor. Since 1999, he has also been associated with Riverside Research Institute, Lexington, Massachusetts, as a consultant.

He has broad research experience and has taught courses on topics including signal and image processing, stochastic processes, and estimation theory. Dr. Ingle is coauthor of the books *Digital Signal Processing Laboratory Using the ADSP-2101 Microprocessor* (Englewood Cliffs, NJ: Prentice-Hall, 1991), *Digital Signal Processing Using MATLAB V.4* (Boston: Brooks/Cole–Thomson Learning, 1997, 1st ed.; 2000, 2nd ed.), *Discrete Systems Laboratory* (Boston: Brooks/Cole–Thomson Learning, 2000), and *Statistical and Adaptive Signal Processing* (New York: McGraw–Hill, 2000).

Chapter 2

Discrete-Time Signals and Systems

The purpose of this chapter is to introduce the reader to the characterization of discrete-time signals and systems in the time domain and the implementation of such signals and systems by using MATLAB. The reader should study material in Chapter 2 of the DSP book:

- Discrete-time Signals — Sections 2.1 and 2.6
- Discrete-time Systems — Section 2.2
- Convolution Operation — Section 2.3
- Difference Equations — Section 2.4

The notation, material, and MATLAB exercises covered in this chapter are essential for the study of the remaining chapters.

Learning Objectives

- Understand the concept of a sequence and its representation in MATLAB.
- Learn basic sequences and operations to construct more complicated sequences.
- Become familiar with various characteristics of sequences, such as power, energy, even–odd decomposition, and how to compute these quantities.
- Understand the concepts of linearity, shift-invariance, stability, and causality.
- Comprehend the time-domain representation of an LTI system in terms of its impulse response and the resulting convolution summation.
- Become familiar with the properties of convolution.
- Learn the computation of convolution, using the summation and matrix operations.
- Learn the difference-equation representation of linear systems and its various solutions.
- Master the solution of difference equations by using MATLABs `filter` function.
- Become familiar with FIR and IIR filters in terms of impulse response, the difference equation, and the `filter` function.

2.1 Discrete-Time Signals

A signal is a vehicle that conveys information. Without signals, any equipment, appliance, or system is meaningless. In this section, we will study discrete-time signals (or sequences) and their representation, generation, operations, and synthesis.

Study Topics: Signals and Their Characteristics

1. Discrete-time signals are called (number) sequences and are denoted by one of the following notations:

$$x(n) = \{x(n)\} = \{\cdots, x(-1), \underset{\uparrow}{x(0)}, x(1), \cdots\} \tag{2.1}$$

   where the *up-arrow* indicates the signal value at $n = 0$. Analog signals will be denoted by $x_a(t)$.

2. The discrete-time signal sequence can be of either finite or infinite length. It is described over the interval $N_1 \le n \le N_2$ with $N_1 < N_2$. Such a sequence has a duration of $(N_2 - N_1 + 1)$ samples.

3. *Unit-sample sequence*: This sequence has only one nonzero sample (at $n = 0$) and is given by

$$\delta(n) = \begin{cases} 1, & n = 0 \\ 0, & n \neq 0 \end{cases} = \left\{\cdots, 0, 0, \underset{\uparrow}{1}, 0, 0, \cdots\right\} \tag{2.2}$$

   The book toolbox function `[x,n] = impseq(n0,n1,n2)` generates

$$\delta(n - n_0) = \begin{cases} 1, & n = n_0 \\ 0, & n \neq n_0 \end{cases}$$

   over the interval $n_1 \le n_0 \le n_2$.

4. *Unit-step sequence*:

$$u(n) = \begin{cases} 1, & n \ge 0 \\ 0, & n < 0 \end{cases} = \left\{\cdots, 0, 0, \underset{\uparrow}{1}, 1, 1, \cdots\right\} \tag{2.3}$$

   The function `[x,n] = stepseq(n0,n1,n2)` generates

$$u(n - n_0) = \begin{cases} 1, & n \ge n_0 \\ 0, & n < n_0 \end{cases}$$

   over the interval $n_1 \le n_0 \le n_2$.

5. *Real-valued exponential sequence*:

$$x(n) = a^n, \forall n;\ a \in \mathbb{R} \tag{2.4}$$

   For example, the sequence $x(n) = (0.9)^n,\ 0 \le n \le 10$ is a finite-duration real exponential sequence of length 11.

6. *Complex-valued exponential sequence*:

$$x(n) = e^{(\sigma + j\omega_0)n}, \forall n \tag{2.5}$$

where σ is called an attenuation factor ($\sigma < 0$) and ω_0 is the frequency in radians. For example, the sequence $x(n) = \exp\left[(2 + j3)\,n\right],\ 0 \le n \le 10$ is an eleven-point complex exponential sequence.

7. *Sinusoidal sequence*:

$$x(n) = \cos(\omega_0 n + \theta), \forall n \tag{2.6}$$

where θ is the phase in radians. The sequence $x(n) = 3\cos(0.1\pi n + \pi/3),\ 0 \le n \le 10$ is an example of a sinusoidal sequence. In discrete time, not all sinusoids are periodic (see Problem P2.3).

8. *Periodic sequence*: A sequence $x(n)$ is periodic if $x(n) = x(n + N), \forall n$. The smallest integer N that satisfies the above relation is called the *fundamental* period. We will use $\tilde{x}(n)$ to denote a periodic sequence. Study MATLAB's powerful indexing capabilities that can be used in generating periodic sequences.

9. *Signal energy*: The energy of a sequence $x(n)$ is defined as

$$\mathcal{E}_x = \sum_{-\infty}^{\infty} x(n)x^*(n) = \sum_{-\infty}^{\infty} |x(n)|^2 \tag{2.7}$$

where the superscript * denotes the operation of complex conjugation. A signal that has finite energy, that is, $\mathcal{E}_x < \infty$, is called an energy-type signal.

10. *Signal power*: The average power of a sequence $x(n)$ is given by

$$\text{P}_x = \lim_{N \to \infty} \frac{1}{2N + 1} \sum_{-N}^{N} |x(n)|^2 \tag{2.8}$$

A signal that has non-zero, finite average power is called a power-type signal. For a periodic sequence $\tilde{x}(n)$ with fundamental period N, the average power is given by

$$\text{P}_x = \frac{1}{N} \sum_{0}^{N-1} |\tilde{x}(n)|^2 \tag{2.9}$$

11. Any arbitrary sequence $x(n)$ can be *synthesized* as a weighted sum of delayed and scaled unit-sample sequences:

$$x(n) = \sum_{k=-\infty}^{\infty} x(k)\delta(n - k) \tag{2.10}$$

12. A real-valued sequence $x_e(n)$ is called *even* (*symmetric*) if

$$x_e(-n) = x_e(n) \tag{2.11}$$

Similarly, a real-valued sequence $x_o(n)$ is called *odd (antisymmetric)* if

$$x_o(-n) = -x_o(n) \tag{2.12}$$

Then any arbitrary real-valued sequence $x(n)$ can be decomposed into its even and odd components,

$$x(n) = x_e(n) + x_o(n) \tag{2.13}$$

where the even and odd parts are given by

$$x_e(n) = \frac{1}{2}\left[x(n) + x(-n)\right] \text{ and } x_o(n) = \frac{1}{2}\left[x(n) - x(-n)\right] \tag{2.14}$$

respectively.

13. A one-sided exponential sequence of the form

$$\alpha^n,\ n \geq 0;\ \alpha\text{: arbitrary constant}$$

is called a *geometric series*. The series converges for $|\alpha| < 1$, and the sum of its components converges to

$$\sum_{n=0}^{\infty} \alpha^n \longrightarrow \frac{1}{1-\alpha}, \quad \text{for } |\alpha| < 1 \tag{2.15}$$

The sum of a finite number N of terms of the series is given by

$$\sum_{n=0}^{N-1} \alpha^n = \frac{1-\alpha^N}{1-\alpha}, \ \forall\alpha \tag{2.16}$$

14. Given two real-valued sequences $x(n)$ and $y(n)$ of finite energy, the *crosscorrelation* of $x(n)$ and $y(n)$ is a sequence $r_{xy}(\ell)$ defined as

$$r_{xy}(\ell) = \sum_{n=-\infty}^{\infty} x(n)y(n-\ell) \tag{2.17}$$

The index ℓ is called the shift or lag parameter. The special case of (2.17) when $y(n) = x(n)$ is called *autocorrelation*, defined by

$$r_{xx}(\ell) = \sum_{n=-\infty}^{\infty} x(n)x(n-\ell) \tag{2.18}$$

2.2 Discrete-Time Systems

A discrete-time system (or *discrete system* for short) takes an input sequence $x(n)$ and produces an output sequence $y(n)$ that has a desired characteristic. A system is characterized through an input–output transformation defined as

$$y(n) = \mathrm{T}\left[x(n)\right] \text{ or } x(n) \longrightarrow \boxed{\mathrm{T}[\cdot]} \longrightarrow y(n) \tag{2.19}$$

where $\mathrm{T}[\cdot]$ is a system operator. Discrete systems are classified into *linear* and *nonlinear* systems or *time-varying* and *time-invariant* systems. In the time domain, linear and time-invariant systems can be characterized by the convolution representation or by the difference-equation representation. These topics are discussed in this section.

Study Topics: LTI Systems and Their Properties

1. A discrete system is linear if and only if $\mathrm{T}[\cdot]$ satisfies the generalized superposition principle, that is,

$$\mathrm{T}[a_1x_1(n) + a_2x_2(n)] = a_1\mathrm{T}[x_1(n)] + a_2\mathrm{T}[x_2(n)], \forall a_1, a_2, x_1(n), x_2(n) \tag{2.20}$$

2. A linear system in which an input–output pair, $x(n)$ and $y(n)$, is invariant to a shift in n is called a linear time-invariant system. Hence, the following is true:

$$\begin{gathered} x(n) \longrightarrow \boxed{\mathrm{T}[\cdot]} \longrightarrow y(n) \longrightarrow \boxed{\text{Shift by } k} \longrightarrow y(n-k) \\ x(n) \longrightarrow \boxed{\text{Shift by } k} \longrightarrow x(n-k) \longrightarrow \boxed{\mathrm{T}[\cdot]} \longrightarrow y(n-k) \end{gathered} \tag{2.21}$$

   We will denote a linear, time-invariant (LTI) system by the operator $\mathrm{LTI}\,[\cdot]$.

3. Let $x(n)$ and $y(n)$ be the input–output pair of an LTI system. Then the output is given by

$$y(n) = \mathrm{LTI}\,[x(n)] = \sum_{k=-\infty}^{\infty} x(k)\,\{\mathrm{LTI}\,[\delta]\,(n-k)\} \triangleq \sum_{k=-\infty}^{\infty} x(k)h(n-k) \tag{2.22}$$

   where the response of an LTI system to the unit-sample sequence $\delta(n)$ is called the *impulse response* and is denoted by $h(n)$.

4. The mathematical operation in (2.22) is called a *linear convolution sum* and is denoted by

$$y(n) \triangleq x(n) * h(n) = \sum_{k=-\infty}^{\infty} x(k)h(n-k) \tag{2.23}$$

5. An LTI system is completely characterized in the time domain by the impulse response $h(n)$, given by

$$x(n) \longrightarrow \boxed{h(n)} \longrightarrow y(n) = x(n) * h(n) \tag{2.24}$$

6. If sequences $x(n)$ and $h(n)$ are of finite duration, then `y = conv(x,h)` computes the convolution between $x(n)$ and $y(n)$. Study various approaches given in the text to compute the convolution sum.

7. A system is said to be *bounded-input bounded-output (BIBO) stable* if every bounded input sequence produces a bounded output sequence — that is, if

$$|x(n)| < \infty \Rightarrow |y(n)| < \infty, \forall x, y, n \tag{2.25}$$

   An LTI system is BIBO stable if and only if its impulse response is *absolutely summable*:

$$\text{BIBO stability} \Longleftrightarrow \sum_{-\infty}^{\infty} |h(n)| < \infty \tag{2.26}$$

8. A system is said to be causal if the output at index n_0 depends only on the input up to and including the index n_0 — that is; if the output does not depend on the future values of the input.

   An LTI system is causal if and only if the impulse response

$$h(n) = 0, \ n < 0 \tag{2.27}$$

   Such a sequence is termed a *causal sequence*.

9. An LTI discrete system can also be described by a linear constant-coefficient difference equation of the form

$$\sum_{k=0}^{N} a_k y(n-k) = \sum_{m=0}^{M} b_m x(n-m), \quad \forall n \tag{2.28}$$

   where $\{a_k\}$ and $\{b_k\}$ are the system coefficients. If $a_N \neq 0$, then the difference equation is of order N.

10. Another form of this equation is (with $a_0 = 1$)

$$y(n) = \sum_{m=0}^{M} b_m x(n-m) - \sum_{k=1}^{N} a_k y(n-k) \tag{2.29}$$

   which is used for recursive computation of $y(n)$ from $n = -\infty$ to $n = \infty$.

11. A solution to this equation can be obtained in the form

$$y(n) = y_H(n) + y_P(n) \tag{2.30}$$

   where $y_H(n)$ is called the *homogeneous part* of the solution and $y_P(n)$ is known as the *particular part* of the solution. The solution of difference equations is deferred until Chapter 4.

12. The MATLAB function `filter` is available to solve difference equations numerically when given the input and the difference-equation coefficients. In its simplest form, this routine is invoked by
   `y = filter(b,a,x)`
   where
   `b = [b0, b1, ..., bM]; a = [a0, a1, ..., aN];`
   are the coefficient arrays of the equation given in (2.28) and `x` is the input sequence array. The output `y` has the same length as the input `x`.

13. If the unit-impulse response of an LTI system is of finite duration,

$$h(n) = \begin{cases} \text{Some non-zero value} & n_1 \leq n \leq n_2 \\ 0, & \text{otherwise} \end{cases} \tag{2.31}$$

   then the system is called a *finite-duration impulse response* (or FIR) filter. FIR filters are also called *nonrecursive* or *moving-average (MA)* filters.

14. The following part of the difference equation (2.28) describes a causal FIR filter:

$$y(n) = \sum_{m=0}^{M} b_m x(n-m) \tag{2.32}$$

The impulse response for this filter is

$$h(n) = \begin{cases} b_n & 0 \le n \le M \\ 0, & \text{otherwise} \end{cases} \tag{2.33}$$

15. In MATLAB, FIR filters are represented either as impulse-response values $\{h(n)\}$ or as difference-equation coefficients $\{b_m\}$ and $\{a_0 = 1\}$. Therefore, to implement FIR filters, we can use either the `conv(x,h)` function or the `filter(b,1,x)` function. In practice (and especially for processing signals), the use of the `filter` function is encouraged.

16. If the impulse response of an LTI system is of infinite duration, then the system is called an *infinite-duration impulse response* (or IIR) filter.

17. The part

$$\sum_{k=0}^{N} a_k y(n-k) = x(n) \tag{2.34}$$

of the difference equation (2.28) obtained by setting $b_o = 1$ and $b_m = 0$, $m \neq 0$, describes a *recursive* filter in which the output $y(n)$ is recursively computed from its previously computed values and is called an *autoregressive (AR)* filter. The impulse response of such a filter is of infinite duration and, hence, it is an IIR filter.

18. The general equation (2.28) with some non-zero coefficients $\{a_k\}$ and $\{b_k\}$ also describes an IIR filter. It has two parts: an AR part and an MA part. Such an IIR filter is called an *autoregressive-moving-average*, or ARMA, filter.

19. In MATLAB, IIR filters are described by the difference-equation coefficients $\{b_m\}$ and $\{a_k\}$ and are implemented by the `filter(b,a,x)` function.

2.3 Summary

MATLAB Functions

Discrete-Time Signals

- `[x,n]=impseq(n0,n1,n2)`: Unit-impulse sequence generation
- `[x,n]=stepseq(n0,n1,n2)`: Unit-step sequence generation
- `[y,n]=sigadd(x1,n1,x2,n2)`: Addition of two sequences
- `[y,n]=sigmult(x1,n1,x2,n2)`: Multiplication of two sequences
- `[y,n]=sigshift(x,m,n0)`: Shifting operation on a sequence
- `[y,n]=sigfold(x,n)`: Folding operation on a sequence
- `[xe,xo,m]=evenodd(x,n)`: Decomposition of real sequence into even and odd components

Discrete-Time Systems

- `[y,ny]=conv_m(x,nx,h,nh)`: Modified convolution operation for signal processing
- `[y]=filter(b,a,x)`: Implementation of a filtering operation

2.4 Problems

P2.1 Generate and plot the samples (use the `stem` function) of the following sequences, using MATLAB.

(a) $x_1(n) = \sum_{m=0}^{10} (m+1)\left[\delta(n-2m) - \delta(n-2m-1)\right], \ 0 \le n \le 25$

(b) $x_2(n) = n^2\left[u(n+5) - u(n-6)\right] + 10\delta(n) + 20(0.5)^n\left[u(n-4) - u(n-10)\right]$

(c) $x_3(n) = (0.9)^n \cos(0.2\pi n + \pi/3), \ 0 \le n \le 20$

(d) $x_4(n) = 10\cos(0.0008\pi n^2) + w(n), \ 0 \le n \le 100$ where $w(n)$ is a random sequence uniformly distributed between $[-1, 1]$. How do you characterize this sequence?

(e) $\tilde{x}_5(n) = \{\ldots, 1, 2, 3, \underset{\uparrow}{2}, 1, 2, 3, 2, 1, \ldots\}_{\text{PERIODIC}}$. Plot five periods.

P2.2 Let $x(n) = \{1, -2, 4, 6, \underset{\uparrow}{-5}, 8, 10\}$. Generate and plot the samples (use the `stem` function) of the following sequences.

(a) $x_1(n) = 3x(n+2) + x(n-4) - 2x(n)$

(b) $x_2(n) = 5x(5+n) + 4x(n+4) + 3x(n)$

(c) $x_3(n) = x(n+4)x(n-1) + x(2-n)x(n)$

(d) $x_4(n) = 2e^{0.5n}x(n) + \cos(0.1\pi n)\, x(n+2), \ -10 \le n \le 10$

(e) $x_5(n) = \sum_{k=1}^{5} nx(n-k)$

P2.3 The complex exponential sequence $e^{j\omega_0 n}$ (or the sinusoidal sequence $\cos(\omega_0 n)$) is periodic if the *normalized* frequency $f_0 \stackrel{\triangle}{=} \dfrac{\omega_0}{2\pi}$ is a rational number (i.e., $f_0 = \dfrac{K}{N}$ where K and N are integers).

(a) Prove the above result.

(b) Generate and plot $\cos(0.3\pi n), \ -20 \le n \le 20$. Is this sequence periodic? If it is, what is its fundamental period? From the examination of the plot, what interpretation can you give to the integers K and N above?

(c) Generate and plot $\cos(0.3n), \ -20 \le n \le 20$. Is this sequence periodic? What do you conclude from the plot? If necessary, examine the values of the sequence in MATLAB to arrive at your answer.

P2.4 A complex-valued sequence $x_e(n)$ is called *conjugate-symmetric* if

$$x_e(n) = x_e^*(-n)$$

Similarly, a complex-valued sequence $x_o(n)$ is called *conjugate-antisymmetric* if

$$x_o(n) = -x_o^*(-n)$$

Then any arbitrary complex-valued sequence $x(n)$ can be decomposed into

$$x(n) = x_e(n) + x_o(n)$$

where $x_e(n)$ and $x_o(n)$ are given by

$$x_e(n) = \tfrac{1}{2}\left[x(n) + x^*(-n)\right] \text{ and } x_o(n) = \tfrac{1}{2}\left[x(n) - x^*(-n)\right] \tag{2.35}$$

respectively.

(a) Modify the evenodd function discussed in the text so that it accepts an arbitrary sequence and decomposes it into its symmetric and anti-symmetric components by implementing (2.35).

(b) Decompose the sequence

$$x(n) = 10e^{-j(0.4\pi n)}, \; 0 \le n \le 10$$

into its conjugate-symmetric and conjugate-antisymmetric components. Plot their real and imaginary parts to verify the decomposition. (Use the `subplot` function.)

P2.5 Here are three systems:

$$T_1[x(n)] = 2^{x(n)}; \quad T_2[x(n)] = 3x(n) + 4; \quad T_3[x(n)] = x(n) + 2x(n-1) - x(n-2)$$

(a) Using (2.20), test analytically for whether these systems are linear.

(b) Let $x_1(n)$ be a uniformly distributed random sequence between [0, 1] over $0 \le n \le 100$, and let $x_2(n)$ be a Gaussian random sequence with mean 0 and variance 10 over $0 \le n \le 100$. Using these sequences, test the linearity of the three systems. Choose any values for constants a_1 and a_2 in (2.20). You should use several realizations of the above sequences to arrive at your answers.

P2.6 Here are three systems:

$$T_1[x(n)] = \sum_{0}^{n} x(k); \quad T_2[x(n)] = \sum_{n-10}^{n+10} x(k); \quad T_3[x(n)] = x(-n)$$

(a) Using (2.21), test analytically for whether these systems are shift-invariant.

(b) Let $x(n)$ be a Gaussian random sequence with mean 0 and variance 10 over $0 \le n \le 100$. Using this sequence, test the shift-invariance of the three systems. Choose any values for sample shift k in (2.21). You should use several realizations of the above sequence to arrive at your answers.

P2.7 For the systems given in Problems P2.5 and P2.6, determine analytically their BIBO stability and causality.

P2.8 The linear convolution defined in (2.23) has several properties:

$$\begin{aligned} x_1(n) * x_2(n) &= x_1(n) * x_2(n) && \text{: Commutation} \\ [x_1(n) * x_2(n)] * x_3(n) &= x_1(n) * [x_2(n) * x_3(n)] && \text{: Association} \\ x_1(n) * [x_2(n) + x_3(n)] &= x_1(n) * x_2(n) + x_1(n) * x_3(n) && \text{: Distribution} \\ x(n) * \delta(n - n_0) &= x(n - n_0) && \text{: Identity} \end{aligned} \tag{2.36}$$

(a) Prove these properties analytically.

(b) Using the following three sequences, verify the preceding properties.

$$\begin{aligned} x_1(n) &= n\left[u(n+10) - u(n-20)\right] \\ x_2(n) &= \cos(0.1\pi n)\left[u(n) - u(n-30)\right] \\ x_3(n) &= (1.2)^n\left[u(n+5) - u(n-10)\right] \end{aligned}$$

Use the `conv_m` function.

P2.9 When the sequences $x(n)$ and $h(n)$ are of finite duration N_x and N_h respectively, then their linear convolution (2.23) can also be implemented, by using *matrix-vector multiplication*. If elements of $y(n)$ and $x(n)$ are arranged in column vectors **x** and **y** respectively, then, from (2.23), we obtain

$$\mathbf{y} = \mathbf{H}\,\mathbf{x}$$

where linear shifts in $h(n-k)$ for $n = 0, \ldots, N_h - 1$ are arranged as rows in the matrix **H**. This matrix has an interesting structure and is called a *Toeplitz* matrix. To investigate this matrix, consider the sequences

$$x(n) = \{\underset{\uparrow}{1}, 2, 3, 4\} \text{ and } h(n) = \{\underset{\uparrow}{3}, 2, 1\}$$

(a) Produce the linear convolution $y(n) = h(n) * x(n)$.

(b) Express $x(n)$ as a 4×1 column vector **x** and $y(n)$ as a 6×1 row vector **y**. Now compute the 6×4 matrix **H** so that $\mathbf{y} = \mathbf{H}\,\mathbf{x}$.

(c) Characterize the matrix **H**. From this characterization, can you give a definition of a Toeplitz matrix? How does this definition compare with that of time invariance?

(d) What can you say about the first column and the first row of **H**?

P2.10 MATLAB provides a function called `toeplitz` that generates a Toeplitz matrix when given the first row and the first column.

(a) Using this function and your answer to Problem P2.9 part (d), develop an alternate MATLAB function to implement linear convolution. The format of the function should be

```
function [y,H]=conv\_tp(h,x)
% Linear Convolution using Toeplitz Matrix
% ----------------------------------------
% [y,H] = conv_tp(h,x)
% y = output sequence in column vector form
% H = Toeplitz matrix corresponding to sequence h so that y = Hx
% h = Impulse response sequence in column vector form
% x = input sequence in column vector form
```

(b) Verify your function on the sequences given in Problem P2.9.

P2.11 Let $x(n) = (0.8)^n \, u(n)$.

(a) Determine $x(n) * x(n)$ analytically.

(b) Using the `filter` function, determine first 50 samples of $x(n) * x(n)$. Compare your results with part (a) above.

P2.12 A linear and shift-invariant system is described by the difference equation

$$y(n) - 0.5y(n-1) + 0.25y(n-2) = x(n) + 2x(n-1) + x(n-3)$$

(a) Determine the stability of the system.

(b) Determine and plot the impulse response of the system over $0 \le n \le 100$. Determine the stability from this impulse response.

(c) If the input to this system is $x(n) = [5 + 3\cos(0.2\pi n) + 4\sin(0.6\pi n)]\, u(n)$, calculate the response $y(n)$ over $0 \le n \le 200$.

Chapter 3

The z-Transform

The purpose of this chapter is to introduce the reader to the z-transform and to illustrate its use in the characterization of discrete-time signals and systems. We shall also demonstrate the use of the z-transform in determining the response of an LTI system to an arbitrary excitation.

The reader should study the following material in Chapter 3 of the DSP book:

- The bilateral z-transform and its important properties — Sections 3.1, 3.2, and 3.3
- Inversion of the z-transform — Section 3.4
- Solution of difference equations — Sections 3.5 and 3.6

Learning Objectives

- Become familiar with the computation of the z-transform of various signals and with the determination of the ROC.
- Learn and understand the properties of the z-transform and how these properties can be used to simplify the computations.
- Learn the process of inverting the z-transform by using the method of partial-fraction expansion.
- Understand how LTI systems are represented in the z-domain and the relationship to the frequency response.
- Learn how to solve difference equations that describe LTI systems with initial conditions.
- Understand the difference between the bilateral and the unilateral z-transform.

3.1 The Bilateral z-Transform

In this section we focus on the bilateral z-transform and its properties. The bilateral z-transform of a sequence $x(n)$ is defined as

$$X(z) \triangleq \mathcal{Z}[x(n)] = \sum_{n=-\infty}^{\infty} x(n)\, z^{-n} \tag{3.1}$$

where z is a complex variable. For any given sequence $x(n)$, the set of values of z for which $X(z)$ exists is called the *region of convergence* (ROC) and is given by

$$R_{x^-} < |z| < R_{x^+}$$

for some positive numbers R_{x^-} and R_{x^+}.

Study Topics: Important Properties of the z-Transform

1. Linearity:

$$\mathcal{Z}[a_1\, x_1(n) + a_2\, x_2(n)] = a_1\, X_1(z) + a_2\, X_2(z); \quad \text{ROC: at least } \text{ROC}_{x_1} \cap \text{ROC}_{x_2} \tag{3.2}$$

2. Time sample shifting:

$$\mathcal{Z}[x(n-n_0)] = z^{-n_0} X(z); \quad \text{ROC: ROC}_x \tag{3.3}$$

3. Frequency shifting:

$$\mathcal{Z}[a^n x(n)] = X\left(\frac{z}{a}\right); \quad \text{ROC: ROC}_x \text{ scaled by } |a| \tag{3.4}$$

4. Folding:

$$\mathcal{Z}[x(-n)] = X\left(\frac{1}{z}\right); \quad \text{ROC: Inverted ROC}_x \tag{3.5}$$

5. Complex conjugation:

$$\mathcal{Z}[x^*(n)] = X^*(z^*); \quad \text{ROC: ROC}_x \tag{3.6}$$

6. Differentiation in the z-domain:

$$\mathcal{Z}[n\, x(n)] = -z\,\frac{\mathrm{d}X(z)}{\mathrm{d}z}; \quad \text{ROC: ROC}_x \tag{3.7}$$

7. Multiplication of two sequences:

$$\mathcal{Z}[x_1(n)\, x_2(n)] = \frac{1}{j2\pi}\oint_C X_1(v) X_2(z/v)\, v^{-1}\, \mathrm{d}v; \quad \text{ROC: ROC}_{x_1} \cap \text{Inverted ROC}_{x_2} \tag{3.8}$$

where C is a closed contour in the z-plane that encloses the origin and lies in the common region of convergence.

8. Convolution of two sequences:

$$\mathcal{Z}[x_1(n) * x_2(n)] = X_1(z)\, X_2(z); \quad \text{ROC: ROC}_{x_1} \cap \text{ROC}_{x_2} \tag{3.9}$$

3.2 Inversion of the z-Transform

The inverse z-transform of a complex function $X(z)$ is given by the contour integral

$$x(n) \triangleq \mathcal{Z}^{-1}[X(z)] = \frac{1}{j2\pi} \oint_C X(z) z^{n-1}\, dz \tag{3.10}$$

where C is a counter-clockwise closed contour encircling the origin and lying in the ROC.

From the definition of the inverse z-transform, we observe that its computation requires the evaluation of a contour integral. In general, the computation is at best tedious and often very complicated. However, in practice, $X(z)$ is usually a ratio of two polynomials in z, that is, $X(z) = B(z)/A(z)$, where $A(z)$ and $B(z)$ are polynomials. When $X(z)$ is a rational function, it can be expressed as a sum of simple first-order factors by using a partial-fraction expansion. By taking advantage of the linearity property of the z-transform and its inverse, the sequence $x(n)$ corresponding to $X(z)$ is then expressed as a sum of the individual sequences corresponding to the factors of $X(z)$. Typically, a small z-transform table is used to find the individual sequences corresponding to each of the factors of $X(z)$.

The procedure for performing the partial-fraction expansion of $X(z)$ is described in the DSP book.

3.3 Representation of an LTI System in the z-Domain

In this section, we consider the representation of an LTI discrete-time system in the z-domain. The z-domain of such a system is simply the z-transform of the impulse response $h(n)$ of the system. Thus, we define the system function $H(z)$ as

$$H(z) = \mathcal{Z}[h(n)] = \sum_{n=-\infty}^{\infty} h(n)\, z^{-n}, \quad R_{h^-} < |z| < R_{h^+} \tag{3.11}$$

The system function is an alternative method for determining the response of an LTI system to an input sequence $x(n)$. Instead of performing the convolution $x(n) * h(n) = y(n)$ to obtain the output sequence $y(n)$, we can use the convolution property of the z-transform to find

$$Y(z) = H(z)\, X(z) \tag{3.12}$$

Then we can compute the inverse z-transform to obtain $y(n)$. This approach is much simpler than performing the convolution in the time domain, especially when $H(z)$ and $X(z)$ are rational.

It is demonstrated in the DSP book that $H(z)$ is rational when the LTI system is described by a difference equation of the form

$$y(n) + \sum_{k=1}^{N} a_k\, y(n-k) = \sum_{k=0}^{M} b_k\, x(n-k) \tag{3.13}$$

In this case,

$$H(z) = \frac{\sum_{k=0}^{M} b_k\, z^{-k}}{1 + \sum_{k=1}^{N} a_k\, z^{-k}} = \frac{B(z)}{A(z)} \tag{3.14}$$

The system function $H(z)$ of an LTI system is related to the frequency response of the system. Specifically, if the region of convergence $H(z)$ includes the unit circle, that is, the region in the z-plane for which $|z| = 1$, then the frequency response of the LTI system can be obtained from $H(z)$ by evaluating $H(z)$ at $z = e^{j\omega}$. Thus,

$$H(\omega) = H(z)|_{z=e^{j\omega}} \tag{3.15}$$

When $H(z)$ is rational

$$H(\omega) = \frac{B(\omega)}{A(\omega)} \tag{3.16}$$

3.4 Solution of Difference Equations

In this section, we focus on determining the response of an LTI system characterized by the difference equation

$$y(n) + \sum_{k=1}^{N} a_k y(n-k) = \sum_{k=0}^{M} b_k x(n-k) \tag{3.17}$$

for $n \geq 0$. The response of the system for $n \geq 0$ depends not only on the excitation for $n \geq 0$, but also on the state of the system at $n = 0$. The state of the system at $n = 0$ is completely characterized by the initial conditions $\{y(i),\ i = -1, -2, \ldots, -N\}$.

To construct the solution of the difference equation in (3.17) for $n \geq 0$ by using z-transforms, it is convenient to use a special form of the z-transform, called the *one-sided* or *unilateral* z-transform, of a sequence. This transform is defined as

$$\mathcal{Z}^+[x(n)] \triangleq \mathcal{Z}[x(n)u(n)] \triangleq X^+(z) = \sum_{n=0}^{\infty} x(n)\, z^{-n} \tag{3.18}$$

Note that

$$\begin{aligned}
\mathcal{Z}^+[x(n-k)] &= \mathcal{Z}[x(n-k)u(n)] \\
&= \sum_{n=0}^{\infty} x(n-k)\, z^{-n} = \sum_{m=-k}^{\infty} x(m)\, z^{-(m+k)} \\
&= \sum_{m=-k}^{-1} x(m)\, z^{-(m+k)} + \left[\sum_{m=0}^{\infty} x(m)\, z^{-m}\right] z^{-k}
\end{aligned}$$

Therefore,

$$\mathcal{Z}^+[x(n-k)] = \sum_{m=-k}^{-1} x(m)\, z^{-(m+k)} + z^{-k} X^+(z) \tag{3.19}$$

When we compute the one-sided z-transform of the difference equation in (3.17), we obtain

$$\left[1+\sum_{k=1}^{N} a_k z^{-k}\right] Y^+(z) + \sum_{k=1}^{N} a_k \left[\sum_{m=-k}^{-1} y(m)\, z^{-(m+k)}\right] = \left[\sum_{k=0}^{M} b_k z^{-k}\right] X^+(z) + \sum_{k=1}^{M} b_k \left[\sum_{m=-k}^{-1} x(m)\, z^{-(m+k)}\right] \tag{3.20}$$

Note that the second term on the left-hand side of (3.20) involves the initial condition that constitutes the state of the LTI system at $n = 0$. If the system is initially at rest, that is, $y(n) = 0$ for $n < 0$, this term is zero. The second term on the right-hand side of (3.20) is zero if the input sequence $x(n)$ is causal, that is, $x(n) = 0$ for $n < 0$. Hence, if the system is initially at rest and the input sequence $x(n)$ is causal, (3.20) reduces to

$$\begin{aligned} Y^+(z) &= \frac{\sum_{k=0}^{M} b_k z^{-k}}{1+\sum_{k=1}^{N} a_k z^{-k}} X^+(z) \\ &= H(z)\, X^+(z) \end{aligned} \tag{3.21}$$

which is exactly the same result that is obtained with a bilateral z-transform. On the other hand, if the system is not at rest initially, but the input sequence still is causal, there is an additional term in the output sequence due to the presence of the nonzero initial conditions. In this case, the unilateral z-transform of the output sequence can be expressed as

$$Y^+(z) = H(z)\, X^+(z) + \frac{C(z)}{1+\sum_{k=1}^{N} a_k z^{-k}} \tag{3.22}$$

where $C(z)$ is a defined as the polynomial

$$C(z) = \sum_{k=1}^{N} a_k \left[\sum_{m=-k}^{-1} y(m)\, z^{-(m+k)}\right] \tag{3.23}$$

Hence, the response of the system corresponding to the term $H(z)\, X^+(z)$ is usually called the *zero-state response* of the LTI system, whereas the response corresponding to the second term on the right-hand side of (3.22) is usually called the *zero-input response* of the LTI system. The latter response is due to nonzero initial conditions.

3.5 Summary

MATLAB Functions

The following MATLAB functions are used in this chapter:

The Bilateral z-Transform

- `[y,ny] = conv_m(x1,n1,x2,n2)`: Modified routine for convolution
- `[x2,r] = deconv(x3,x1)`: Deconvolution of `x3/x1` with remainder `r`

Inversion of the z-Transform

- `[R,p,C] = residuez(b,a)`: Partial-fraction expansion from rational function
- `[b,a] = residuez(R,p,C)`: Rational function from partial-fraction expansion
- `[b] = roots(a)`: Roots of polynomial `a`
- `[a] = poly(b)`: Polynomial `a` from roots in `b`

Representation of the LTI System in the z-Domain

- `[H,w] = freqz(b,a,N)`: Frequency response `H` at `N` frequencies around the unit circle
- `[H] = freqz(b,a,w)`: Frequency response `H` at frequencies `w`
- `magH = abs(H)`: Magnitude of `H`
- `phaH = angle(H)`: Phase of `H` in radians

Solution of Difference Equations

- `[y] = filter(b,a,x,xic)`: Implementation of `filter` with equivalent initial conditions `xic`
- `[xic] = filtic(b,a,Y,X)`: Computation of equivalent initial conditions `xic` from the initial conditions `Y` and `X`.

3.6 Problems

P3.1 Determine the z-transform of the following sequences, using the definition (3.1). Indicate the region of convergence for each sequence, and verify the z-transform expression by using MATLAB.

(a) $x(n) = \left(\frac{4}{3}\right)^n u(1-n)$

(b) $x(n) = 2^{-|n|} + \left(\frac{1}{3}\right)^{|n|}$

P3.2 Determine the z-transform of the following sequences, using the z-transform table and the z-transform properties. Express $X(z)$ as a rational function in z^{-1}. Verify your results, using MATLAB. Indicate the region of convergence in each case, and provide a pole–zero plot.

(a) $x(n) = \left(\frac{1}{3}\right)^n u(n-2) + (0.9)^{n-3} u(n)$

(b) $x(n) = \left(\frac{1}{2}\right)^n \cos\left(\frac{\pi n}{4} - 45°\right) u(n-1)$

P3.3 The z-transform of $x(n)$ is $X(z) = \left(1 + 2z^{-1}\right)$, $|z| \neq 0$. Find the z-transforms of the following sequences, and indicate their region of convergence.

(a) $x_2(n) = \left(1 + n + n^2\right) x(n)$

(b) $x_3(n) = \left(\frac{1}{2}\right)^n x(n-2)$

P3.4 The inverse z-transform of $X(z)$ is $x(n) = \left(\frac{1}{2}\right)^n u(n)$. Using the z-transform properties, determine the sequence

$$X_1(z) = zX\left(z^{-1}\right)$$

P3.5 Determine the following inverse z-transforms, using the partial-fraction expansion method.

(a) $X_2(z) = \left(1 - z^{-1} - 4z^{-2} + 4z^{-3}\right) / \left(1 - \frac{11}{4}z^{-1} + \frac{13}{8}z^{-2} - \frac{1}{4}z^{-3}\right)$. The sequence is absolutely summable.

(b) $X_4(z) = z / \left(z^3 + 2z^2 + 1.25z + 0.25\right)$, $|z| > 1$

P3.6 Suppose $X(z)$ is given as follows:

$$X(z) = \frac{2 + 3z^{-1}}{1 - z^{-1} + 0.81z^{-2}}, \quad |z| > 0.9$$

(a) Present $x(n)$ in a form that contains no complex numbers.

(b) Using MATLAB, find the first 20 samples of $x(n)$, and compare them with your answer in part (a).

P3.7 For the linear and shift-invariant systems described by the following impulse responses, determine (i) the system-function representation, (ii) the difference-equation representation, (iii) the pole–zero plot, and (iv) the output $y(n)$ if the input is $x(n) = \left(\frac{1}{4}\right)^n u(n)$:

(a) $h(n) = 2\left(\frac{1}{2}\right)^n u(n)$

(b) $h(n) = n\,[u(n) - u(n-10)]$

P3.8 A stable system has the following pole–zero locations:

$$z_1 = j, \quad z_2 = -j, \quad p_1 = -\frac{1}{2} + j\frac{1}{2}, \quad p_2 = -\frac{1}{2} - j\frac{1}{2}$$

It is known that the frequency response function $H(\omega)$ evaluated at $\omega = 0$ is equal to 0.8 — that is,

$$H(0) = 0.8$$

(a) Determine the system function $H(z)$, and indicate its region of convergence.

(b) Determine the difference-equation representation.

(c) Determine the steady-state response $y_{ss}(n)$ if the input is $x(n) = \frac{1}{\sqrt{2}} \sin\left(\frac{\pi n}{2}\right) u(n)$.

(d) Determine the transient response $y_{tr}(n)$ if the input is $x(n) = \frac{1}{\sqrt{2}} \sin\left(\frac{\pi n}{2}\right) u(n)$.

P3.9 A digital filter is described by the difference equation

$$y(n) = x(n) + x(n-1) - 0.9y(n-1) + 0.81y(n-2)$$

(a) Using the `freqz` function, plot the magnitude and phase of the frequency response of the filter. Note the magnitude and phase at $\omega = \pi/3$ and at $\omega = \pi$.

(b) Generate 200 samples of the signal $x(n) = \sin(\pi n/3) + 5\cos(\pi n)$ and process them through the filter. Compare the steady-state portion of the output to $x(n)$. How are the amplitudes and phases of two sinusoids affected by the filter?

P3.10 Solve the following difference equation for $y(n)$, using the one-sided z-transform approach.

$$\begin{aligned} y(n) &= 0.5y(n-1) + 0.25y(n-2) + x(n), \; n \geq 0; \quad y(-1) = 1, \; y(-2) = 2 \\ x(n) &= (0.8)^n u(n) \end{aligned}$$

Generate the first 20 samples of $y(n)$, using MATLAB, and compare them with your answer.

P3.11 A causal, linear, shift-invariant system is given by the following difference equation:

$$y(n) = y(n-1) + y(n-2) + x(n-1)$$

(a) Find the system function $H(z)$ for this system.

(b) Plot the poles and zeros of $H(z)$ and indicate the region of convergence (ROC).

(c) Find the unit-sample response $h(n)$ of this system.

(d) Is this system stable? If the answer is yes, justify it. If the answer is no, find a stable unit-sample response that satisfies the difference equation.

P3.12 Determine the zero-state response of the system

$$y(n) = \tfrac{1}{4}y(n-1) + x(n) + 3x(n-1), \; n \geq 0; \quad y(-1) = 2$$

to the input

$$x(n) = e^{j\pi n/4} u(n)$$

What is the steady-state response of the system?

Chapter 4

Discrete-Time Fourier Analysis

The purpose of this chapter is to characterize discrete-time signals and systems in the frequency domain. First, we demonstrate that a discrete-time signal can be represented as a linear combination of sequences from the complex exponential signal set $\{e^{j\omega n}\}$. The representation is called the Discrete-Time Fourier Transform (DTFT). Secondly, we show that the DTFT not only is useful in characterizing signals in the frequency domain but also is invaluable in analyzing and designing linear, time-invariant (LTI) systems. Finally, we treat the analog-to-digital conversion (ADC) or sampling operation in the frequency domain using the DTFT.

The reader should study the following material in Chapters 4, 5, and 6 of the DSP book:

- Discrete-time Fourier transform — Sections 4.2.1–4.2.4 and Section 4.4
- Frequency-domain representation of LTI systems — Section 5.1 and 5.2
- Sampling and reconstruction of analog signals — Section 6.1 and 6.2

Learning Objectives

- Understand the concept of the frequency-domain signal representation (spectrum) and the operation of the DTFT.
- Learn the computational aspects of the DTFT, both the analytical ones and those obtained via MATLAB.
- Study the various properties of the DTFT, and develop skills to apply those properties in computations.
- Comprehend why the response to a complex exponential leads to the frequency response $H(\omega)$ of a system and how it describes the system in the frequency domain.
- Understand the LTI system response to a sinusoidal signal and how to use it for steady-state response calculations.
- Master the skills to compute the frequency response $H(\omega)$ from the impulse response and from the difference equation, using MATLAB's `freqz` function.

- Learn the mathematical concepts in the sampling operation and its effect in the frequency domain on the spectra of discrete-time signals.
- Understand the concept of aliasing and how to minimize it.
- Study the conversion of discrete-time signals to analog signals (or the reconstruction operation) and the use of MATLAB's `sinc`, `stairs`, `plot`, and `spline` functions.

4.1 The Discrete-Time Fourier Transform

In this section, we will study the representation of sequences in the frequency domain via the complex exponential and the resulting DTFT, its properties, and its computation with MATLAB.

Study Topics: Definitions and Properties

1. The discrete-time Fourier transform of an arbitrary signal $x(n)$ is a complex-valued continuous function $X(\omega)$ given by

$$X(\omega) \triangleq \mathcal{F}[x(n)] = \sum_{n=-\infty}^{\infty} x(n)e^{-j\omega n} \tag{4.1}$$

   where $\mathcal{F}[x(n)]$ is an operator notation. It can be expressed as

$$X(\omega) = X_{\mathrm{R}}(\omega) + jX_{\mathrm{I}}(\omega)$$

   where $X_{\mathrm{R}}(\omega)$ and $X_{\mathrm{I}}(\omega)$ are the real and imaginary parts of $X(\omega)$ respectively, or

$$X(\omega) = |X(\omega)|\, e^{j\measuredangle X(\omega)}$$

   where $|X(\omega)|$ and $\measuredangle X(\omega)$ are the magnitude and angle parts of $X(\omega)$ respectively.

2. The inverse discrete-time Fourier transform of $X(\omega)$ [or the synthesis of $x(n)$] is given by

$$x(n) \triangleq \mathcal{F}^{-1}[X(\omega)] = \frac{1}{2\pi}\int_{-\pi}^{\pi} X(\omega)e^{j\omega n}d\omega \tag{4.2}$$

   where $\mathcal{F}^{-1}[X(\omega)]$ is an operator notation.

3. The discrete-time Fourier transform $X(\omega)$ exists if $x(n)$ is absolutely summable, that is, if

$$\sum_{-\infty}^{\infty} |x(n)| < \infty \tag{4.3}$$

   This is a sufficient but not necessary condition for the existence of the discrete-time Fourier transform.

4. *Periodicity Property*: The discrete-time Fourier transform $X(\omega)$ is periodic in ω with period 2π:

$$X(\omega) = X(\omega + 2\pi)$$

We need only one period of $X(\omega)$ (e.g., $\omega \in [0, 2\pi]$ or $[-\pi, \pi]$) for analysis and not the whole domain $-\infty < \omega < \infty$.

5. *Symmetry Property*: For real-valued $x(n)$, $X(\omega)$ is conjugate symmetric:

$$X(-\omega) = X^*(\omega)$$

or

$$\begin{aligned} X_R(-\omega) &= X_R(\omega) && \text{(even symmetry)} \\ X_I(-\omega) &= -X_I(\omega) && \text{(odd symmetry)} \\ |X(-\omega)| &= |X(\omega)| && \text{(even symmetry)} \\ \measuredangle X(-\omega) &= -\measuredangle X(\omega) && \text{(odd symmetry)} \end{aligned}$$

6. From the above two properties note that to plot $X(\omega)$ for a real-valued sequence we need to consider only a half period of $X(\omega)$. Generally, in practice, this period is chosen to be $\omega \in [0, \pi]$.

7. *Linearity Property*: The DTFT is a linear operation — that is,

$$\mathcal{F}[\alpha x_1(n) + \beta x_2(n)] = \alpha\mathcal{F}[x_1(n)] + \beta\mathcal{F}[x_2(n)] \tag{4.4}$$

8. *Time-Shifting Property*: The time-shift in a sequence results in a change in the angle (or phase) part of its DTFT, that is,

$$\mathcal{F}[x(n-k)] = X(\omega)e^{-j\omega k} \tag{4.5}$$

9. *Frequency-Shifting Property*: Multiplying a sequence by a complex exponential results in a frequency shifting of its DTFT — that is,

$$\mathcal{F}\left[x(n)e^{j\omega_0 n}\right] = X(\omega - \omega_0) \tag{4.6}$$

10. *Conjugation Property*: Complex conjugation of a sequence results in a complex conjugation and folding of the DTFT — that is,

$$\mathcal{F}\left[x^*(n)\right] = X^*(-\omega) \tag{4.7}$$

11. *Folding Property*: Folding of a sequence results in the folding of its DTFT — that is,

$$\mathcal{F}[x(-n)] = X(-\omega) \tag{4.8}$$

12. *Symmetries in Real Signals*: The real signal can be decomposed into its even and odd parts in the way that we discussed in Chapter 2.

$$x(n) = x_e(n) + x_o(n)$$

Then

$$\begin{aligned} \mathcal{F}[x_e(n)] &= X_R(\omega) \\ \mathcal{F}[x_o(n)] &= jX_I(\omega) \end{aligned} \tag{4.9}$$

13. *Convolution Property*: This most useful property makes system analysis convenient in the frequency domain:

$$\mathcal{F}[x_1(n) * x_2(n)] = \mathcal{F}[x_1(n)]\,\mathcal{F}[x_2(n)] = X_1(\omega)X_2(\omega) \tag{4.10}$$

14. *Multiplication Property*: This is a dual of the convolution property:

$$\mathcal{F}[x_1(n) \cdot x_2(n)] = \mathcal{F}[x_1(n)] \circledast \mathcal{F}[x_2(n)] = \frac{1}{2\pi}\int X_1(\theta)X_2(\omega - \theta)d\theta \tag{4.11}$$

The foregoing convolutionlike operation is called a *periodic convolution* and is denoted by $\circledast$. It is discussed (in its discrete form) in Chapter 5.

15. *Energy*: The energy of the signal $x(n)$ can be written as

$$\begin{aligned} \mathcal{E}_x &= \sum_{-\infty}^{\infty} |x(n)|^2 = \frac{1}{2\pi}\int_{-\pi}^{\pi} |X(\omega)|^2\, d\omega \\ &= \int_0^{\pi} \frac{|X(\omega)|^2}{\pi} d\omega \quad \text{(for real sequences)} \end{aligned} \tag{4.12}$$

This is also known as Parseval's theorem. From (4.12), the *energy-density spectrum* of $x(n)$ is defined as

$$\Phi_x(\omega) \triangleq \frac{|X(\omega)|^2}{\pi} \tag{4.13}$$

Then the energy of $x(n)$ in the $[\omega_1, \omega_2]$ band is given by

$$\int_{\omega_1}^{\omega_2} \Phi_x(\omega)d\omega, \quad 0 \le \omega_1 < \omega_2 \le \pi$$

4.2 The Frequency-Domain Representation of LTI Systems

In this section, you will study how the DTFT is used for the analysis of LTI systems. The computation of the frequency-response function $H(\omega)$ and of the steady-state response will be emphasized.

Study Topics: The Transfer Function of LTI Systems

1. The DTFT of the impulse response $h(n)$ of an LTI system is called the *Frequency Response* (or *Transfer Function*) and is denoted by

$$H(\omega) \triangleq \mathcal{F}[h(n)] = \sum_{-\infty}^{\infty} h(n)e^{-j\omega n} \tag{4.14}$$

2. *Response to a complex exponential* $e^{j\omega_0 n}$: Let $x(n) = e^{j\omega_0 n}$ be the input to an LTI system represented by the impulse response $h(n)$. Then

$$\begin{aligned} y(n) = h(n) * e^{j\omega_0 n} &= \textstyle\sum_{-\infty}^{\infty} h(k) e^{j\omega_0 (n-k)} \\ &= \left[\textstyle\sum_{-\infty}^{\infty} h(k) e^{-j\omega_0 k}\right] e^{j\omega_0 n} = H(\omega_0)\, e^{j\omega_0 n} \end{aligned} \tag{4.15}$$

Thus we have

$$x(n) = e^{j\omega_0 n} \longrightarrow \boxed{H(\omega)} \longrightarrow y(n) = H(\omega_0) \cdot e^{j\omega_0 n} \tag{4.16}$$

This justifies the definition of $H(\omega)$ as a frequency response, because it is what the complex exponential is multiplied by to obtain the output $y(n)$.

3. The result, in topic 2 can be extended to a linear combination of complex exponentials by using the linearity of LTI systems.

$$\sum_k A_k e^{j\omega_k n} \longrightarrow \boxed{h(n)} \longrightarrow \sum_k A_k H(\omega_k)\, e^{j\omega_k n}$$

4. The magnitude $|H(\omega)|$ of $H(\omega)$ is called the *magnitude (or gain) response* function, and the angle $\theta(\omega)$ is called the *phase response* function.

5. *Response to sinusoidal sequences*: Let $x(n) = A\cos(\omega_0 n + \theta_0)$ be an input to an LTI system $h(n)$. Then the response $y(n)$ is another sinusoid of the same frequency ω_0, with its amplitude *scaled* by $|H(\omega_0)|$ and its phase *shifted* by $\measuredangle H(\omega_0)$; that is,

$$y(n) = A\,|H(\omega_0)| \cos\left(\omega_0 n + \theta_0 + \measuredangle H(\omega_0)\right) \tag{4.17}$$

This response is called the *steady-state response* and is denoted by $y_{ss}(n)$. It can be extended to a linear combination of sinusoidal sequences.

$$\sum_k A_k \cos(\omega_k n + \theta_k) \longrightarrow \boxed{H(\omega)} \longrightarrow \sum_k A_k\, |H(\omega_k)| \cos(\omega_k n + \theta_k + \measuredangle H(\omega_k))$$

6. *Response to arbitrary sequences*: Let $X(\omega) = \mathcal{F}[x(n)]$ and $Y(\omega) = \mathcal{F}[y(n)]$; then, using the convolution property (4.10), we have

$$Y(\omega) = H(\omega)\, X(\omega) \tag{4.18}$$

Therefore, the response of an LTI system can be represented in the frequency domain by

$$X(\omega) \longrightarrow \boxed{H(\omega)} \longrightarrow Y(\omega) = H(\omega)\, X(\omega)$$

7. *Frequency-response function from difference equations*: Given the difference-equation representation of an LTI system,

$$y(n) + \sum_{\ell=1}^{N} a_\ell y(n-\ell) = \sum_{m=0}^{M} b_m x(n-m) \tag{4.19}$$

the frequency response is given by

$$H(\omega) = \frac{\sum_{m=0}^{M} b_m \, e^{-j\omega m}}{1 + \sum_{\ell=1}^{N} a_\ell \, e^{-j\omega \ell}} \tag{4.20}$$

In MATLAB, this response can be computed by using the `freqz` function.

4.3 Sampling and Reconstruction of Analog Signals

In this section, we study the use of a sampling operation to convert an analog signal into a discrete-time sequence. Similarly, an inverse operation, called reconstruction, is required to convert the discrete-time sequence back to an analog signal. The emphasis of this section will be on the use of Fourier analysis and of MATLAB to study these operations and their effects.

Study Topics: Sampling and Reconstruction

1. *Sampling*: Let $x_a(t)$ be an analog (absolutely integrable) signal. Its continuous-time Fourier transform (CTFT) is given by

$$X_a(\Omega) \triangleq \int_{-\infty}^{\infty} x_a(t) e^{-j\Omega t} dt \tag{4.21}$$

where Ω is an analog frequency in radians/sec. The inverse CTFT is given by

$$x_a(t) = \frac{1}{2\pi} \int_{-\infty}^{\infty} X_a(\Omega) e^{j\Omega t} d\Omega \tag{4.22}$$

We now sample $x_a(t)$ at times T_s seconds apart (called the sampling interval) to obtain the discrete-time signal $x(n)$, that is,

$$x(n) \triangleq x_a(\,nT_s)$$

Let $X(\omega)$ be the discrete-time Fourier transform of $x(n)$. Then $X(\omega)$ is a sum of an infinite number of amplitude-scaled, frequency-scaled, and translated versions of the Fourier transform $X_a(\Omega)$:

$$X(\omega) = \frac{1}{T_s} \sum_{\ell=-\infty}^{\infty} X_a\left[\left(\frac{\omega}{T_s} - \frac{2\pi T_s}{\ell}\right)\right] \tag{4.23}$$

This relation is known as the *aliasing formula*. The graphical illustration of (4.23) is shown in Figure 4.1.

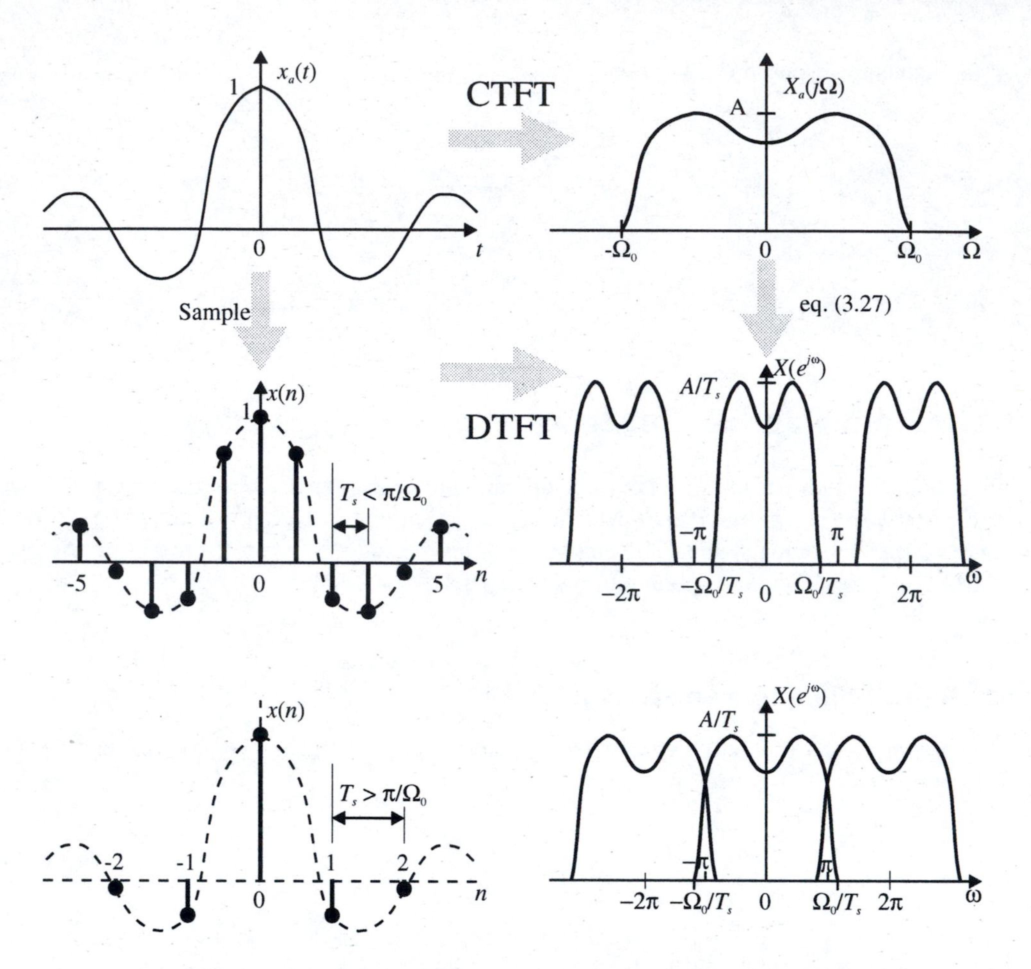

Figure 4.1: Sampling operations in the time and in the frequency domains

2. The frequencies of the analog and discrete-time signals are also related through T_s:

$$\omega = \Omega\, T_s \tag{4.24}$$

3. *Band-limited signal*: A signal is band-limited if there exists a finite-radian frequency Ω_0 such that $X_a(j\Omega)$ is zero for $|\Omega| > \Omega_0$ The frequency $F_0 = \Omega_0/2\pi$ is called the signal bandwidth in Hz.

4. *Sampling principle*: A band-limited signal $x_a(t)$ with bandwidth F_0 can be reconstructed from its sample values $x(n) = x_a(nT_s)$ if the sampling frequency $F_s = 1/T_s$ is greater than twice the bandwidth F_0 of $x_a(t)$ — that is, if

$$F_s > 2F_0$$

Otherwise, aliasing would result in $x(n)$. The sampling rate $2F_0$ for an analog band-limited signal is called the *Nyquist rate*.

5. *Reconstruction*: If we sample a band-limited signal $x_a(t)$ above its Nyquist rate then we can reconstruct $x_a(t)$ from its samples $x(n)$. This reconstruction is given by the *interpolation formula*

$$x_a(t) = \sum_{n=-\infty}^{\infty} x(n)\, \mathrm{sinc}\,[F_s(t - nT_s)] \tag{4.25}$$

where $\mathrm{sinc}(x) = \frac{\sin \pi x}{\pi x}$ is an interpolating function. In MATLAB, the `sinc` function is used for this ideal interpolation.

6. This *ideal* interpolation is not practically feasible, because the system is noncausal. Hence, we use practical nonideal converters.

7. *Zero-order hold (ZOH) interpolation*: In this interpolation, a given sample value is held for the sample interval until the next sample is received.

$$\hat{x}_a(t) = x(n), \ \ nT_s \le t < (n+1)T_s$$

The resulting signal is a piecewise-constant (staircase) waveform which requires an appropriately designed analog post-filter for accurate waveform reconstruction.

$$x(n) \longrightarrow \boxed{\text{ZOH}} \longrightarrow \hat{x}_a(t) \longrightarrow \boxed{\text{Post-Filter}} \longrightarrow x_a(t)$$

In MATLAB, the `stairs` function is used for the ZOH.

8. *First-order hold (FOH) interpolation*: In this case, the adjacent samples are joined by straight lines. This can be obtained by filtering the impulse train through a filter with impulse response

$$h_1(t) = \begin{cases} 1 + \dfrac{t}{T}, & 0 \le t \le T_s \\ 1 - \dfrac{t}{T}, & T_s \le t \le 2T_s \\ 0, & \text{otherwise} \end{cases}$$

The `plot` function implements the FOH.

9. *Cubic spline interpolation*: This approach uses spline interpolants for a smoother, but not necessarily more accurate, estimate of the analog signals between samples. Hence this interpolation does not require an analog post-filter. The smoother reconstruction is obtained by using a set of piecewise continuous third-order polynomials called *cubic splines*. In MATLAB, the `spline` function is available for this interpolation.

4.4 Summary

MATLAB Functions

Sampling and Reconstruction of Analog Signals

- `sinc(x)`: Ideal interpolation function
- `stairs(n,x)`: ZOH interpolation
- `plot(n,x)`: FOH interpolation
- `[xa]=spline(n,x,t)`: Spline interpolation

4.5 Problems

P4.1 Write a MATLAB function to compute the DTFT of a finite-duration sequence. The format of the function should be the following:

```
function [X] = dtft(x,n,w)
% Computes Discrete-time Fourier Transform
% [X] = dtft(x,n,w)
%
% X = DTFT values computed at w frequencies
% x = finite duration sequence over n
% n = sample position vector
% w = frequency location vector
```

P4.2 For each of the following sequences, determine the DTFT $X(\omega)$. Plot the magnitude and angle of $X(\omega)$.

(a) $x(n) = \{\underset{\uparrow}{4}, 3, 2, 1, 2, 3, 4\}$. Comment on the angle plot.

(b) $x(n) = \{\underset{\uparrow}{4}, 3, 2, 1, 1, 2, 3, 4\}$. Comment on the angle plot.

P4.3 Determine analytically the DTFT of each of the following sequences. Plot the magnitude and angle of $X(\omega)$, using MATLAB.

(a) $x(n) = 3\,(0.9)^n\,u(n)$

(b) $x(n) = 2\,(0.8)^{n+2}\,u(n-2)$

(c) $x(n) = n\,(0.5)^u\,u(n)$

(d) $x(n) = (n+2)\,(-0.7)^{n-1}\,u(n-2)$

(e) $x(n) = 5\,(-0.9)^n \cos(0.1\pi n)\,u(n)$

P4.4 A symmetric rectangular pulse is given by

$$\mathcal{R}_N(n) = \begin{cases} 1, & -N \le n \le N \\ 0, & \text{otherwise} \end{cases}$$

Determine the DTFT for $N = 5, 15, 25, 100$. Scale the DTFT so that $X(0) = 1$. Plot the normalized DTFT over $[-\pi, \pi]$. Study these plots, and comment on their behavior as a function of N.

P4.5 Determine the DTFT for $N = 5, 15, 25, 100$ for the symmetric triangular pulse that is given by

$$\mathcal{T}_N(n) = \left[1 - \frac{|n|}{N}\right] \mathcal{R}_N(n)$$

P4.6 A complex-valued sequence $x(n)$ can be decomposed into a conjugate symmetric part $x_e(n)$ and a conjugate antisymmetric part $x_o(n)$, as was discussed in Chapter 2. Show that

$$\mathcal{F}[x_e(n)] = X_R(\omega) \quad \text{and} \quad \mathcal{F}[x_o(n)] = jX_I(\omega)$$

where $X_R(\omega)$ and $X_I(\omega)$ are the real and imaginary parts of the DTFT $X(\omega)$ respectively. Verify this property on

$$x(n) = e^{j0.1\pi n}\left[u(n) - u(n-20)\right]$$

by using the MATLAB functions developed in Chapter 2.

P4.7 Determine $H(\omega)$, and plot its magnitude and phase for each of the following systems:

(a) $y(n) = x(n) + 2x(n-1) + x(n-2) - 0.5y(n-1) - 0.25y(n-2)$

(b) $y(n) = 2x(n) + x(n-1) - 0.25y(n-1) + 0.25y(n-2)$

P4.8 We have the following analog filter, which is realized by using a discrete filter:

$$x_a(t) \longrightarrow \boxed{\text{A/D}} \xrightarrow{x(n)} \boxed{h(n)} \xrightarrow{y(n)} \boxed{\text{D/A}} \longrightarrow y_a(t)$$

The sampling rate in the A/D and D/A is 100 samples per second and the impulse response is $h(n) = (0.5)^n u(n)$.

(a) What is the digital frequency in $x(n)$ if $x_a(t) = 3\cos(20\pi t)$?

(b) Find the steady-state output $y_a(t)$ if $x_a(t) = 3\cos(20\pi t)$.

(c) Find the steady-state output $y_a(t)$ if $x_a(t) = 3u(t)$.

(d) Find two other analog signals $x_a(t)$, with different analog frequencies, that will give the same steady-state output $y_a(t)$ when $x_a(t) = 3\cos(20\pi t)$ is applied.

(e) To prevent aliasing, a prefilter would be required to process $x_a(t)$ before it passes to the A/D converter. What type of filter should be used, and what should be the largest cutoff frequency that would work for the given configuration?

P4.9 Consider an analog signal $x_a(t) = \cos(20\pi t)$, $0 \leq t \leq 1$. It is sampled at $T_s = 0.01$-, 0.05-, and 0.1-sec intervals to obtain, $x(n)$.

(a) For each T_s, plot $x(n)$.

(b) Reconstruct the analog signal $y_a(t)$ from the samples $x(n)$ by means of the sinc interpolation. (Use $\Delta t = 0.001$.) Estimate the frequency in $y_a(t)$ from your plot. (Ignore the end effects.)

(c) Reconstruct the analog signal $y_a(t)$ from the samples $x(n)$, using the cubic spline interpolation, and estimate the frequency in $y_a(t)$ from your plot. (Again, ignore the end effects.)

(d) Comment on your results.

Chapter 5

The Discrete Fourier Transform

The purpose of this chapter is to introduce the reader to the discrete Fourier transform (DFT) and its applications in practice. The DFT is an extremely important tool in the digital processing of discrete-time signals by discrete-time systems.

We recall that the discrete-time Fourier transform (DTFT) and the z-transform are two transform-domain representations for discrete-time signals and LTI systems. The z-transform provides a generalized frequency-domain (z-plane) representation for arbitrary discrete-time signals and systems (sequences), whereas the DTFT provides a frequency domain (ω) representation for absolutely summable sequences. These two transforms are defined for infinite-duration sequences in general and are continuous functions of their arguments. The DFT is a discrete-frequency representation of finite-duration sequences and, as a consequence, it becomes a powerful tool in the digital processing of discrete-time signals.

The reader should study the material in Chapters 4, 7, and 8 of the DSP book:

- The Discrete Fourier Series (DFS) — Section 4.2
- The Discrete Fourier Transform (DFT) — Sections 7.1 and 7.2
- Linear Convolution Using the DFT — Section 7.3
- The Fast Fourier Transform — Sections 8.1 and 8.2

Learning Objectives

- Become familiar with the computation of the DFS coefficients for a periodic sequence.
- Understand the relationship of the DFS coefficients to the DFT of an aperiodic sequence $x(n)$ that equals a single period of a periodic sequence.
- Understand the basic relationships between samples of the frequency-domain characteristics of signals and the corresponding time-domain representation.
- Become familiar with the computation of the DFT.
- Learn the properties of the DFT, and use them to simplify its computation.

- Understand the differences and the relationship between circular convolution and linear convolution of two sequences.
- Learn how to use the DFT to perform the linear convolution of two sequences.
- Learn how to perform block convolution with the DFT by employing the overlap-add and overlap-save methods.
- Understand the basic principles underlying the fast computation of the DFT.
- Learn the basic decimation operations involved in the decimation-in-time and decimation-in-frequency FFT algorithms.

5.1 The Discrete Fourier Series

In this section, we focus on a Fourier-series representation of a periodic discrete-time signal. A periodic sequence $\tilde{x}(n)$ with fundamental period N, where N is the smallest positive integer for which (5.1) holds, is defined as

$$\tilde{x}(n) = \tilde{x}(n + kN) \ , \quad \forall n, k \tag{5.1}$$

The frequency range for discrete-time signals is $-\pi \leq \omega \leq \pi$ or $0 \leq \omega \leq 2\pi$, and the fundamental frequency is $2\pi/N$, so there are N frequency harmonics in the range $0 \leq \omega \leq 2\pi$. As in the Fourier series representation of a periodic continuous-time signal, a periodic discrete-time sequence may be expressed as a linear combination of complex exponentials whose frequencies are harmonics of the fundamental frequency. Hence, the discrete Fourier series (DFS) for a periodic sequence $\tilde{x}(n)$ is given as

$$\tilde{x}(n) = \frac{1}{N} \sum_{k=0}^{N-1} \tilde{X}(k) e^{j\frac{2\pi}{N}kn} \ , \quad n = 0, \pm 1, \pm 2, \ldots \tag{5.2}$$

where N is the fundamental period and $\tilde{X}(k)$, $k = 0, \pm 1, \pm 2, \ldots$ are the coefficients of the Fourier series. These coefficients are given by the formula

$$\tilde{X}(k) = \sum_{n=0}^{N-1} \tilde{x}(n) e^{-j\frac{2\pi}{N}nk} \ , \quad k = 0, \pm 1, \pm 2, \ldots \tag{5.3}$$

In general, the Fourier-series coefficients $\tilde{X}(k)$ are complex-valued and periodic with fundamental period N, (i.e., $\tilde{X}(K + N) = \tilde{X}(k)$).

Given one period of a periodic sequence $\tilde{x}(n)$, one may use MATLAB to compute the DFS coefficients $\tilde{X}(k)$. The function for performing this computation is `dfs(xn,N)`. Conversely, given the DFS coefficients $\tilde{X}(k)$, the MATLAB function `idfs(Xk,N)` may be used to compute the one period of the periodic sequence.

The DFS coefficients are related in a simple way to the DTFT and to the z-transform of a finite-duration sequence that is equal to a single period of the periodic sequence. To be specific, let us define a finite-duration sequence of length N as

$$x(n) = \begin{cases} \tilde{x}(n) & , 0 \leq n \leq N-1 \\ 0 & , \text{otherwise} \end{cases} \tag{5.4}$$

where $\tilde{x}(n)$ is a periodic sequence of fundamental period N. The z-transform of $x(n)$ is

$$X(z) = \sum_{n=0}^{N-1} x(n)z^{-n} \tag{5.5}$$

and the DTFT of $x(n)$ is

$$X(\omega) = X(z)|_{z=e^{j\omega}} = \sum_{n=0}^{N-1} x(n)e^{-j\omega n} \tag{5.6}$$

Then, because $x(n) = \tilde{x}(n)$ for $0 \le n \le N-1$ and $x(n) = 0$ otherwise, it follows that the coefficients in the DFS representation of $\tilde{x}(n)$ are equal to the values of $X(z)$ evaluated at $z_k = e^{j2\pi k/N}, 0 \le k \le N-1$, and to the values of $X(\omega)$ evaluated at $\omega_k = 2\pi k/N, 0 \le k \le N-1$ — that is,

$$\begin{aligned} \tilde{X}(k) &= X(z)\,\big|_{z_k=e^{j2\pi k/N}} \quad , \ 0 \le k \le N-1 \\ &= X(\omega)\,\big|_{\omega_k=2\pi k/N} \ , \ 0 \le k \le N-1 \end{aligned} \tag{5.7}$$

5.2 Frequency Sampling and Reconstruction

Suppose we have a finite-duration sequence $x(n)$, $0 \le n \le N-1$. Its z-transform is

$$X(z) = \sum_{n=0}^{N-1} x(n)z^{-n} \tag{5.8}$$

and its DTFT is

$$X(\omega) = \sum_{n=0}^{N-1} x(n)e^{-j\omega n} \tag{5.9}$$

From Section 5.1, we observed that if we evaluate $X(z)$ or $X(\omega)$ at N equally spaced frequencies

$$z_k = e^{j\omega_k} = e^{j2\pi k/N} \ , \ k = 0, 1, ..., N-1 \tag{5.10}$$

then the values $\tilde{X}(k)$ that we obtain can be associated with the DFS coefficients of a periodic sequence $\tilde{x}(n)$ of period N, which is equal to $x(n)$ over a single period. It is demonstrated in the DSP book that the samples $\tilde{X}(k) = X(z_k), 0 \le k \le N-1$, completely determine $X(z)$, (i.e., $X(z)$ can be reconstructed from these N samples). The reconstruction formula is shown to be

$$X(z) = \frac{1-z^{-N}}{N} \sum_{k=0}^{N-1} \frac{\tilde{X}(k)}{1 - W_N^{-k}\, z^{-1}} \tag{5.11}$$

where W_N is defined as $W_N = \exp(-j2\pi/N)$. By evaluating (5.11) on the unit circle, (i.e., $z = \exp(j\omega)$), we obtain the corresponding reconstruction (interpolation) formula for the DTFT as

$$\begin{aligned} X(\omega) &= \frac{1-e^{-j\omega N}}{N} \sum_{k=0}^{N-1} \frac{\tilde{X}(k)}{1-e^{-j(\omega-2\pi k/N)}} \\ &= \sum_{k=0}^{N-1} \tilde{X}(k)\Phi\left(\omega - \frac{2\pi k}{N}\right) \end{aligned} \tag{5.12}$$

where the interpolation function $\Phi(\omega)$ is defined as

$$\Phi(\omega) \triangleq \frac{\sin(\omega N/2)}{N\sin(\omega/2)} e^{-j\omega(N-1)/2} \tag{5.13}$$

On the other hand, if the sequence $x(n)$ is of length M, where M may be finite or infinite, the coefficients $\tilde{X}(k), 0 \le k \le N$ obtained by sampling the z-transform of $x(n)$ at N equally spaced points on the unit circle, where $N < M$, specify a periodic sequence $\tilde{x}(n)$ that is related to the sequence $x(n)$ by the formula

$$\tilde{x}(n) = \sum_{r=-\infty}^{\infty} x(n - rN) \tag{5.14}$$

Hence, $\tilde{x}(n)$ is a linear combination of the original sequence and its infinite number of replicas each shifted by multiples of $\pm N$. In this case, the DFS coefficients do not represent uniquely the z-transform or the DTFT of the original sequence. Instead, the DFS coefficients represent the periodic sequence $\tilde{x}(n)$ given by (5.14).

5.3 The Discrete Fourier Transform and Its Properties

In Section 5.2, we observed that if we have a finite-duration sequence $x(n)$ of length N, its DTFT $X(\omega)$ evaluated at N equally spaced frequencies $\omega_k = 2\pi k/N$, $0 \le k \le N-1$, uniquely represents the DTFT. That is, the DTFT can be reconstructed from the N samples $\tilde{X}(k), 0 \le k \le N-1$, by using the interpolation formula in (5.12). By associating the N samples $\tilde{X}(k), 0 \le k \le N-1$, as the coefficients in a DFS representation of a periodic sequence $\tilde{x}(n)$, we observed that

$$x(n) = \begin{cases} \tilde{x}(n), & 0 \le n \le N-1 \\ 0, & \text{otherwise} \end{cases} \tag{5.15}$$

Thus, the N frequency samples $\tilde{X}(k)$ uniquely determine the finite duration sequence.

These relationships lead us to define the Discrete Fourier Transform (DFT) as the N equally spaced frequency samples of the DTFT of a finite-duration sequence of length N. That is, the DFT of $x(n)$ is defined as

$$\begin{aligned} X(\omega_k) \triangleq X(k) &= \sum_{n=0}^{N-1} x(n)e,^{-j2\pi nk/N} \;,\; k = 0, 1, ..., N-1 \\ &= \sum_{n=0}^{N-1} x(n)\, W_N^{nk} \end{aligned} \tag{5.16}$$

From our previous discussion, it follows that

$$X(k) = \begin{cases} \tilde{X}(k), & 0 \le k \le N-1 \\ 0, & \text{otherwise} \end{cases} \tag{5.17}$$

As a result of the relationships given by (5.15) and (5.17), it follows that the inverse DFT is given as

$$x(n) = \frac{1}{N} \sum_{k=0}^{N-1} X(k) W_N^{-nk} \;,\; 0 \le n \le N-1 \tag{5.18}$$

The MATLAB functions `dft` and `idft` compute the DFT and inverse DFT, respectively.

The DFT and the inverse DFT given by (5.16) and (5.18), respectively, can be expressed in matrix form. If we express the sequence $x(n)$ as an N-point vector $\mathbf{x}$ and its DFT by the N-point vector $\mathbf{X}$, then

$$\begin{aligned} \mathbf{X} &= \mathbf{W}_N \mathbf{x} \\ \mathbf{x} &= \frac{1}{N}\mathbf{W}_N^* \mathbf{X} \end{aligned} \tag{5.19}$$

where $\mathbf{W}_N$ is an $N \times N$ matrix whose elements are powers of $\exp(-j2\pi/N)$, as defined in the DSP book, and $\mathbf{W}_N^*$ is the complex conjugate of $\mathbf{W}_N$.

As we have observed above, the DFT of an N-point sequence $x(n)$ is the minimum number of frequency samples that are required to represent the sequence $x(n)$ uniquely in the frequency domain. However, the DFT does not provide a sufficiently detailed view of the frequency content of the sequence $x(n)$. Such a detailed view can be obtained by oversampling the DTFT of $x(n)$ by some multiple of N. For example, if we oversample the DTFT by a factor of two, at frequencies $\omega_k = \pi k/N$, $k = 0, 1, ..., 2N - 1$, the corresponding frequency samples $X(k)$, $0 \le k \le 2N - 1$, are associated with a sequence of length $2N$, where the first N elements of the sequence are identical with $x(n)$ and the remaining elements are zero. Therefore, if we wish to obtain a more detailed view of the spectrum of a sequence $x(n)$ of length N, we first construct a longer sequence by padding $x(n)$ with zeros. Thus, we obtain a new sequence $x_1(n)$, of length M, defined as

$$x_1(n) = \begin{cases} x(n), & 0 \le n \le N-1 \\ 0, & N < n \le M-1 \end{cases}$$

Then, we compute the M-point DFT of the sequence $x_1(n)$. This M-point DFT $X(k)$, $0 \le k \le M - 1$, represents a more dense sampling of the DTFT of $x(n)$ than does the N-point DFT.

Study Topics: Important Properties of the DFT

1. *Linearity*: If $x_1(n)$ and $x_2(n)$ are two N-point sequences,

   $$\text{DFT}[ax_1(n) + bx_2(n)] = a\,\text{DFT}[x_1(n)] + b\,\text{DFT}[x_2(n)]$$

   where a and b are constants.

2. *Circular folding*: If $x(n)$ is an N-point sequence with DFT $X(k)$, the DFT of the circularly folded sequence, defined as

   $$x((-n))_N = \begin{cases} x(0), & n = 0 \\ x(N-n), & 1 \le n \le N-1 \end{cases}$$

   is

   $$X((-k))_N = \begin{cases} X(0), & k = 0 \\ X(N-k), & 1 \le k \le N-1 \end{cases}$$

   $X((-k))_N$ is called the circularly folded version of $X(k)$.

3. *Conjugation*:

   $$\text{DFT}[x^*(n)] = X^*((-k))_N$$

4. *Circular shift of a sequence*: A circular shift of an N-point sequence $x(n)$ is denoted as $x((N - m))_N R_N(n)$, where $R_N(n)$ is defined by

$$R_N(n) = \begin{cases} 1, & 0 \leq n \leq N-1 \\ 0, & \text{otherwise} \end{cases}$$

The DFT of $x((n - m))_N R_N(n)$ is

$$\text{DFT}[x((n-m))_N R_N(n)] = W_N^{km} X(k)$$

5. *Circular shift in frequency*:

$$\text{DFT}[W_N^{-ln} x(n)] = X((k-l))_N R_N(n)$$

6. *Circular convolution*: If $x_1(n)$ and $x_2(n)$ are two finite-duration sequences of length N, the N-point circular convolution of $x_1(n)$ and $x_2(n)$ is defined as

$$x_1(n) \circledN x_2(n) \triangleq \sum_{m=0}^{N-1} x_1(m) x_2((n-m))_N \ , \ 0 \leq 1 \leq N-1$$

The resulting sequence $x_3(n) = x_1(n) \circledN x_2(n)$ is an N-point sequence whose N-point DFT $x_3(k)$ is the product of the DFTs of $x_1(n)$ and $x_2(n)$. That is,

$$\text{DFT}[x_1(n) \circledN x_2(n)] = X_1(k) X_2(k)$$

Therefore, the multiplication of two N-point DFTs corresponds to the circular convolution of their respective sequences in the time domain. The function `circonvt` may be used to perform circular convolution directly in the time domain.

7. *Multiplication of two sequences*: If $x_1(n)$ and $x_2(n)$ are two N-point sequences, the DFT of their product is given as

$$\text{DFT}[x_1(n)\, x_2(n)] = \frac{1}{N} X_1(k) \circledN X_2(k)$$

This property is basically the dual of the preceding property.

8. *Parseval's relation*: This property relates the energy of a finite-duration sequence to its DFT. It is given as

$$\sum_{n=0}^{N-1} |x(n)|^2 = \frac{1}{N} \sum_{k=0}^{N-1} |X(k)|^2$$

5.4 Linear Convolution Using the DFT

In practice, we are often faced with the problem of processing some signal sequence $x(n)$ by an LTI system (filter). Usually, this processing is performed on some general-purpose digital signal processor or on a digital computer. In many cases, it is computationally more efficient to perform this processing by using the DFT. We know that the output of an LTI system to some input sequence $x(n)$ is simply the linear convolution of the input sequence with the impulse response of the system. If the LTI system has a finite-duration impulse response (FIR filter) and the input sequence has a finite duration, we would like to compute the DFT of the two sequences, multiply the two DFTs, and perform the inverse DFT to obtain the system response. However, we have observed in the preceding section that the result of multiplying two DFTs and performing the inverse DFT is equivalent to performing a circular convolution. The question is this: How can we modify these computations so that the result obtained from circular convolution is identical to the result from linear convolution?

The answer is relatively simple. If the impulse response of the FIR filter is a sequence of length N_1, and the input sequence has length N_2, the output sequence will have a length $N = N_1 + N_2 - 1$. Then, as shown in the DSP book, if we increase the length of both the impulse response and the input sequence to N points by padding each sequence with zeros, and perform the N-point DFTs of the resulting sequences, the N-point inverse DFT of their product will be the output sequence that would result from the linear convolution of the two sequences. Thus, the result of performing the N-point circular convolution of the two sequences padded with zeros is identical to the result obtained from linear convolution.

In the case where the input sequence $x(n)$ is extremely long and exceeds the memory of the digital processor that stores $x(n)$, we can exploit the linearity property of the filter. That is, we subdivide the input sequence into smaller blocks and process each block separately, using the DFTs of these smaller blocks, multiplying each DFT by the DFT of the filter impulse, and performing the inverse DFT. Then, we linearly combine the filter responses from each block of input data. There are two ways of dealing with this linear combination of blocks, as described in the DSP book. One is called the overlap-save method and the other is called the overlap-add method. The MATLAB functions for performing these computations are `ovrlsav` and `ovrladd`.

5.5 The Fast Fourier Transform (FFT) Algorithms

The DFT and inverse DFT of an N-point sequence $x(n)$ are given as

$$X(k) = \sum_{n=0}^{N-1} x(n) W_N^{nk} \ , \ 0 \le k \le N-1 \tag{5.20}$$

$$x(n) = \frac{1}{N} \sum_{k=0}^{N-1} X(k) W_N^{-nk} \ , \ 0 \le n \le N-1 \tag{5.21}$$

where $W_N = \exp(-j2\pi/N)$. To compute the DFT via the formula in (5.20) requires N^2 complex multiplications and $N(N-1) \approx N^2$ complex additions. However, it is possible to reduce the number of arithmetic operations by exploiting the following properties of the exponential factors W_N^{nk}.

$$\text{periodicity property:} \quad W_N^{nk} = W_N^{k(n+N)} = W_N^{(k+N)n}$$

$$\text{symmetry property:} \quad W_N^{kn+N/2} = W_N^{N/2} W_N^{kn} = -W_N^{kn}$$

By exploiting these periodicity properties, it is possible to perform the computation of the DFT with only $(N/2)\log_2 N$ complex multiplications, where N is a power of 2. Two types of relevant algorithms are described in the DSP book; one type is called *decimation-in-time* algorithm, and the second type is called *decimation-in-frequency* algorithm. These two types of algorithms are also applicable to the computation of the inverse DFT in (5.21).

An FFT algorithm is especially efficient in processing signals through FIR filters. Let us suppose that we have an FIR filter of length M and an input sequence of length N_1. If the output of the filter is computed directly by using the convolution formula

$$y(n) = \sum_{m=0}^{M-1} h(m)x(n-m)\ ,\ n = 0, 1, ...$$

each output point $y(n)$ requires M (real) multiplications and $M - 1$ (real) additions. If M and N_1 are large, say $M,\ N_1 > 100$, it is computationally more efficient to use the DFT method to compute the output sequence $y(n)$, where the FFT algorithm is used to compute the DFT. Thus, with appropriate zero padding of $h(m)$ and $x(m)$, we can use an FFT algorithm to compute the N-point $(N \geq M + N_1 - 1)$ DFTs of $h(m)$ and $x(m)$, multiply the two DFTs together and perform the inverse DFT, also by using an FFT algorithm. If the length N_1 of the input sequence is very large, we can subdivide the input sequence into blocks, perform the above computations for each block and, finally, use either the overlap-save or the overlap-add method to obtain the filter output for the entire sequence.

The MATLAB functions `fft` and `ifft` implement the efficient computation of the DFT and inverse DFT, using an FFT algorithm.

5.6 Summary

MATLAB Functions

The following MATLAB functions are used in this chapter:

The Discrete Fourier Series

- `[Xk] = dfs(xn,N)`: Computes the DFS coefficients of the periodic sequence `xn` of period `N`
- `[xn] = idfs(Xk,N)`: Computes the periodic sequence from the DFS coefficients

The Discrete Fourier Transform and Its Properties

- `m = rem(n,N)`: Computes the remainder after dividing `n` by `N`
- `[Xk] = dft(xn,N)`: Computes the DFT of the `N`-point sequence `xn`
- `[xn] = idft(Xk,N)`: Computes the inverse DFT to yield the `N`-point sequence `xn`
- `[y] = x(mod(-n,N)+1)`: Circular folding of an `N` -point sequence
- `[xec,xoc] = circevod(x)`: Signal decomposition into circular-even and circular-odd parts
- `[y] = cirshftt(x,m,N)`: Circular shift of `m` samples with size `N` in sequence `x`
- `[y] = circonvt(x1,x2,N)`: `N`-point circular convolution between `x1` and `x2`

Linear Convolution Using the DFT

- `[y] = ovrlpsav(x,k,N)`: Overlap-Save Method of block convolution
- `[y] = ovrlpadd(x,k,N)`: Overlap-Add Method of block convolution

The Fast Fourier Transform Algorithm

- `[X] = fft(x,N)`: Computation of the N-point DFT via the FFT Algorithms
- `[x] = ifft(X,N)`: Computation of the N-point IDFT via the FFT Algorithm

5.7 Problems

P5.1 Determine the DFS coefficients of the following periodic sequences by using the DFS definition, and verify them with MATLAB.

(a) $\tilde{x}_1(n) = \{2, 0, 2, 0\}$, $N = 4$

(b) $\tilde{x}_2(n) = \{0, 0, 1, 0, 0\}$, $N = 5$

(c) $\tilde{x}_3(n) = \{3, -3, 3, -3\}$, $N = 4$

(d) $\tilde{x}_4(n) = \{j, j, -j, -j\}$, $N = 4$

(e) $\tilde{x}_5(n) = \{1, j, j, 1\}$, $N = 4$

P5.2 Use the IDFS definition to determine the periodic sequences, given the following periodic DFS coefficients, then verify your results by using MATLAB:

(a) $\tilde{X}_1(k) = \{5, -2j, 3, 2j\}$, $N = 4$

(b) $\tilde{X}_2(k) = \{4, -5, 3, -5\}$, $N = 4$

(c) $\tilde{X}_3(k) = \{1, 2, 3, 4, 5\}$, $N = 5$

(d) $\tilde{X}_4(k) = \{0, 0, 2, 0\}$, $N = 4$

(e) $\tilde{X}_5(k) = \{0, j, -2j, -j\}$, $N = 4$

P5.3 Let $\tilde{x}_1(n)$ be periodic with fundamental period $N = 50$, where one period is given by

$$\tilde{x}_1(n) = \begin{cases} ne^{-0.3n}, & 0 \leq n \leq 25 \\ 0, & 26 \leq n \leq 49 \end{cases}$$

and let $\tilde{x}_2(n)$ be periodic with fundamental period $N = 100$, where one period is given by

$$\tilde{x}_2(n) = \begin{cases} ne^{-0.3n}, & 0 \leq n \leq 25 \\ 0, & 26 \leq n \leq 99 \end{cases}$$

These two periodic sequences differ in their periodicity, but otherwise have equal nonzero samples.

(a) Find the DFS $\tilde{X}_1(k)$ of $\tilde{x}_1(n)$, and plot samples (using the `stem` function) of its magnitude and angle versus k.

(b) Find the DFS $\tilde{X}_2(k)$ of $\tilde{x}_2(n)$, and plot samples of its magnitude and angle versus k.

(c) What is the difference between the two DFS plots?

P5.4 Consider the periodic sequence $\tilde{x}_1(n)$ given in Problem P5.3. Let $\tilde{x}_3(n)$ be periodic with period 100, obtained by concatenating two periods of $\tilde{x}_1(n)$; that is,

$$\tilde{x}_3(n) = [\tilde{x}_1(n), \tilde{x}_2(n)]_{\text{PERIODIC}}$$

Clearly, $\tilde{x}_3(n)$ is different from $\tilde{x}_2(n)$ of Problem P5.4 even though both of them are periodic with period 100.

(a) Find the DFS $\tilde{X}_3(k)$ of $\tilde{x}_3(n)$, and plot samples of its magnitude and angle versus k.

(b) What effect does the periodicity doubling have on the DFS?

(c) Generalize the above result to M-fold periodicity. In particular, show that if

$$\tilde{x}_M(n) = \Bigg[\underbrace{\tilde{x}_1(n), \ldots, \tilde{x}_2(n)}_{M \text{ times}}\Bigg]_{\text{PERIODIC}}$$

then

$$\begin{aligned} \tilde{X}_M(Mk) &= M\tilde{X}_1(k), & k &= 0, 1, \ldots, N-1 \\ \tilde{X}_M(k) &= 0, & k &\neq 0, M, \ldots, MN \end{aligned}$$

P5.5 Let $X(\omega)$ be the DTFT of a 10-point sequence

$$x(n) = \{2, 5, 3, -4, -2, 6, 0, -3, -3, 2\}$$

(a) Let

$$y_1(n) = \overset{\text{3-point}}{\text{IDFT}} \, [X(0), X(2\pi/3), X(4\pi/3)]$$

Determine $y_1(n)$ by using the frequency-sampling theorem. Verify your answer with MATLAB.

(b) Let

$$y_2(n) = \overset{\text{20-point}}{\text{IDFT}} \left[X(0), X(e^{2\pi/20}), X(e^{4\pi/20}), \ldots, X(e^{2\pi(19)/20})\right]$$

Determine $y_2(n)$ by using the frequency-sampling theorem. Verify your answer with MATLAB.

P5.6 A 12-point sequence is $x(n)$ defined as

$$x(n) = \{1, 2, 3, 4, 5, 6, 6, 5, 4, 3, 2, 1\}$$

(a) Determine the DFT $X(k)$ of $x(n)$. Plot (using the `stem` function) its magnitude and phase.

(b) Plot the magnitude and phase of the DTFT $X(\omega)$ of $x(n)$, using MATLAB.

(c) Verify that this DFT is the sampled version of $X(\omega)$. It might be helpful to combine the two plots in one graph by using the `hold` function.

(d) Is it possible to reconstruct the DTFT $X(\omega)$ from the DFT $X(k)$? If it is possible, give the necessary interpolation formula for reconstruction. If not, state why this reconstruction cannot be done.

P5.7 Plot the DTFT magnitudes of the following sequences, using the DFT as a computation tool. Make an educated guess about the length N so that your plots will be meaningful.

(a) $x_1(n) = 2\cos(0.2\pi n)\,[u(n) - u(n-10)]$

(b) $x_2(n) = \sin(0.45\pi n)\sin(0.55\pi n)\,,\ 0 \le n \le 50$

(c) $x_3(n) = 3\,(2)^n\,,\ -10 \le n \le 10$

(d) $x_5(n) = 5\left(0.9e^{j\pi/4}\right)^n u(n)$

P5.8 Let $H(\omega)$ be the frequency response of a real, causal, discrete-time LSI system.

(a) If

$$\Re e\{H(\omega)\} = \sum_{k=0}^{5}(0.5)^k\cos(k\omega)$$

determine the impulse response $h(n)$ analytically. Verify your answer by using IDFT as a computation tool. Choose the length N judiciously.

(b) If

$$\Im m\{H(\omega)\} = \sum_{\ell=0}^{5}2\ell\sin(\ell\omega)\,,\ \text{and}\ \int_{-\pi}^{\pi}H(\omega)d\omega = 0$$

determine the impulse response $h(n)$ analytically. Verify your answer by using IDFT as a computation tool. Again, choose the length N judiciously.

P5.9 Let $X(k)$ denote the N-point DFT of an N-point sequence $x(n)$. The DFT $X(k)$ itself is an N-point sequence.

(a) If the DFT of $X(k)$ is computed to obtain another N-point sequence $x_1(n)$, show that

$$x_1(n) = Nx\left((-n)\right)_N\,,\quad 0 \le n \le N-1$$

(b) Using the property derived in part (a), design a MATLAB function to implement an N-point circular folding operation $x_2(n) = x_1\left((-n)\right)_N$. The format should be

```
x2 = circfold(x1,N)
% Circular folding using DFT
% x2 = circfold(x1,N)
% x2 = circularly folded output sequence
% x1 = input sequence of length <= N
%  N = circular buffer length
```

(c) Determine the circular folding of the following sequence:

$$x_1(n) = \{1, 2, 3, 4, 5, 6, 6, 5, 4, 3, 2, 1\}$$

P5.10 Complex-valued N-point sequences are decomposed into N-point even and odd sequences by using the following relations

$$\begin{aligned} x_{ec}(n) &\triangleq \tfrac{1}{2}\left[x(n) + x^*((-n))_N\right] \\ x_{oc}(n) &\triangleq \tfrac{1}{2}\left[x(n) - x^*((-n))_N\right] \end{aligned}$$

Then

$$\begin{aligned} \text{DFT}\left[x_{ec}(n)\right] &= \Re e\left[X(k)\right] = \Re e\left[X((-k))_N\right] \\ \text{DFT}\left[x_{oc}(n)\right] &= \Im m\left[X(k)\right] = \Im m\left[X((-k))_N\right] \end{aligned}$$

(a) Prove the preceding property analytically.

(b) Modify the `circevod` function developed in the chapter so that it can be used for complex-valued sequences.

(c) Verify the foregoing symmetry property and your MATLAB function on the following sequence:

$$x(n) = \left(0.9e^{j\pi/3}\right)^n \left[u(n) - u(n-20)\right]$$

P5.11 The first five values of the eight-point DFT of a real-valued sequence $x(n)$ are given by

$$\{0.25, 0.125 - j0.3, 0, 0.125 - j0.06\}$$

Compute the DFT of each of the following sequences, using properties.

(a) $x_1(n) = x((2-n))_8$

(b) $x_2(n) = x((n+5))_{10}$

(c) $x_3(n) = x^2(n)$

(d) $x_4(n) = x(n) \textcircled{8} x((-n))_8$

P5.12 If $X(k)$ is the DFT of an N-point complex-valued sequence

$$x(n) = x_R(n) + jx_I(n)$$

where $x_R(n)$ and $x_I(n)$ are the real and imaginary parts of $x(n)$, then

$$\begin{aligned} X_R(k) &\triangleq \text{DFT}\left[x_R(n)\right] = X_{ec}(k) \\ X_I(k) &\triangleq \text{DFT}\left[x_I(n)\right] = X_{oc}(k) \end{aligned}$$

where $X_{ec}(k)$ and $X_{oc}(k)$ are the circular-even and circular-odd components of $X(k)$, as defined in Problem P5.10.

(a) Prove this property analytically.

(b) The property can be utilized to compute the DFTs of two real-valued N-point sequences by using one N-point DFT operation. Specifically, let $x_1(n)$ and $x_2(n)$ be two N-point sequences. Then we can form a complex-valued sequence

$$x(n) = x_1(n) + jx_2(n)$$

and use the foregoing property. Develop a MATLAB function to implement this approach with the following format:

```
function [X1,X2] = real2dft(x1,x2,N)
% DFTs of two real sequences
% [X1,X2] = real2dft(x1,x2,N)
%  X1 = n-point DFT of x1
%  X2 = n-point DFT of x2
%  x1 = sequence of length <= N
%  x2 = sequence of length <= N
%   N = length of DFT
```

(c) Compute the DFTs of the following two sequences:

$$x_1(n) = \cos(0.25\pi n), \; x_2(n) = \sin(0.75\pi n); \quad 0 \le n \le 63$$

P5.13 Using the frequency-domain approach, develop a MATLAB function to determine a circular shift $x((n-m))_N$, given an N_1-point sequence $x(n)$ where $N_1 \le N$. Your function should have the following format:

```
function y = cirshftf(x,m,N)
%
%function y=cirshftf(x,m,N)
%
% Circular shift of m samples wrt size N in sequence x: (freq domain)
% -----------------------------------------------------------------
%       y : output sequence containing the circular shift
%       x : input sequence of length <= N
%       m : sample shift
%       N : size of circular buffer
%
% Method: y(n) = idft(dft(x(n))*WN^(mk))
%
%  If m is a scalar then y is a sequence (row vector)
%  If m is a vector then y is a matrix, each row is a circular shift
%      in x corresponding to entries in vector m
%  M and x should not be matrices
```

Verify your function on the sequence

$$x_1(n) = 11 - n, \; 0 \le n \le 10$$

with $m = 10$ and $N = 15$.

P5.14 Using the analysis and synthesis equations of the DFT, show that

$$\sum_{n=0}^{N-1} |x(n)|^2 = \frac{1}{N} \sum_{k=0}^{N-1} |X(k)|^2$$

This is commonly referred to as a Parseval's relation for the DFT. Verify this relation by using MATLAB on the sequence Problem P5.9.

P5.15 Using the frequency-domain approach, develop a MATLAB function to implement the circular convolution operation between two sequences. The format of the sequence should be

```
function x3 = circonvf(x1,x2,N)
% Circular convolution in the frequency domain
%  x3 = circonvf(x1,x2,N)
%  x3 = convolution result of length N
%  x1 = sequence of length <= N
%  x2 = sequence of length <= N
%   N = length of circular buffer
```

P5.16 The `circonvt` function developed in this chapter implements the circular convolution as a matrix-vector multiplication. The matrix corresponding to the circular shifts $\{x\left((n-m)\right)_N ; 0 \leq n \leq N-1\}$ has an interesting structure. This matrix is called a *circulant* matrix, which is a special case of the Toeplitz matrix (introduced in Chapter 2).

(a) Consider the following sequences:

$$x_1(n) = \{1, 2, 2\} \text{ and } x_2(n) = \{1, 2, 3, 4\}$$

Express $x_1(n)$ as a column vector $\mathbf{x}_1$ and $x_2\left((n-m)\right)_N$ as a matrix $\mathbf{X}_2$ with rows corresponding to $n = 0, 1, 2, 3$. Characterize this matrix $\mathbf{X}_2$. Can it completely be described by its first row (or column)?

(b) Evaluate the circular convolution as $\mathbf{X}_2\mathbf{x}_1$, and verify your calculations.

P5.17 Develop a MATLAB function to construct a circulant matrix $\mathbf{C}$ when given an N-point sequence $x(n)$. Use the `cirshftf` function developed in Problem P5.15. Your subroutine function should have the following format:

```
function [C] = circulnt(x,N)
% Circulant Matrix from a sequence
% [C] = circulnt(x,N)
% C = circulant matrix of size NxN
% x = sequence of length <= N
% N = size of circulant matrix
```

Using this function, modify the circular convolution function `circonvt` discussed in the chapter.

P5.18 Compute the N-point circular convolution for the following sequences:

(a) $x_1(n) = \{1, 1, 1, 1\}, \; x_2(n) = \cos(\pi n/4)\,\mathcal{R}_N(n); \quad N = 8$

(b) $x_1(n) = n\mathcal{R}_N(n), \; x_2(n) = (N-n)\,\mathcal{R}_N(n); \quad N = 10$

P5.19 For the following sequences, compute (i) the N-point circular convolution $x_3(n) = x_1(n) \circledN x_2(n)$, (ii) the linear convolution $x_4(n) = x_1(n) * x_2(n)$, and (iii) the error sequence $e(n) = x_3(n) - x_4(n)$.

(a) $x_1(n) = \cos(2\pi n/N)\,\mathcal{R}_{16}(n)$, $x_2(n) = \sin(2\pi n/N)\,\mathcal{R}_{16}(n)$; $\quad N = 32$

(b) $x_1(n) = \{1, -1, 1, -1\}$, $x_2(n) = \{1, 0, -1, 0\}$; $\quad N = 5$

In each case, verify that $e(n) = x_4(n + N)$.

5.20 Let

$$\begin{aligned} x(n) &= \sum_{m=0}^{10} (0.9)^m \cos(0.1\pi\, m\, n), \; 0 \le n \le 10^6, \\ h(n) &= \cos(0.5\pi n), \; 0 \le n \le 100 \end{aligned}$$

(a) Using the conv function determine the output sequence $y(n) = x(n) * y(n)$.

(b) Use the overlap-and-save method of block convolution along with the FFT algorithm to implement high-speed block convolution. Using this approach determine $y(n)$ with FFT sizes of 1024 and 2048.

(d) Compare the above three approaches in terms of the convolution results and their execution times.

Chapter 6

Realization of Digital Filters

Before embarking on the digital-filter design, we first discuss the structural representation of digital filters as interconnected basic building blocks. This issue of filter structures is important because the design of digital filters is influenced not only by such factors as the type of the filter (i.e., IIR or FIR) but also by the form of its hardware or software implementation (or structures). Different filter structures dictate different design strategies.

The reader should study the following material in Chapter 9 of the DSP book:

- IIR Filter Structures — Sections 9.1 and 9.3
- FIR Filter Structures — Section 9.2
- Lattice Filter Structures — Sections 9.3.4 and 9.3.5

Several basic building blocks that are used in the filter implementation are described with various forms of IIR, FIR, and lattice filter structures. Study carefully the MATLAB functions that are discussed in the text to implement these structures. Review topics are given below.

Learning Objectives

- Learn the characteristics and the input/output relations of the basic building blocks.
- Develop skills to transform difference equations into direct form structures for both IIR and FIR filters.
- Understand the analytical steps needed to transform a system function $H(z)$ into cascade, parallel, and lattice/ladder forms, and vice versa, for IIR filters.
- Understand the analytical steps needed to transform a system function $H(z)$ into cascade, frequency-sampling, and lattice forms, and vice versa, for FIR filters.
- Learn how to use MATLAB's conversion functions — that is, how to map array values into block-diagram structures.
- Learn how to use MATLAB's `filter` implementation functions.

6.1 IIR Filter Structures

IIR filters are characterized by infinite-duration impulse responses. In this section, we focus on IIR filters that are characterized by rational system functions or (equivalently) by difference equations.

Study Topics: IIR Filter Realizations

1. We need three elements to describe digital filter structures. These elements are (a) the adder that has two inputs and one output, shown in Figure 6.1(a); (b) the multiplier that is a single-input single-output element, shown in Figure 6.1(b); and (c) the unit that delays the signal passing through it by one sample, shown in Figure 6.1(c).

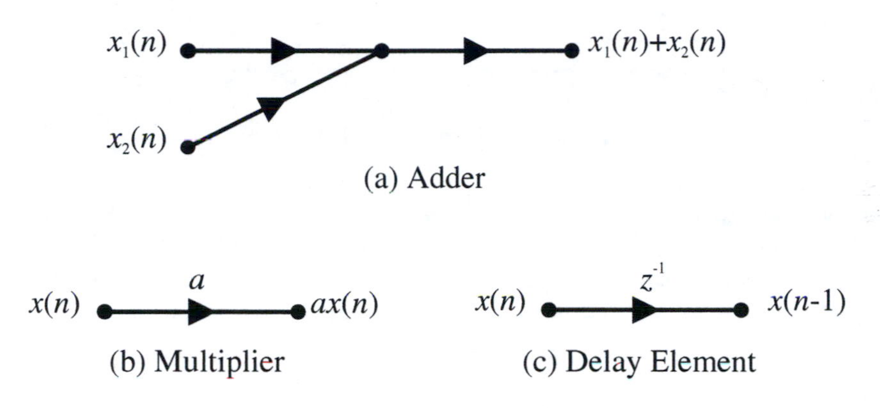

Figure 6.1: Three basic elements

2. The system function of a causal infinite-duration impulse response (IIR) filter is given by

$$H(z) = \frac{B(z)}{A(z)} = \frac{\sum_{n=0}^{M} b_n z^{-n}}{\sum_{n=0}^{N} a_n z^{-n}} = \frac{b_0 + b_1 z^{-1} + ... + b_M z^{-M}}{1 + a_1 z^{-1} + ... + a_N z^{-N}}; \; a_0 = 1 \quad (6.1)$$

   where b_n and a_n are the coefficients of the filter. We have assumed, without loss of generality, that $a_0 = 1$. The order of such an IIR filter is called N if $a_N \neq 0$. The difference equation representation of an IIR filter is expressed as

$$y(n) = \sum_{m=0}^{M} b_m x(n-m) - \sum_{m=1}^{N} a_m y(n-m) \quad (6.2)$$

3. When the difference equation (6.2) is implemented by using delays, multipliers, and adders in which the multiplier coefficients are actually the coefficients of the system function, we obtain the *direct-form structure*. Figure 6.2 shows one form, called the *Direct Form I* structure, for $M = N = 4$.

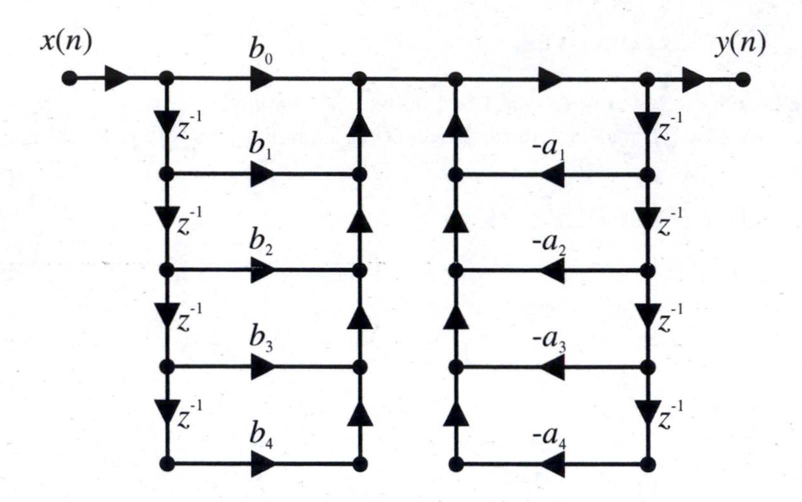

Figure 6.2: Direct Form I structure

4. There are two separate delay lines in the direct-form structure in Figure 6.2. We can eliminate one delay line by a simple block-diagram manipulation that leads to a canonical structure, called the *Direct Form II* structure. It is shown in Figure 6.3. In MATLAB, the direct-form structure is implemented by the `filter` function that was discussed in Chapter 2.

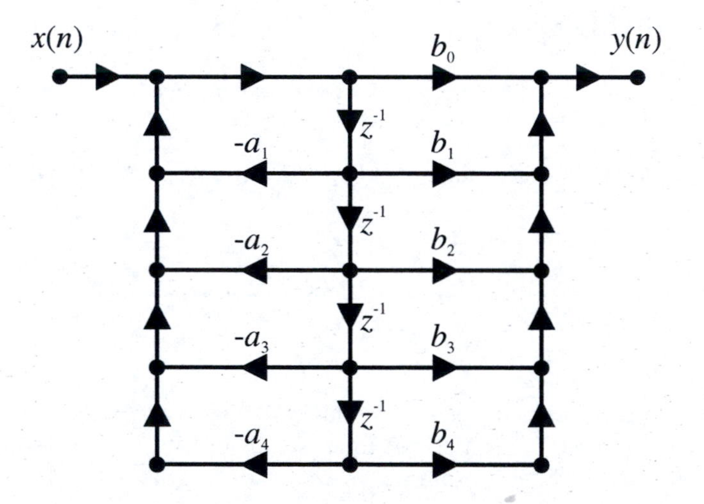

Figure 6.3: Direct Form II structure

5. By expressing the system function $H(z)$ as a product of second-order sections with real coefficients, we can realize a filter as a cascade of these lower-order sections. In this case, $H(z)$ is expressed as (assuming $N \geq M$)

$$H(z) = b_0 \prod_{k=1}^{\lceil \frac{N}{2} \rceil} \frac{1 + B_{k,1} z^{-1} + B_{k,2} z^{-2}}{1 + A_{k,1} z^{-1} + A_{k,2} z^{-2}} \tag{6.3}$$

where $B_{k,1}$, $B_{k,2}$, $A_{k,1}$, and $A_{k,2}$ are real numbers representing the coefficients of the second-order sections and $\lceil\cdot\rceil$ is the `ceil` operation. The second-order section

$$H_k(z) = \frac{1 + B_{k,1}z^{-1} + B_{k,2}z^{-2}}{1 + A_{k,1}z^{-1} + A_{k,2}z^{-2}};\ k = 1, \dots, K$$

is called the k th *biquad* section and can be implemented in Direct Form II. The entire filter is then implemented as a cascade of biquads called the *Cascade Form* structure. Figure 6.4 shows a cascade-form structure for a fourth-order IIR filter. The conversion of a direct-form to a cascade-form structure is obtained by the function `dir2cas`. Similarly, `cas2dir` can be used to obtain the direct form from the cascade form.

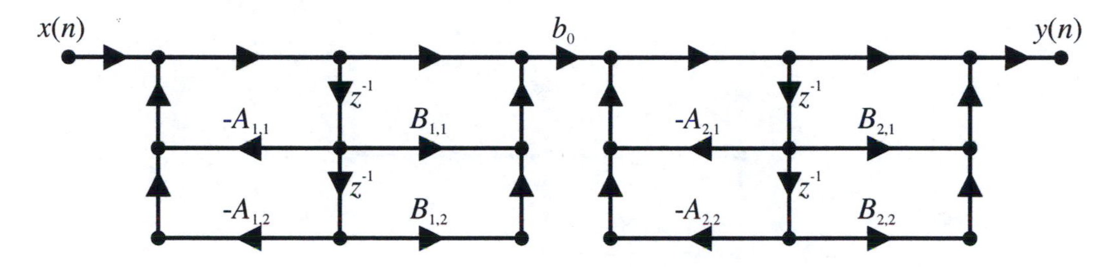

Figure 6.4: Cascade-form structure

6. When the system function $H(z)$ is expressed as a sum of second-order sections by using Partial Fraction Expansion (PFE), the resulting $H(z)$ can be implemented as a *Parallel Form* structure. In this case, $H(z)$ is expressed as

$$H(z) = \sum_{k=1}^{\lceil \frac{N}{2} \rceil} \frac{B_{k,0} + B_{k,1}z^{-1}}{1 + A_{k,1}z^{-1} + A_{k,2}z^{-2}} + \underbrace{\sum_{0}^{M-N} C_k z^{-k}}_{\text{only if } M \geq N} \tag{6.4}$$

where $B_{k,0}$, $B_{k,1}$, $A_{k,1}$, and $A_{k,2}$ are real numbers representing the coefficients of second-order sections. The second-order section

$$H_k(z) = \frac{B_{k,0} + B_{k,1}z^{-1}}{1 + A_{k,1}z^{-1} + A_{k,2}z^{-2}};\ k = 1, \cdots, K$$

is the kth proper-rational biquad section that can be implemented in Direct Form II. By summation of subsections, a parallel structure can be built to realize $H(z)$, as shown in Figure 6.5 for a fourth-order IIR filter.

The conversion of a direct-form to a parallel-form structure is obtained by the function `dir2par`. Similarly, `par2dir` can be used to obtain the direct form from the parallel form.

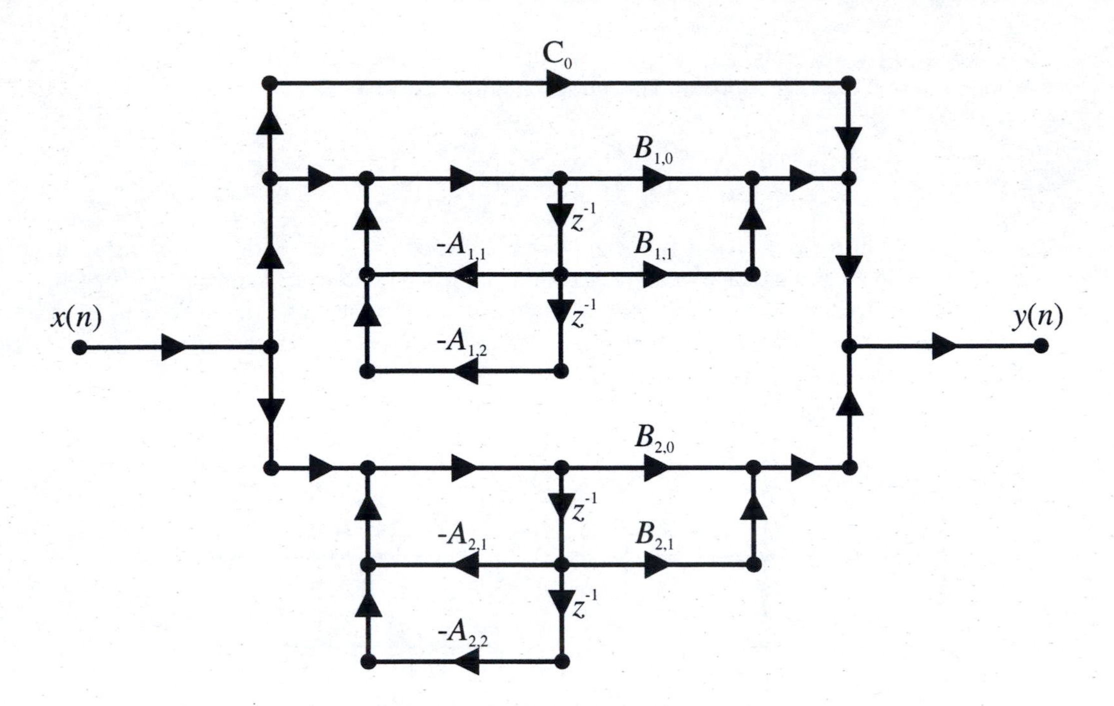

Figure 6.5: Parallel-form structure

6.2 FIR Filter Structures

FIR filters are characterized by finite-duration impulse responses. Signal processing with an FIR filter can be accomplished in the frequency domain by using the DFT as described in the preceding chapter or directly in the time domain. In this section, we focus on FIR filters that perform the signal processing in the time domain.

Study Topics: FIR Filter Realizations

1. The system function of a causal finite-duration impulse response (FIR) filter is given by

$$H(z) = b_0 + b_1 z^{-1} + \cdots + b_{M-1} z^{1-M} = \sum_{n=0}^{M-1} b_n z^{-n} \tag{6.5}$$

   The order of the filter is $M - 1$, and the *length* of the filter (which is equal to the number of coefficients) is M. The difference-equation representation is

$$y(n) = b_0 x(n) + b_1 x(n-1) + \cdots + b_{M-1} x(n - M + 1) \tag{6.6}$$

   Hence, the impulse response $h(n)$ is

$$h(n) = \begin{cases} b_n & 0 \le n \le M-1 \\ 0, & \text{else} \end{cases} \tag{6.7}$$

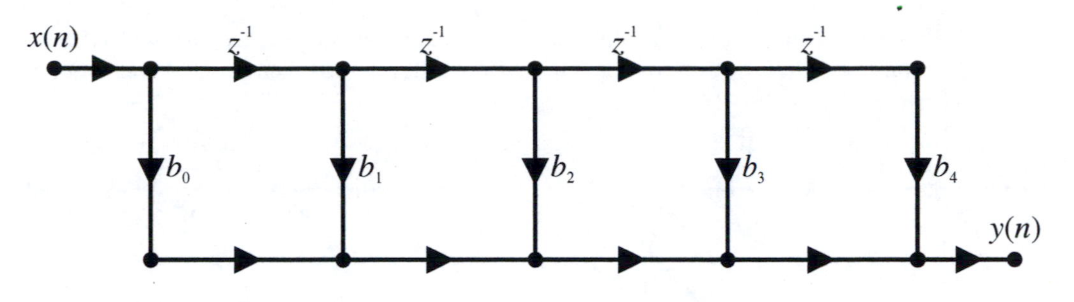

Figure 6.6: Direct-form FIR structure

2. The difference equation (6.6) can be implemented as a tapped delay line, because there are no feedback paths resulting in a direct-form structure for the FIR filter. This structure is shown in Figure 6.6 for $M = 5$ (i.e., a fourth-order FIR filter).

3. The higher-order FIR system function $H(z)$ can be expressed as a product of second-order sections with real coefficients — that is,

$$H(z) = b_0 \prod_{k=1}^{\lfloor \frac{M}{2} \rfloor} \left(1 + B_{k,1}z^{-1} + B_{k,2}z^{-2}\right) \tag{6.8}$$

where $B_{k,1}$ and $B_{k,2}$ are real numbers representing the coefficients of second-order sections and $\lfloor \cdot \rfloor$ is the `floor` operation. Each section (or biquad) is implemented in direct form, and the entire filter is implemented as a cascade of second-order sections. This cascade-form structure is shown in Figure 6.7 for $M = 7$. The conversion of a direct-form to a cascade-form structure is obtained by using the function `dir2cas`, in which the denominator vector `a` is set equal to 1. Similarly, `cas2dir` can be used to obtain the direct form from the cascade form.

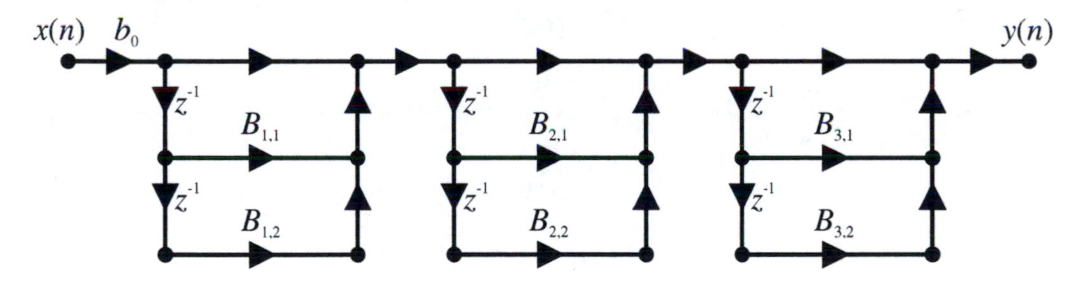

Figure 6.7: Cascade-form FIR structure

4. A causal and *linear-phase* FIR filter with impulse response over $[0, \ M-1]$ interval can be characterized either by a symmetric impulse response

$$h(n) = h(M-1-n); \quad 0 \le n \le M-1 \tag{6.9}$$

or by an antisymmetric impulse response

$$h(n) = -h(M-1-n); \quad 0 \le n \le M-1 \tag{6.10}$$

These symmetry conditions can now be exploited in a structure, called the *linear-phase form*, that reduces multiplications by 50% over the direct form. The block-diagram implementation of this form is shown in Figure 6.8 for odd and for even M.

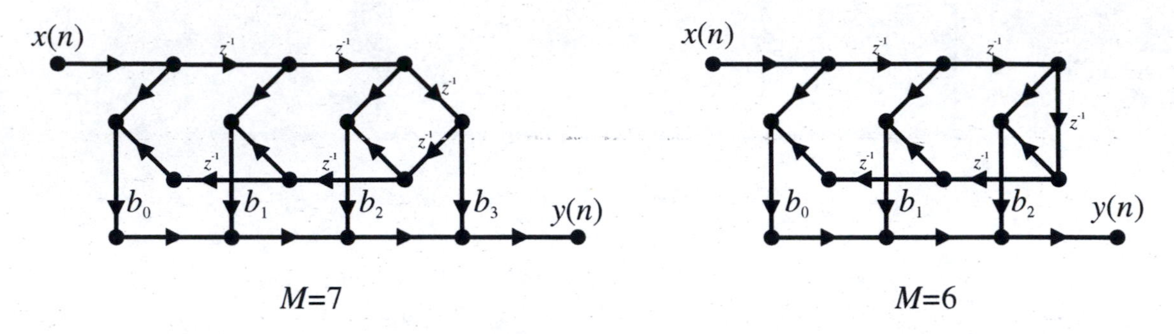

Figure 6.8: Linear-phase-form FIR structures (symmetric impulse response)

5. A structure that is akin to the parallel-form structure of the IIR filter is the *frequency-sampling form*. The system function $H(z)$ of an FIR filter can be reconstructed from its samples, $\{H(k),\ 0 \le k \le M-1\}$, on the unit circle. (See Chapter 5). That is, we have

$$H(z) = \left(\frac{1 - z^{-M}}{M}\right) \sum_{k=0}^{M-1} \frac{H(k)}{1 - W_M^{-k} z^{-1}} \tag{6.11}$$

The system function in (6.11) leads to a parallel structure, as shown in Figure 6.9 for $M = 4$.

6. To eliminate the complex arithmetic required in Figure 6.9 for a real-valued filter, we use the symmetry properties of the DFT and the W_M^{-k} factor. Then (6.11) can be expressed as

$$H(z) = \frac{1 - z^{-M}}{M} \left\{ \sum_{k=1}^{L} 2\left|H(k)\right| H_k(z) + \frac{H(0)}{1 - z^{-1}} + \frac{H(M/2)}{1 + z^{-1}} \right\} \tag{6.12}$$

where $L = \frac{M-1}{2}$ for M odd and $L = \frac{M}{2} - 1$ for M even and where the $\{H_k(z),\ k = 1, \ldots, L\}$ are second-order sections given by

$$H_k(z) = \frac{\cos\left[\angle H(k)\right] - z^{-1}\cos\left[\angle H(k) - \frac{2\pi k}{M}\right]}{1 - 2z^{-1}\cos\left(\frac{2\pi k}{M}\right) + z^{-2}} \tag{6.13}$$

A frequency-sampling structure for $M = 4$ containing real coefficients is shown in Figure 6.10. The conversion of a direct-form to a frequency-sampling form structure is obtained by the function `dir2fs`.

6.3 Lattice Filter Structures

Lattice and lattice-ladder filter structures are basically alternative realizations of LTI systems that are characterized by rational system functions. In this section, we consider such filter structures that exhibit either finite-duration or infinite-duration impulse responses.

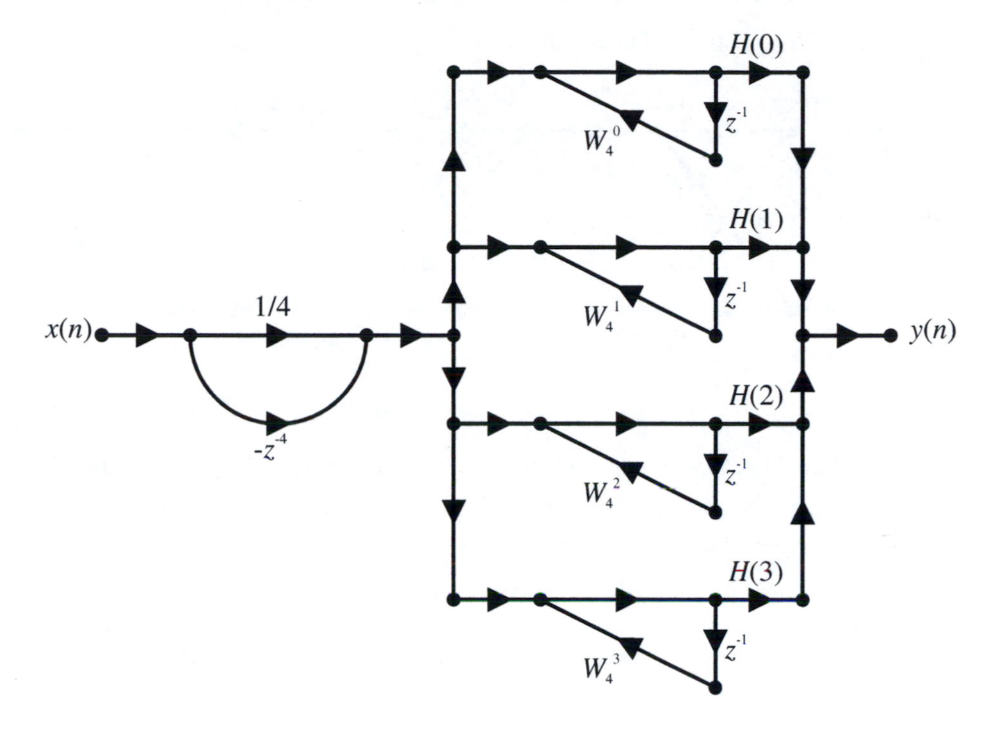

Figure 6.9: Frequency-sampling structure for $M = 4$

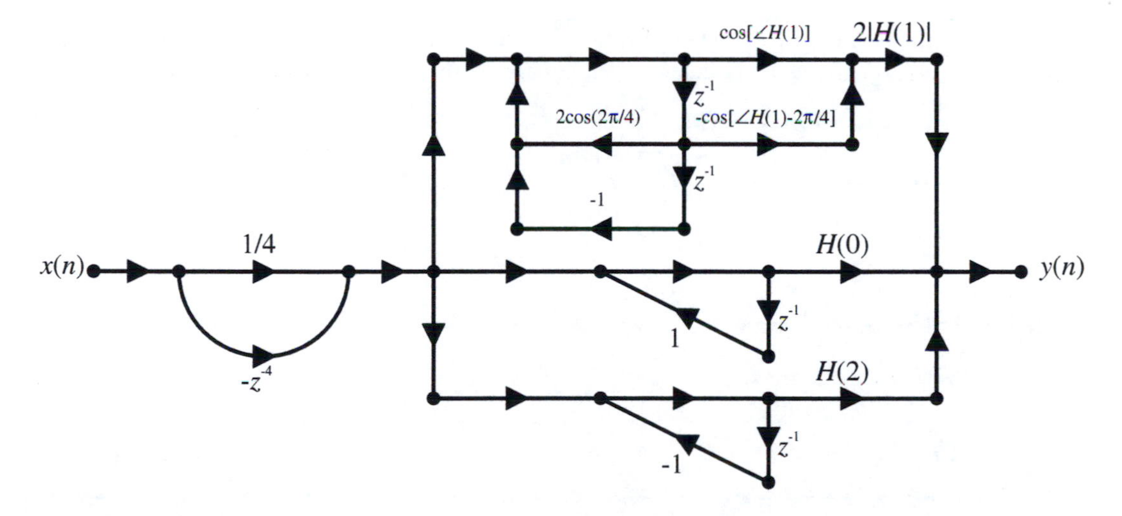

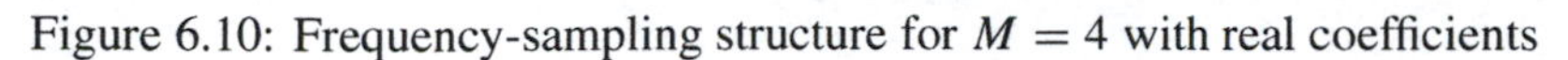
Figure 6.10: Frequency-sampling structure for $M = 4$ with real coefficients

Study Topics: Lattice and Lattice-Ladder Filter Realizations

1. An FIR filter of length M has a lattice structure with $M-1$ stages, as shown in Figure 6.11.

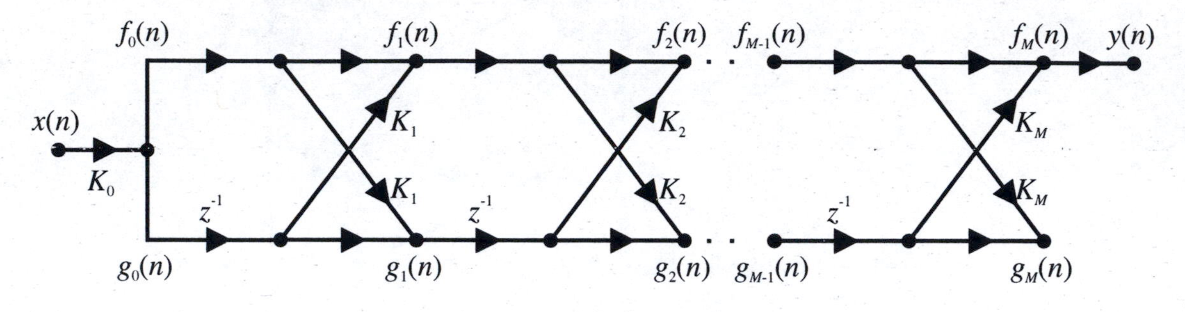

Figure 6.11: All-zero lattice filter

Each stage of the filter has an input and output that are related by the order-recursive equations

$$\begin{aligned} f_m(n) &= f_{m-1}(n) + K_m g_{m-1}(n-1), \quad m = 1, 2, \dots, M-1 \\ g_m(n) &= K_m f_{m-1}(n) + g_{m-1}(n-1), \quad m = 1, 2, \dots, M-1 \end{aligned} \tag{6.14}$$

where the parameters K_m, $m = 1, 2, \dots, M-1$, called the *reflection coefficients*, are the lattice-filter coefficients. The conversion of a direct-form to a lattice structure is obtained by the function `dir2latc` that is described in the text. Similarly, `latc2dir` can be used to obtain a direct-form structure from a lattice structure.

2. A lattice structure for an IIR filter is restricted to an all-pole system function given by

$$H(z) = \frac{1}{1 + \sum\limits_{m=1}^{N} a_N(m) z^{-m}} \tag{6.15}$$

This IIR filter of order N has a lattice structure with N stages, as shown in Figure 6.12. Each stage of the filter has an input and output that are related by the order-recursive equations

$$\begin{aligned} f_N(n) &= x(n) \\ f_{m-1}(n) &= f_m(n) - K_m g_{m-1}(n-1), \quad m = N, N-1, \dots, 1 \\ g_m(n) &= K_m f_{m-1}(n) + g_{m-1}(n-1), \quad m = N, N-1, \dots, 1 \\ y(n) &= f_0(n) = g_0(n) \end{aligned} \tag{6.16}$$

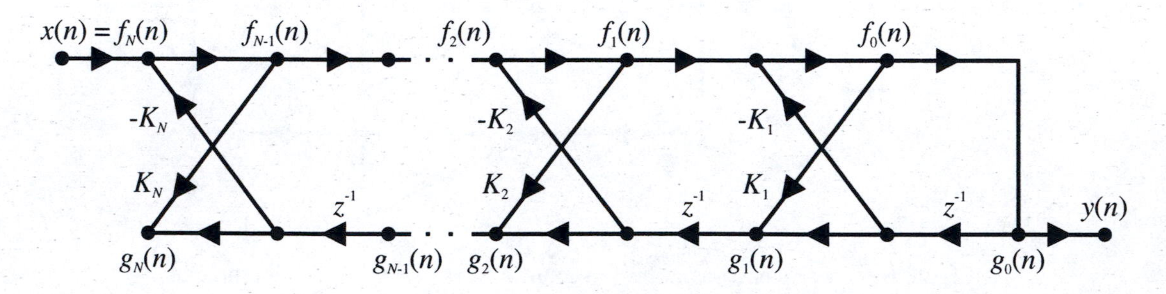

Figure 6.12: All-pole lattice filter

where the parameters, K_m, $m = 1, 2, \ldots, M-1$, are the reflection coefficients of the all-pole lattice. The conversion of a direct-form to an all-pole lattice structure is obtained by the function `dir2latc`. (Care must be taken to ignore the K_0 coefficient in the `K` array.) Similarly, `latc2dir` can be used to obtain the direct form from the all-pole form, provided that $K_0 = 1$ is used as the first element of the `K` array.

3. A general IIR filter containing both poles and zeros can be realized as a lattice-type structure by using an all-pole lattice as the basic building block. This results in a *lattice-ladder* structure, exemplified in Figure 6.13. The conversion of the direct-form to the lattice-ladder structure is obtained by the function `dir2ladr`. Similarly, `ladr2dir` can be used to obtain the direct form from the lattice-ladder form.

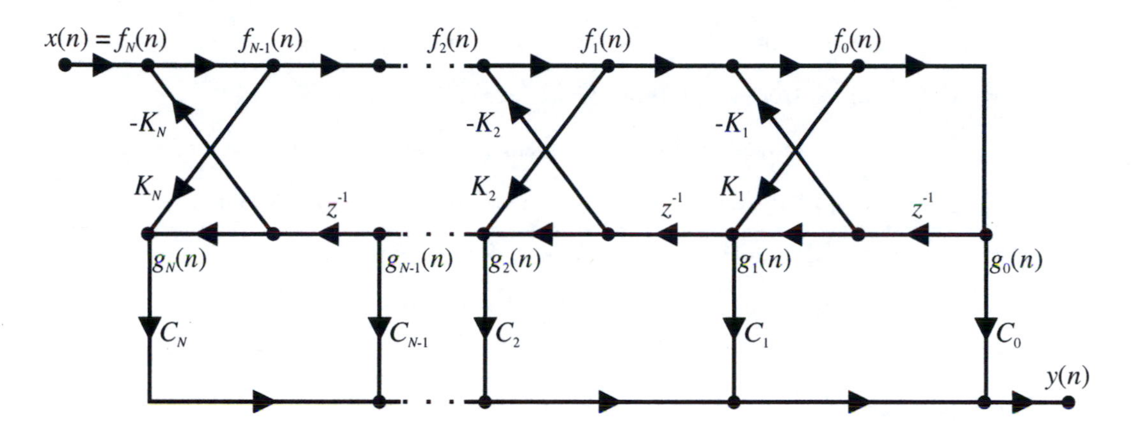

Figure 6.13: Lattice-ladder structure for realizing a pole–zero IIR filter

6.4 Summary

MATLAB Functions

The following MATLAB functions are used in this chapter:

IIR Filter Structures

- `y = filter(b,a,x)`: Implementation of Direct Form I and II
- `[b0,B,A] = dir2cas(b,a)`: Direct-form to cascade-form conversion
- `[b,a] = cas2dir(b0,B,A)`: Cascade-form to direct-form conversion
- `y = casfiltr(b0,B,A,x)`: Implementation of cascade form
- `[C,B,A] = dir2par(b,a)`: Direct-form to parallel-form conversion
- `[b,a] = par2dir(C,B,A)`: Parallel-form to direct-form conversion
- `y = parfiltr(C,B,A,x)`: Implementation of parallel form

FIR Filter Structures

- `y = filter(b,1,x)`: Implementation of direct form
- `[b0,B] = dir2cas(b,1)`: Direct-form to cascade-form conversion
- `[b,a] = cas2dir(b0,B,ones(K,3))`: Cascade-form to direct-form conversion
- `y = casfiltr(b0,B,ones(K,3),x)`: Implementation of cascade form
- `[C,B,A] = dir2fs(h)`: Direct-form to sampling-form conversion

Lattice Filter Structures 049

- `[K] = dir2latc(b)`: FIR direct-form to all-zero lattice-form conversion
- `[b] = latc2dir(K)`: All-zero lattice-form to FIR direct-form conversion
- `[y] = latcfilt(K,x)`: All-zero lattice-form implementation of FIR filters
- `[K,C] = dir2ladr(b,a)`: IIR direct-form to pole–zero lattice-ladder conversion
- `[b,a] = ladr2dir(K,C)`: Lattice-ladder to IIR direct-form conversion
- `[y] = ladrfilt(K,C,x)`: Implementation of lattice-ladder form

6.5 Problems

P6.1 A causal, linear, time-invariant system is described by

$$y(n) = \sum_{k=0}^{5} \left(\frac{1}{2}\right)^k x(n-k) + \sum_{\ell=1}^{5} \left(\frac{1}{3}\right)^\ell y(n-\ell)$$

Determine and draw the block diagrams of the following structures. Compute the response of the system to

$$x(n) = u(n),\ 0 \le n \le 100$$

using the corresponding structures in each case.

(a) Direct Form I

(b) Direct Form II

(c) Cascade form containing second-order Direct Form II sections

(d) Parallel form containing second-order Direct Form II sections

P6.2 An IIR filter is described by the following system function:

$$H(z) = 2\left(\frac{1+0z^{-1}+z^{-2}}{1-0.8z^{-1}+0.64z^{-2}}\right)\left(\frac{2-z^{-1}}{1-0.75z^{-1}}\right)\left(\frac{1+2z^{-1}+z^{-2}}{1+0.81z^{-2}}\right)$$

Determine and draw the following structures.

(a) Direct Form I

(b) Direct Form II

(c) Cascade form containing second-order Direct Form II sections

(d) Parallel form containing second-order Direct Form II sections

(e) Lattice-ladder form

P6.3 An IIR filter is described by the following system function:

$$H(z) = \left(\frac{-14.75 - 12.9z^{-1}}{1 - \frac{7}{8}z^{-1} + \frac{3}{32}z^{-2}}\right) + \left(\frac{24.5 + 26.82z^{-1}}{1 - z^{-1} + \frac{1}{2}z^{-2}}\right)$$

Determine and draw the following structures:

(a) Direct Form I

(b) Direct Form II

(c) Cascade form containing second-order Direct Form II sections

(d) Parallel form containing second-order Direct Form II sections

P6.4 A linear, time-invariant system having the system function

$$H(z) = \frac{0.5\left(1 + z^{-1}\right)^6}{\left(1 - \frac{3}{2}z^{-1} + \frac{7}{8}z^{-2} - \frac{13}{16}z^{-3} - \frac{1}{8}z^{-4} - \frac{11}{32}z^{-5} + \frac{7}{16}z^{-6}\right)}$$

is to be implemented to match a flow graph of the form shown in Figure 6.14.

(a) Fill in all the coefficients in the diagram.

(b) Is your solution unique? Explain.

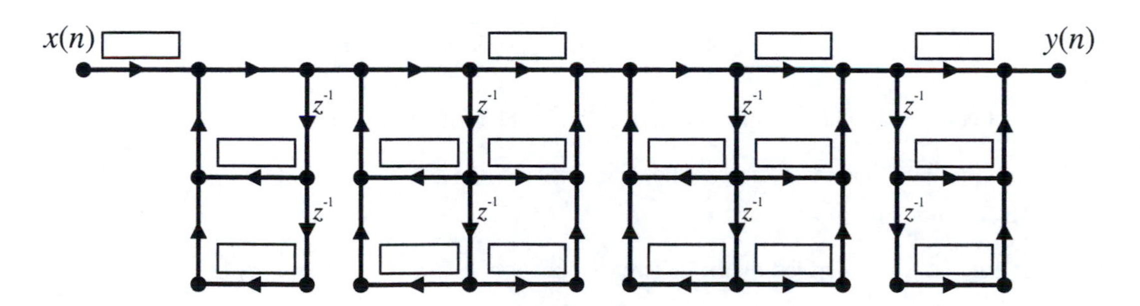

Figure 6.14: Structure for P6.4

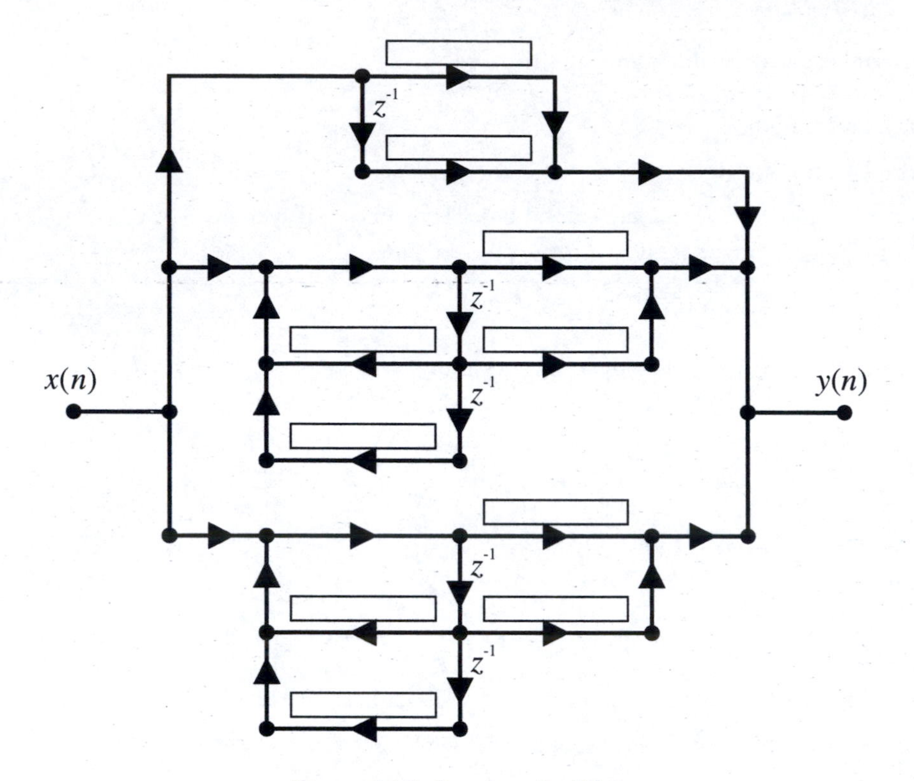

Figure 6.15: Structure for P6.5

P6.5 A linear, time-invariant system having the system function

$$H(z) = \frac{5 + 11.2z^{-1} + 5.44z^{-2} - 0.384z^{-3} - 2.3552z^{-4} - 1.2288z^{-5}}{1 + 0.8z^{-1} - 0.512z^{-3} - 0.4096z^{-4}}$$

is to be implemented to match a flow graph of the form shown in Figure 6.15. Fill in all the coefficients in the diagram.

P6.6 Consider the linear, time-invariant system given in P6.4.

$$H(z) = \frac{0.5\left(1 + z^{-1}\right)^6}{\left(1 - \frac{3}{2}z^{-1} + \frac{7}{8}z^{-2} - \frac{13}{16}z^{-3} - \frac{1}{8}z^{-4} - \frac{11}{32}z^{-5} + \frac{7}{16}z^{-6}\right)}$$

It is to be implemented to match a flow graph of the form shown in Figure 6.16.

(a) Fill in all the coefficients in the diagram.

(b) Is your solution unique? Explain.

P6.7 An FIR filter is described by the difference equation

$$y(n) = \sum_{k=0}^{10} \left(\frac{1}{2}\right)^{|5-k|} x(n-k)$$

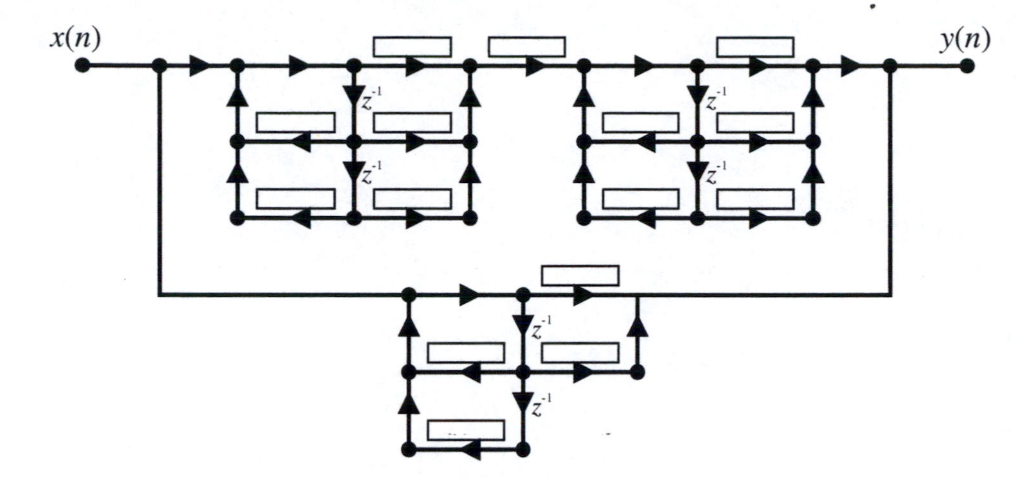

Figure 6.16: Structure for P6.6

Determine and draw the block diagrams of the following structures:

(a) Direct form

(b) Linear-phase form

(c) Cascade form

(d) Frequency-sampling form

P6.8 A linear, time-invariant system is given by the system function

$$H(z) = \sum_{k=0}^{10} (2z)^{-k}$$

Determine and draw the block diagrams of the following structures:

(a) Direct form

(b) Cascade form

(c) Frequency-sampling form

(d) Lattice form

Chapter 7

FIR Filter Design

In this chapter and the next, we treat the problem of designing filters that approximate a given set of specifications. The design of FIR filters will be treated in this chapter; IIR filters will be treated in the next chapter. FIR filters are generally used in applications that require that the filter have linear phase. A linear-phase filter has the desirable characteristic that it filters out undesirable signals without distorting the desired signal that it passes. Multiband FIR frequency-selective filters will be considered, and three main design techniques will be discussed — namely, the window, the frequency-sampling, and the optimal-design techniques. The emphasis will be on the understanding of design concepts and on the use of various MATLAB functions.

The reader should study the following material in Chapter 10 of the DSP book:

- Properties of Linear-phase FIR Filters — Sections 10.1 and 10.2.1
- Window Design Technique — Section 10.2.2
- Frequency-Sampling Design Technique — Section 10.2.3
- Optimal-Filter Design Technique — Sections 10.2.4–10.2.7

Learning Objectives

- Understand how digital-filter design requirements are specified and what parameters are needed to specify each frequency band.
- Learn the advantages of linear-phase FIR filters and how four different types of linear-phase filters are obtained.
- Study the properties of the four filter types in terms of the impulse response, the amplitude response, and the placement of zeros.
- Understand the restrictions placed on the kind of frequency-selective filter that can be designed with each filter type.
- Understand the basic idea behind the window design technique, that is, the roles played by the ideal filter $h_d(n)$ and the window function $w(n)$ in the overall approximation.

- Study various fixed- and adjustable-performance window functions — in particular, how their parameters are related to filter specifications.
- Develop skill in using MATLAB window functions and the ideal-filter function to design linear-phase FIR filters.
- Learn the concept behind the frequency-sampling design technique and the role played by the interpolation of the frequency response in the approximation process.
- Practice frequency-sampling design techniques, using the transition value tables and the DFT.
- Understand the concept of spreading the approximation error uniformly over the frequency bands in order to minimize the filter order and the resulting equiripple FIR filters.
- Learn how the design requirements in terms of approximation error are transformed into a uniform set of equations across all four types of linear-phase FIR filters.
- Understand how the maxima and minima of a trigonometric polynomial and the results of the alternation theorem are related to the order of the FIR filter.
- Understand the basic philosophy behind the Parks–McClellan algorithm and how the Remez exchange idea is used in the algorithm to obtain the optimum filter.
- Develop skill in using MATLAB's `remez` function to iteratively obtain the minimum-order linear-phase FIR filter via the Parks–McClellan algorithm.

7.1 Specifications and Properties of FIR Filters

In this section, we present the specifications and properties of linear-phase FIR filters that are subsequently used in the design. Study the review topics, and then solve the practice problems.

Study Topics: Filter specifications and FIR Filter Types

1. The magnitude specifications on an FIR filter are given in one of two ways. The first approach, called *Absolute specifications*, provides a set of requirements on the magnitude response function $|H(\omega)|$. The second approach, called *Relative specifications*, provides requirements in *decibels* (dB), given by

$$\text{dB scale} = -20\log_{10}\frac{|H(\omega)|_{\max}}{|H(\omega)|} \geq 0$$

   These specifications are shown in Figure 7.1 for a lowpass filter $H(\omega)$. The band of frequencies that is allowed to pass through the filter is called *passband* and is given by $0 \leq |\omega| \leq \omega_p$. The band of frequencies that is suppressed by the filter is called *stopband* and is given by $\omega_s \leq |\omega| \leq \pi$. The band $\omega_p \leq |\omega| \leq \omega_s$ is called the *transition band.*

2. In the absolute specifications, we require that the passband and stopband ripples δ_1 and δ_2, respectively, satisfy

$$\begin{array}{lll} \text{passband:} & 1-\delta_1 \leq |H(\omega)| \leq 1+\delta_1, & \text{for} \quad |\omega| \leq \omega_p \\ \text{stopband:} & |H(\omega)| \leq \delta_2, & \text{for} \quad \omega_s \leq |\omega| \leq \pi \end{array}$$

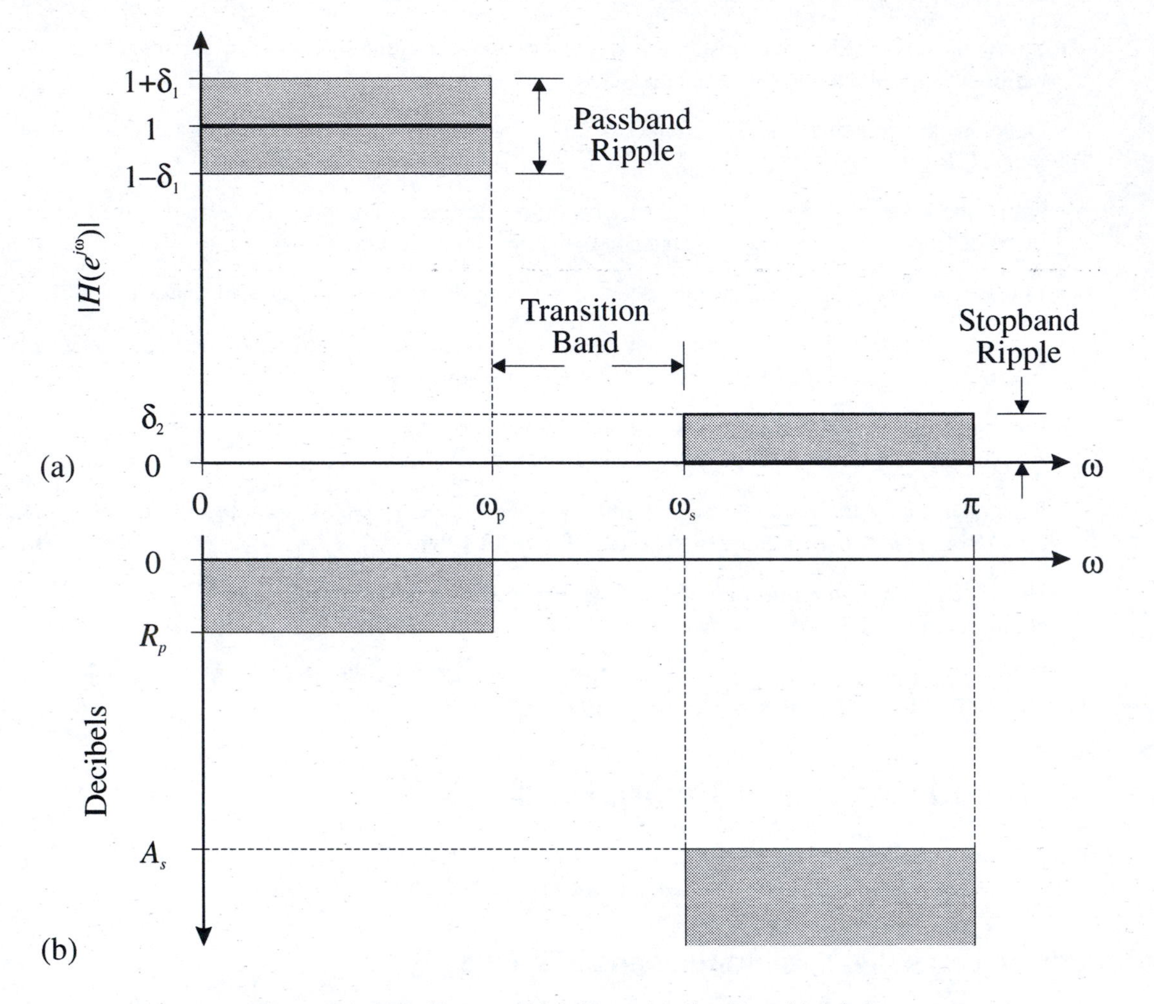

Figure 7.1: FIR filter specifications: (a) absolute, (b) relative

3. In the relative specifications, we require that the passband ripple R_p and the stopband attenuation A_s satisfy

$$\begin{array}{lll} \text{passband:} & 0 \le \text{dB scale} \le R_p, & \text{for} \quad |\omega| \le \omega_p \\ \text{stopband:} & \text{dB scale} > A_s, & \text{for} \quad \omega_s \le |\omega| \le \pi \end{array}$$

4. The relationship between the two sets of parameters is given by

$$R_p = -20\log_{10}\frac{1-\delta_1}{1+\delta_1} \tag{7.1}$$

and

$$A_s = -20\log_{10}\frac{\delta_2}{1+\delta_1} \tag{7.2}$$

5. FIR filters can be designed to have *exact* linear phase. There are two types of linear-phase conditions: (a) *constant phase delay*, in which

$$\sphericalangle H(\omega) = -\alpha\,\omega$$

for which the impulse response $\{h(n)\}_0^{M-1}$ is *symmetric*, that is,

$$h(n) = h(M-1-n),\ 0 \le n \le M-1$$

and (b) *constant group delay*, in which

$$\tau(\omega) \triangleq -\frac{\mathrm{d}}{\mathrm{d}\omega}[\sphericalangle H(\omega)] = -\alpha$$

for which the impulse response is *antisymmetric*, that is,

$$h(n) = -h(M-1-n),\ 0 \le n \le M-1$$

where $\sphericalangle H(\omega)$ is an unwrapped phase response.

6. There are four types of linear-phase FIR filters:

   Type-1: Symmetric impulse response, odd M
   Type-2: Symmetric impulse response, even M
   Type-3: Antisymmetric impulse response, odd M
   Type-4: Antisymmetric impulse response, even M

7. The frequency response of a linear-phase FIR filter can be written as

$$H(\omega) \triangleq H_r(\omega)\,e^{j\sphericalangle H(\omega)} = H_r(\omega)\,e^{j(\beta-\alpha\omega)}$$

where $H_r(\omega)$ is called an *amplitude response* and $\beta = 0$ for constant-phase-delay response and $\beta = \pm\frac{\pi}{2}$ for constant-group-delay response.

8. The amplitude response, $H_r(\omega)$, of Type-2 FIR filters is zero at $\omega = \pi$; hence, they cannot be used for highpass or bandstop filters. For the Type-3 filter, amplitude response, $H_r(\omega)$, is zero at $\omega = 0$ and at $\omega = \pi$; hence, it cannot be used in the design of lowpass, highpass, or bandstop filters. It is most suitable for designing bandpass filters and Hilbert transformers. Similarly, the amplitude response, $H_r(\omega)$, of a Type-4 FIR filter is zero at $\omega = 0$; hence, it cannot be used for lowpass or bandstop filters. It is most suitable for designing digital differentiators and highpass filters. Finally, the Type-1 filter has no restrictions; it can be used for designing any type of frequency-selective filter. Study the MATLAB functions `Hr_type1` through `Hr_type4` that compute amplitude responses of the respective types.

9. The system function $H(z)$ of a Type-2 filter must have a zero at $z = -1$, that of a Type-3 filter must have zeros at $z = 1$ and $z = -1$, and that of a Type-4 filter must have a zero at $z = 1$. The rest of the zeros have mirror symmetries with respect to the real axis and the unit circle.

7.2 Window Design Technique

In this section, we treat in detail the most straightforward (and often the easiest) design technique. It is based on the use of various tapered windows to truncate the impulse responses of ideal digital filters. Study the review topics and related MATLAB functions to solve the chapter problems given at the end of the section.

Study Topics: Window Design Method and Window Types

1. The natural approach to designing an FIR filter via the window design is to choose a proper ideal (or desired) frequency-selective filter (which always has a noncausal, infinite-duration impulse response, $h_d(n)$) and then truncate (or window) its impulse response to obtain a *linear-phase and causal* FIR filter. Hence, the impulse response of the designed filter is given by

$$h(n) = h_d(n)w(n) \tag{7.3}$$

where

$$w(n) = \begin{cases} \text{some symmetric tapered function over } 0 \le n \le M-1 \\ 0, \text{ otherwise} \end{cases} \tag{7.4}$$

Therefore, the emphasis in this method is on selecting an appropriate *ideal* filter, $h_d(n)$, and an appropriate *windowing* function, $w(n)$.

2. The frequency response $H(\omega)$ of the designed filter is given by the periodic convolution of $H_d(\omega)$ and the window response $W(\omega)$; that is,

$$H(\omega) = H_d(\omega) \circledast W(\omega) = \frac{1}{2\pi}\int_{-\pi}^{\pi} W(\lambda)\, H_d(\omega - \lambda)\, d\lambda \tag{7.5}$$

which is shown pictorially in Figure 7.2. The periodic convolution (7.5) produces a smeared version of the ideal response $H_d(\omega)$. One form of smearing is the creation of a transition band that is due to the width ($\propto 1/M$) of the main lobe of the window response, $W(\omega)$. The other form of smearing is the creation of passband and stopband ripples that are produced by the side lobes of $W(\omega)$.

3. An ideal linear-phase *lowpass* filter with *cutoff* frequency $0 < \omega_c < \pi$ is defined as

$$H_{\text{LP}}(\omega) = \begin{cases} 1 \cdot e^{-j\alpha\omega}, & |\omega| \le \omega_c \\ 0, & \omega_c < |\omega| \le \pi \end{cases} \tag{7.6}$$

The impulse response of this filter is of infinite duration and is given by

$$h_{\text{LP}}(n) = \mathcal{F}^{-1}\left[H_{\text{LP}}(\omega)\right] = \frac{\sin\left[\omega_c(n-\alpha)\right]}{\pi(n-\alpha)} \tag{7.7}$$

Note that $h_{\text{LP}}(n)$ is symmetric with respect to α, which is chosen to be $\alpha = (M-1)/2$ for $h(n)$. It should also be observed that the DTFT of the infinite duration impulse response exists, although the impulse response of the ideal filter is not absolute summable. The MATLAB function `ideal_lp` can be used to obtain $h_{\text{LP}}(n)$ in (7.7).

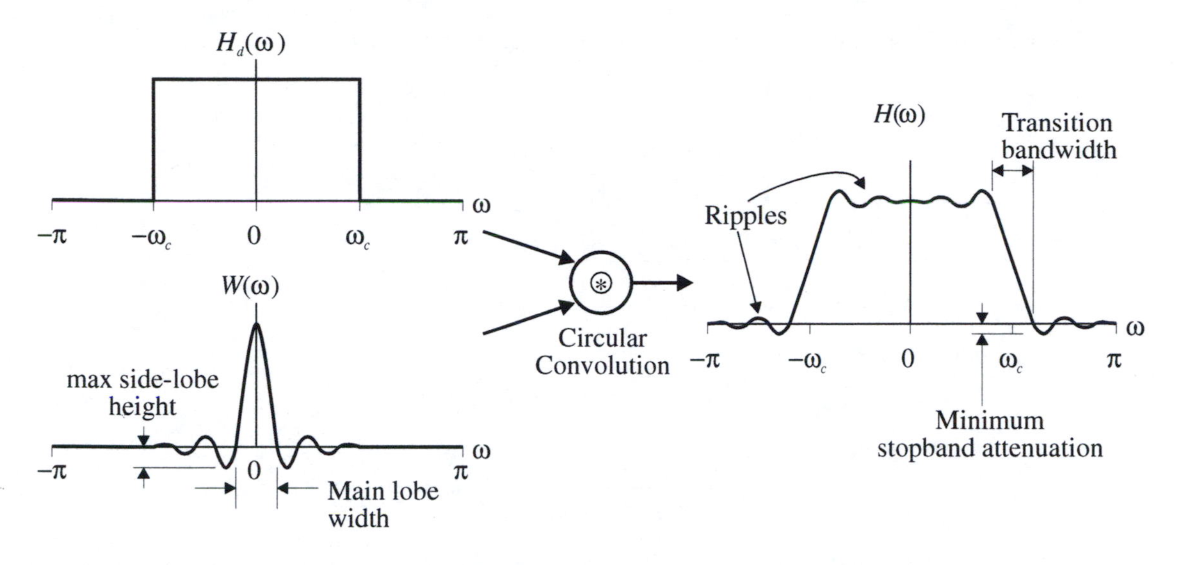

Figure 7.2: Windowing operation in the frequency domain

4. The impulse response of an ideal linear-phase *highpass* filter with cutoff frequency $0 < \omega_c < \pi$ is given by

$$h_{\mathrm{HP}}(n) = \begin{cases} 1 - \omega_c/\pi, & n = \alpha \\ -\dfrac{\sin[\omega_c(n-\alpha)]}{\pi(n-\alpha)}, & |n-\alpha| > 0 \end{cases}, \quad \alpha = \frac{M-1}{2} \tag{7.8}$$

The impulse response of an ideal linear-phase *bandpass* filter with cutoff frequencies $0 < \omega_{c_1} < \omega_{c_2} < \pi$ is given by

$$h_{\mathrm{BP}}(n) = \frac{\sin\left[\omega_{c_2}(n-\alpha)\right]}{\pi(n-\alpha)} - \frac{\sin\left[\omega_{c_1}(n-\alpha)\right]}{\pi(n-\alpha)}, \quad \alpha = \frac{M-1}{2} \tag{7.9}$$

Finally, the impulse response of an ideal linear-phase *bandstop* filter with cutoff frequencies $0 < \omega_{c_1} < \omega_{c_2} < \pi$ is given by

$$h_{\mathrm{BP}}(n) = \begin{cases} 1 - \left(\omega_{c_2} - \omega_{c_1}\right)/\pi, & n = \alpha \\ \dfrac{\sin\left[\omega_{c_1}(n-\alpha)\right]}{\pi(n-\alpha)} - \dfrac{\sin\left[\omega_{c_2}(n-\alpha)\right]}{\pi(n-\alpha)}, & |n-\alpha| \neq 0 \end{cases}, \quad \alpha = \frac{M-1}{2} \tag{7.10}$$

These impulse response values can also be obtained from the `ideal_lp` function by appropriate linear operations.

5. The ideal linear-phase *Hilbert transformer* is a 90° phase shifter given by

$$H_{\mathrm{HT}}(\omega) = \begin{cases} e^{j(\pi/2-\alpha\omega)}, & -\pi < \omega < 0 \\ -e^{j(\pi/2-\alpha\omega)}, & 0 < \omega < \pi \end{cases} \tag{7.11}$$

The impulse response of this Hilbert transformer is given by

$$h_{\text{HT}}(n) = \begin{cases} 0, & n = \alpha \\ \dfrac{\sin^2\left[\pi\left(n-\alpha\right)/2\right]}{\pi\left(n-\alpha\right)/2}, & |n-\alpha| \neq 0 \end{cases}, \quad \alpha = \frac{M-1}{2} \tag{7.12}$$

6. The ideal linear-phase *digital differentiator* is given by

$$H_{\text{DIF}}(\omega) = j\omega, \quad -\pi < \omega < \pi \tag{7.13}$$

with impulse response

$$h_{\text{DIF}}(n) = \begin{cases} 0, & n = \alpha \\ \dfrac{\cos\left[\pi\left(n-\alpha\right)\right]}{\left(n-\alpha\right)}, & |n-\alpha| \neq 0 \end{cases}, \quad \alpha = \frac{M-1}{2} \tag{7.14}$$

7. The basic idea behind the window design technique is to choose the filter length M and a window function $w(n)$ for the narrowest main-lobe width and the smallest side-lobe ripple (or attenuation) possible.

8. The simplest window function is the *rectangular* window, defined by

$$w(n) = \begin{cases} 1, & 0 \leq n \leq M-1 \\ 0, & \text{otherwise} \end{cases} \tag{7.15}$$

with amplitude response

$$W_r(\omega) = \frac{\sin\left(\frac{\omega M}{2}\right)}{\sin\left(\frac{\omega}{2}\right)} \tag{7.16}$$

The *approximate transition bandwidth* is $4\pi/M$; the *exact transition bandwidth* is $1.8\pi/M$, which is computed by using the *integrated* amplitude response. The *peak side-lobe magnitude* is 21.22% (or 13 dB) of the main-lobe amplitude. The integrated amplitude response has its first side-lobe magnitude at 21 dB, which results in a *minimum stopband attenuation* of 21 dB, irrespective of the window length M. This attenuation is insufficient for many applications. Furthermore, the rectangular window suffers from the *Gibbs phenomenon*, which renders it impractical for many applications.

The rectangular window is implemented either by the `boxcar` function or the `rectwin` function.

9. Many other fixed-performance window functions are available. Some of the most popular are the following:

Bartlett: Also known as the triangular window, it is given by

$$w(n) = \begin{cases} \dfrac{2n}{M-1}, & 0 \leq n \leq \dfrac{M-1}{2} \\ 2 - \dfrac{2n}{M-1}, & \dfrac{M-1}{2} \leq n \leq M-1 \\ 0, & \text{otherwise} \end{cases} \tag{7.17}$$

The exact transition width is $\Delta\omega = 6.1\pi/M$, the minimum stopband attenuation is 25 dB, and it is implemented by the `triang` function.

Hanning: This is a raised cosine window function given by

$$w(n) = \begin{cases} 0.5\left[1 - \cos\left(\frac{2\pi n}{M-1}\right)\right], & 0 \le n \le M-1 \\ 0, & \text{otherwise} \end{cases} \tag{7.18}$$

The exact transition width is $\Delta\omega = 6.2\pi/M$, the minimum stopband attenuation is 44 dB, and it is implemented by the `hanning` function.

Hamming: This window is similar to the Hanning window and is given by

$$w(n) = \begin{cases} 0.54 - 0.46\cos\left(\frac{2\pi n}{M-1}\right), & 0 \le n \le M-1 \\ 0, & \text{otherwise} \end{cases} \tag{7.19}$$

The exact transition width is $\Delta\omega = 6.6\pi/M$, the minimum stopband attenuation is 53 dB, and it is implemented by the `hamming` function.

Blackman: This window is similar to the previous two, but it contains a second-harmonic term and is given by

$$w(n) = \begin{cases} 0.42 - 0.5\cos\left(\frac{2\pi n}{M-1}\right) + 0.08\cos\left(\frac{4\pi n}{M-1}\right), & 0 \le n \le M-1 \\ 0, & \text{otherwise} \end{cases} \tag{7.20}$$

The exact transition width is $\Delta\omega = 11\pi/M$, the minimum stopband attenuation is 74 dB, and it is implemented by the `blackman` function.

10. The Kaiser window is an adjustable-performance window and one of the most useful in practice. It is given by

$$w(n) = \frac{I_0\left[\beta\sqrt{1-\left(1-\dfrac{2n}{M-1}\right)^2}\right]}{I_0\left[\beta\right]}, \quad 0 \le n \le M-1 \tag{7.21}$$

where $I_0[\cdot]$ is the *modified zero-order Bessel function* given by

$$I_0[\beta] = 1 + \sum_{k=1}^{\infty}\left[\frac{(\beta/2)^k}{k!}\right] \tag{7.22}$$

and β is a parameter that depends on M and that can be chosen to yield various transition widths and near-optimum stopband attenuation. The design equations for this window follow: Given ω_p,

ω_s, R_p, and A_s,

$$
\begin{aligned}
\text{Norm. transition width} &= \Delta f \triangleq \frac{\omega_s - \omega_p}{2\pi} \\
\text{Filter order } M &\simeq \frac{A_s - 7.95}{14.36\,\Delta f} + 1 \\
\text{Parameter } \beta &= \begin{cases} 0.1102\,(A_s - 8.7)\,, & A_s \geq 50 \\ 0.5842\,(A_s - 21)^{0.4} + 0.07886\,(A_s - 21)\,, & 21 < A_s < 50 \end{cases}
\end{aligned}
\tag{7.23}
$$

The Kaiser window is implemented by the `kaiser` function.

7.3 Frequency-Sampling Design Techniques

In this section, we treat the technique that is based on the frequency-sampling structure discussed in Chapter 6. The design process takes place in the frequency, domain, and the inverse DFT is used for the computation. There are two approaches: a straightforward naive design method and the optimum design method. Review these topics and the many design examples given in the text to solve the problems given at the end of the chapter.

Study Topics: Sampling the Frequency Response

1. Given the samples $\{H(k) = H(2\pi k/M)\,,\ k = 0, \ldots, M-1\}$ of the frequency response $H(\omega)$, we can obtain the interpolated version of $H(\omega)$ as

$$
H(\omega) = \frac{1 - e^{-j\omega M}}{M} \sum_{k=0}^{M-1} \frac{H(k)}{1 - e^{-j\omega} e^{j2\pi k/M}} \tag{7.24}
$$

2. The values of $\{H(k)\}_{k=0}^{M-1}$ are obtained by uniformly sampling the ideal frequency response $H_d(\omega)$ at $\omega_k = 2\pi k/M$. By using the conjugate-symmetry property and the linear-phase property, the amplitude and angle samples can be properly assembled. This is shown in Figure 7.3.

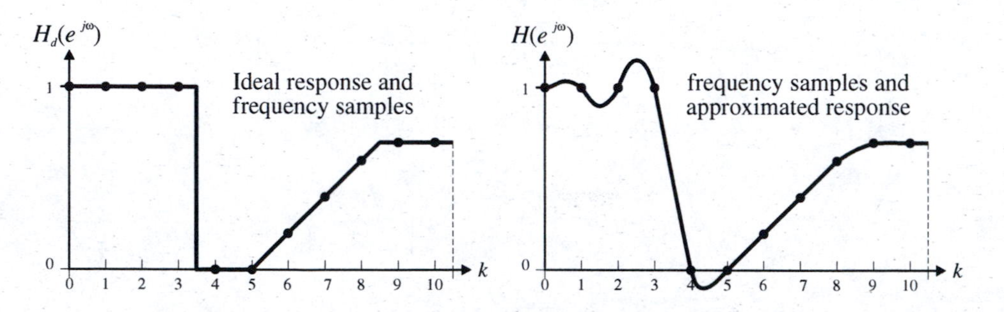

Figure 7.3: Pictorial description of frequency-sampling technique

3. For the naive design method, the amplitude values of the DFT $H(k)$ are assembled as

$$\{\text{amplitude of } H(k)\}_{k=0}^{M-1} = [H_r(0), H_r(2\pi/M), H_r(4\pi/M), \cdots, H_r(4\pi/M), H_r(2\pi/M)] \tag{7.25}$$

and the angle values are assembled as

$$\sphericalangle H(k) = \begin{cases} \beta - \left(\dfrac{M-1}{2}\right)\left(\dfrac{2\pi k}{M}\right) & k = 0, \ldots, \left\lfloor \dfrac{M-1}{2} \right\rfloor \\ -\beta + \left(\dfrac{M-1}{2}\right)\dfrac{2\pi}{M}(M-k) & k = \left\lfloor \dfrac{M-1}{2} \right\rfloor + 1, \ldots, M-1 \end{cases} \tag{7.26}$$

where $\beta = 0$ for Type-1 and Type-2 and $\beta = \pm\pi/2$ for Type-3 and Type-4 linear-phase FIR filters.

4. The impulse response of the designed filter is obtained as

$$\begin{aligned} h(n) &= \text{IDFT}[H(k)] = \text{IDFT}\left[\{\text{amplitude of } H(k)\}\, e^{j\sphericalangle H(k)}\right] \\ &= \pm h(M-1-n),\ n = 0, 1, ..., M-1 \end{aligned}$$

where the positive sign is for the Type-1 and Type-2 linear-phase filters, the negative sign for the Type-3 and Type-4 linear-phase filters.

5. In the naive design method, the frequency samples are fixed either to the passband values or to the stopband values. This can lead to an unacceptable amount of rippling near the band edges, especially for high-order filters. In the optimum design method, the samples in the transition bandwidth (or free samples) are allowed to vary to minimize the stopband ripples or attenuation. This optimization problem is called a *minimax* problem. The solution of it is available as tables of transition values for one and two free samples as described in Section 10.2.3 of the DSP book.

7.4 Optimal Equiripple Design Technique

In this section, we treat the optimal FIR filter design technique that can be used to design any type of linear-phase FIR filter. Compared to other techniques, this technique designs the smallest-order FIR filter for the given set of specifications. It is based on the concept that the filter exhibits ripples of equal height in the passband and the stopband. The design algorithm of these equiripple FIR filters uses iterative polynomial interpolation. This algorithm is known as the Parks–McClellan algorithm, and it incorporates the Remez-exchange routine for polynomial interpolation. Review these topics and the design examples given in the DSP text to solve the chapter problems given at the end of the section.

Study Topics: Design Procedure

1. The first step in understanding this technique is to express the frequency-response function in a common form across all four types of linear-phase FIR filters. Using simple trigonometric identities, we can write

$$\begin{aligned} H(\omega) &= e^{j\beta} e^{-j\frac{M-1}{2}\omega} H_r(\omega) \\ &= e^{j\beta} e^{-j\frac{M-1}{2}\omega} Q(\omega) P(\omega) \end{aligned} \tag{7.27}$$

where $Q(\omega)$ is a fixed function of ω and $P(\omega)$ is a polynomial function in $\cos\omega$ as shown in Table 7.1.

LP FIR Filter Type	$Q(\omega)$	L	$P(\omega)$
Type-1	1	$\frac{M-1}{2}$	$\sum_0^L a(n)\cos\omega n$
Type-2	$\cos\frac{\omega}{2}$	$\frac{M}{2}-1$	$\sum_0^L \tilde{b}(n)\cos\omega n$
Type-3	$\sin\omega$	$\frac{M-3}{2}$	$\sum_0^L \tilde{c}(n)\cos\omega n$
Type-4	$\sin\frac{\omega}{2}$	$\frac{M}{2}-1$	$\sum_0^L \tilde{d}(n)\cos\omega n$

Table 7.1: $Q(\omega)$, L, and $P(\omega)$ for linear-phase FIR filters

2. To formulate the design problem as a minimax optimization problem, we define the desired amplitude response $H_{dr}(\omega)$ and a weighting function $W(\omega)$, both defined over passbands and stopbands. The weighted error is defined as

$$E(\omega) \triangleq W(\omega)\left[H_{dr}(\omega) - H_r(\omega)\right], \quad \omega \in \mathcal{S} \triangleq \left[0, \omega_p\right] \cup \left[\omega_s, \pi\right] \tag{7.28}$$

The weighting function $W(\omega)$ is chosen as

$$W(\omega) = \begin{cases} \delta_2/\delta_1, & \text{in the passband} \\ 1, & \text{in the stopband} \end{cases} \tag{7.29}$$

so that the weighted error $E(\omega)$ has the same maximum error, equal to δ_2, in both the passband and the stopband.

3. Substituting $H_r(\omega)$ from (7.27) into (7.28), we obtain

$$E(\omega) = W(\omega)\left[H_{dr}(\omega) - Q(\omega)P(\omega)\right] = W(\omega)Q(\omega)\left[\frac{H_{dr}(\omega)}{Q(\omega)} - P(\omega)\right], \omega \in \mathcal{S}$$

If we define

$$\hat{W}(\omega) \triangleq W(\omega)Q(w) \text{ and } \hat{H}_{dr}(\omega) \triangleq \frac{H_{dr}(\omega)}{Q(\omega)}$$

then we obtain

$$E(\omega) = \hat{W}(\omega)\left[\hat{H}_{dr}(\omega) - P(\omega)\right], \ \omega \in \mathcal{S} \tag{7.30}$$

Thus we have a common form of $E(\omega)$ for all four cases.

4. The minimax optimization problem is to determine the filter impulse-response coefficients $h(n)$ that will minimize the maximum absolute value of $E(\omega)$ over the passband and stopband; that is,

$$\min_{\text{over coeff.}} \left[\max_{\omega \in \mathcal{S}} |E(\omega)|\right] \tag{7.31}$$

The solution to the preceding problem is given in the approximation theory. It suggests that the optimal equiripple filter has either $(L+2)$ or $(L+3)$ alternations in its error function $E(\omega)$ over S.

5. The most efficient algorithm for obtaining the above solution is the *Parks–McClellan algorithm.* It is an iterative algorithm that uses the Remez-exchange routine to fit $(L+2)$ or $(L+3)$ alternations in the amplitude response. This algorithm is available in MATLAB as the `remez` function.

6. Given the filter specifications, an approximate filter order can be estimated (or in some cases a reasonable value is simply chosen). The `remez` function computes the impulse-response values when given this order. However, the filter response (i.e., the passband and/or stopband specifications) might not be satisfactory. By increasing or decreasing the estimated order, a satisfactory filter design is obtained in an iterative fashion.

7.5 Summary

MATLAB Functions

The following MATLAB functions are used in this chapter:

Properties of the Linear-Phase FIR Filters

- `[Hr,w,a,L]=Hr_Type1(h)`: Computation of the Type-1 filter amplitude response
- `[Hr,w,a,L]=Hr_Type2(h)`: Computation of the Type-2 filter amplitude response
- `[Hr,w,a,L]=Hr_Type3(h)`: Computation of the Type-3 filter amplitude response
- `[Hr,w,a,L]=Hr_Type4(h)`: Computation of the Type-4 filter amplitude response

Window Design Technique

- `hd = ideal_lp(wc,M)`: Ideal lowpass filter computation
- `[db,mag,pha,grd,w] = freqz_m(b,a)`: Modified version of `freqz` function
- `w = boxcar(M)`: Rectangular window of length M
- `w = triang(M)`: Triangular window of length M
- `w = hanning(M)`: Hanning window of length M
- `w = hamming(M)`: Hamming window of length M
- `w = blackman(M)`: Blackman window of length M
- `w = kaiser(M,beta)`: Kaiser window of length M and parameter β

Optimal Equiripple Filter Design Technique

- `[h] = remez(N,f,m,weights,ftype)`: Parks–McClellan algorithm

7.6 Problems

P7.1 The Type-2 linear-phase FIR filter is characterized by

$$h(n) = h(M-1-n), \ 0 \le n \le M-1, \ M \text{ even}$$

(a) Show that its amplitude response $H_r(\omega)$ is given by

$$H_r(\omega) = \sum_{n=1}^{M/2} b(n) \cos\left\{\omega\left(n - \frac{1}{2}\right)\right\}$$

where coefficients $\{a(n)\}$ are obtained from $h(n)$.

(b) Show that the above $H_r(\omega)$ can be further expressed as

$$H_r(\omega) = \cos\frac{\omega}{2} \sum_{n=0}^{L} \tilde{b}(n) \cos\omega n, \ L = \frac{M}{2} - 1$$

where $\tilde{b}(n)$ is derived from $b(n)$.

P7.2 The Type-3 linear-phase FIR filter is characterized by

$$h(n) = -h(M-1-n), \ 0 \le n \le M-1, \ M \text{ odd}$$

(a) Show that its amplitude response $H_r(\omega)$ is given by

$$H_r(\omega) = \sum_{n=1}^{M/2} c(n) \sin\omega n$$

where coefficients $\{c(n)\}$ are obtained from $h(n)$.

(b) Show that the above $H_r(\omega)$ can be further expressed as

$$H_r(\omega) = \sin\omega \sum_{n=0}^{L} \tilde{c}(n) \cos\omega n, \ L = \frac{M-3}{2}$$

where $\tilde{c}(n)$ is derived from $c(n)$.

P7.3 Write a MATLAB function to compute the amplitude response $H_r(\omega)$ when given a linear-phase impulse response $h(n)$. The format of this function should be the following:

```
function [Hr,w,P,L] = Ampl_Res(h);
%
% function [Hr,w,P,L] = Ampl_Res(h)
% Computes Amplitude response Hr(w) and its polynomial P of order L,
% given a linear-phase FIR filter impulse response h.
% The type of filter is determined automatically by the subroutine.
%
```

```
% Hr = Amplitude Response
% w = frequencies between [0 pi] over which Hr is computed
% P = Polynomial coefficients
% L = Order of P
% h = Linear Phase filter impulse response
```

The subroutine should first determine the type of the linear-phase FIR filter and then use the appropriate `Hr_Type#` function discussed in the chapter. It should also check for whether the given $h(n)$ is of a linear-phase type. Check your subroutine on the following sequences:

$$h_1(n) = \{1, 2, 3, 2, 1\}, h_2(n) = \{1, 2, 2, 1\}, h_3(n) = \{1, 2, 0, -2, -1\}, h_4(n) = \{1, 2, -2, -1\}$$

P7.4 If $H(z)$ has zeros at

$$z_1 = re^{j\theta}, \quad z_2 = \frac{1}{r}e^{-j\theta}, \quad z_3 = re^{-j\theta}, \quad z_4 = \frac{1}{r}e^{-j\theta}$$

show that $H(z)$ represents a linear-phase FIR filter.

P7.5 Design a bandstop filter, using the Hanning window design technique. The specifications are:

lower stopband edge:	0.4π	$A_s = 40$ dB
upper stopband edge:	0.6π	
lower passband edge:	0.3π	$R_p = 0.5$ dB
upper passband edge:	0.7π	

Plot the impulse response and the magnitude response (in dB) of the designed filter.

P7.6 Design a bandpass filter, using the Hamming window design technique. The specifications are:

lower stopband edge:	0.3π	$A_s = 50$ dB
upper stopband edge:	0.6π	
lower passband edge:	0.4π	$R_p = 0.5$ dB
upper passband edge:	0.5π	

Plot the impulse response and the magnitude response (in dB) of the designed filter.

P7.7 Design a highpass filter, using the Kaiser window design technique. The specifications are:

$$\text{stopband edge: } 0.4\pi, \quad A_s = 60 \text{ dB}$$
$$\text{passband edge: } 0.6\pi, \quad R_p = 0.5 \text{ dB}$$

Plot the impulse response and the magnitude response (in dB) of the designed filter.

P7.8 We wish to use the Kaiser window method to design a linear-phase FIR digital filter that meets the following specifications:

$$\begin{array}{rcccl} 0 \le & |H(\omega)| & \le 0.01, & 0 \le & \omega \le 0.25\pi \\ 0.95 \le & |H(\omega)| & \le 1.05, & 0.35\pi \le & \omega \le 0.65\pi \\ 0 \le & |H(\omega)| & \le 0.01, & 0.75\pi \le & \omega \le \pi \end{array}$$

Determine the minimum-length impulse response $h(n)$ of such a filter. Provide a plot containing subplots of the amplitude response and the magnitude response in dB.

P7.9 Following the procedure used in this chapter, develop the following MATLAB functions to design FIR filters via the Kaiser window technique. These functions should check for the valid band-edge frequencies and restrict the filter length to 255.

(a) Lowpass filter: The format should be the following:

```
function [h,M] = kai_lpf(wp,ws,As);
% [h,M] = kai_lpf(wp,ws,As);
% Low-Pass FIR filter design using Kaiser window
%
%  h = Impulse response of length M of the designed filter
%  M = Length of h which is an odd number
% wp = Pass-band edge in radians (0 < wp < ws < pi)
% ws = Stop-band edge in radians (0 < wp < ws < pi)
% As = Stop-band attenuation in dB (As > 0)
```

(b) Highpass filter: The format should be the following:

```
function [h,M] = kai_hpf(ws,wp,As);
% [h,M] = kai_hpf(ws,wp,As);
% HighPass FIR filter design using Kaiser window
%
%  h = Impulse response of length M of the designed filter
%  M = Length of h which is an odd number
% ws = Stop-band edge in radians (0 < wp < ws < pi)
% wp = Pass-band edge in radians (0 < wp < ws < pi)
% As = Stop-band attenuation in dB (As > 0)
```

(c) Bandpass filter: The format should be the following:

```
function [h,M] = kai_bpf(ws1,wp1,wp2,ws2,As);
% [h,M] = kai_bpf(ws1,wp1,wp2,ws2,As);
% Band-Pass FIR filter design using Kaiser window
%
%  h = Impulse response of length M of the designed filter
%  M = Length of h which is an odd number
% ws1 = Lower stop-band edge in radians
% wp1 = Lower pass-band edge in radians
% wp2 = Upper pass-band edge in radians
% ws2 = Upper stop-band edge in radians
%       0 < ws1 < wp1 < wp2 < ws2< pi
% As = Stop-band attenuation in dB (As > 0)
```

(d) Bandstop filter: The format should be the following:

```
function [h,M] = kai_bsf(wp1,ws1,ws2,wp2,As);
% [h,M] = kai_bsf(wp1,ws1,ws2,wp2,As);
% Band-Pass FIR filter design using Kaiser window
%
%  h = Impulse response of length M of the designed filter
```

```
%  M = Length of h which is an odd number
% wp1 = Lower stop-band edge in radians
% ws1 = Lower pass-band edge in radians
% ws2 = Upper pass-band edge in radians
% wp2 = Upper stop-band edge in radians
%       0 < wp1 < ws1 < ws2 < wp2 < pi
% As = Stop-band attenuation in dB (As > 0)
```

You can now develop similar functions for other windows discussed in this chapter.

P7.10 Design a staircase filter from the following specifications, using the Blackman window approach:

$$\begin{array}{lrcccll} \text{Band-1:} & 0 & \leq & \omega & \leq & 0.3\pi, & \text{Ideal gain} = 1, & \delta_1 = 0.01 \\ \text{Band-2:} & 0.4\pi & \leq & \omega & \leq & 0.7\pi, & \text{Ideal gain} = 0.5, & \delta_2 = 0.005 \\ \text{Band-3:} & 0.8\pi & \leq & \omega & \leq & \pi, & \text{Ideal gain} = 0, & \delta_3 = 0.001 \end{array}$$

Provide a plot of the magnitude response in dB.

P7.11 The frequency response of an ideal bandpass filter is given by

$$H_d(\omega) = \begin{cases} 0, & 0 \leq |\omega| \leq \pi/3 \\ 1, & \pi/3 \leq |\omega| \leq 2\pi/3 \\ 0, & 2\pi/3 \leq |\omega| \leq \pi \end{cases}$$

(a) Determine the coefficients of a 25-tap filter based on the Parks–McClellan algorithm with stopband attenuation of 50 dB. The designed filter should have the smallest possible transition width.

(b) Plot the amplitude response of the filter, using the function developed in Problem P7.3.

P7.12 Consider the bandstop filter given in Problem P7.5.

(a) Design a linear-phase bandstop FIR filter, using the Parks–McClellan algorithm. Note that the length of the filter must be odd. Provide a plot of the impulse response and the magnitude response in dB of the designed filter.

(b) Plot the amplitude response of the designed filter, and count the total number of extrema in the stopband and the passband. Check this number against the theoretical estimate of the total number of extrema.

(c) Compare the order of this filter with that of the filter in Problem P7.5.

(d) Verify the operation of the designed filter on the following signal:

$$x(n) = 5 - 5\cos\left(\frac{\pi n}{2}\right); \quad 0 \leq n \leq 300$$

P7.13 Using the Parks–McClellan algorithm, design a 25-tap FIR differentiator with slope equal to 1 sample/cycle.

(a) Choose the frequency band of interest between 0.1π and 0.9π. Plot the impulse response and the amplitude response.

(b) Generate 100 samples of the sinusoid

$$x(n) = 3\sin(0.25\pi n),\ n = 0, ..., 100$$

and process them through the FIR differentiator. Compare the result with the theoretical "derivative" of $x(n)$. Note: Don't forget to take the 12-sample delay of the FIR filter into account.

P7.14 Design a lowest-order equiripple linear-phase FIR filter to satisfy the specifications given in Figure 7.4. Provide a plot of the amplitude response and a plot of the impulse response.

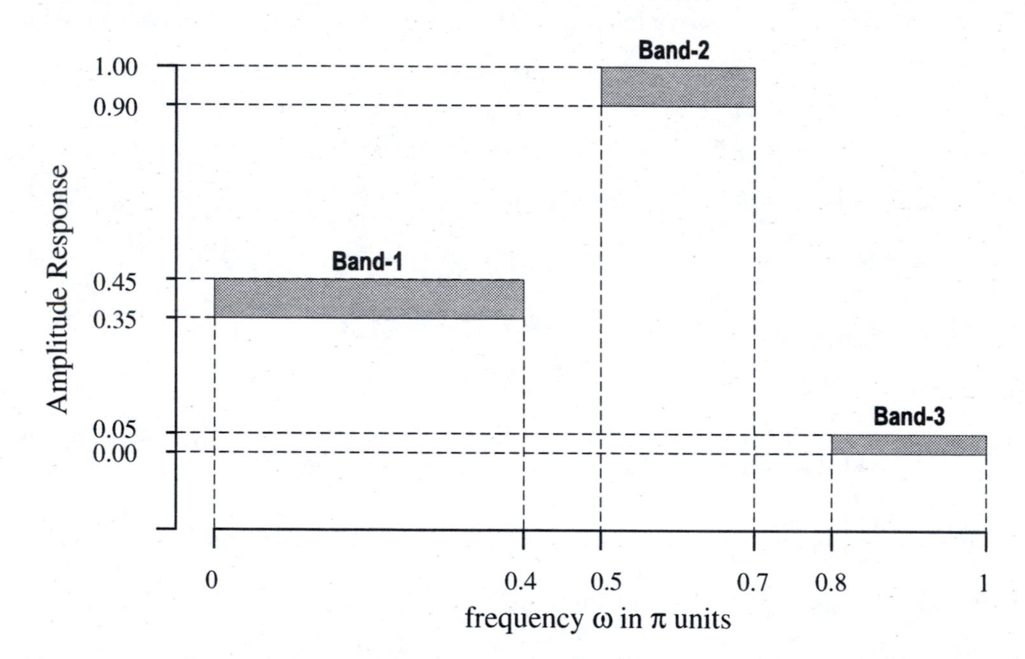

Figure 7.4: Filter specification for Problem P7.14

P7.15 A digital signal $x(n)$ contains a sinusoid of frequency $\pi/2$ and a Gaussian noise $w(n)$ of zero mean and unit variance:

$$x(n) = 2\cos\frac{\pi n}{2} + w(n)$$

We want to filter out the noise component, using a 50th-order, causal, linear-phase FIR filter.

(a) Using the Parks–McClellan algorithm, design a narrow bandpass filter with passband width of no more than 0.02π and stopband attenuation of at least 30 dB. Note that no other parameters are given, so you have to choose the remaining parameters for the `remez` function to satisfy the requirements. Provide a plot of the log-magnitude response in dB of the designed filter.

(b) Generate 200 samples of the sequence $x(n)$, and process them through the filter to obtain the output $y(n)$. Provide subplots of $x(n)$ and $y(n)$ for $100 \le n \le 200$ on one plot, and comment on your results.

P7.16 Design a Bandpass filter using the frequency sampling method. The specifications are

Lower stopband edge:	0.3π	$A_s = 50$ dB
Upper stopband edge:	0.6π	
Lower passband edge:	0.4π	$R_p = 0.5$ dB
Upper passband edge:	0.5π	

Choose the order of the filter appropriately so that there are two samples in the transition band. Use optimum values for these samples.

Chapter 8

IIR Filter Design

The straightforward method of designing IIR digital filters is based on the conversion of analog lowpass filters. In this method, an analog lowpass filter is first designed by using any one of the several analog prototypes available in the literature. Next, the analog lowpass filter is transformed into a digital lowpass filter by using one of several filter-mapping algorithms. Finally, the digital lowpass filter is converted into a desired digital filter by using the appropriate frequency-band transformation. Once again, the emphasis in this chapter will be on the understanding of design concepts and on the use of various MATLAB functions.

The reader should study the following material in Chapter 10 of the DSP book:

- Characteristics and design of prototype analog filters — Section 10.3.4
- Analog-to-digital filter transformations — Sections 10.3.2 and 10.3.3
- Frequency-band transformations — Section 10.4

Learning Objectives

- Understand the concept of spectral factorization of the magnitude-squared response $|H_a(j\Omega)|^2$ into a minimum-phase system function $H_a(s)$ — that is, the separation of $H_a(s)H_a(-s)$ into $H_a(s)$.
- Learn the characteristics and parameter structures of the basic analog Butterworth, Chebyshev, and elliptic filter prototypes.
- Study and learn how to use various MATLAB functions developed for designing analog lowpass filters.
- Learn the basic concept and the limitations of the impulse-invariance transformation.
- Understand the bilinear mapping and frequency prewarping concept.
- Study the basic requirements of frequency-band mapping and the parameter structures for various filter transformations.
- Develop skill in designing digital lowpass filters by using any analog prototype and filter transformations.

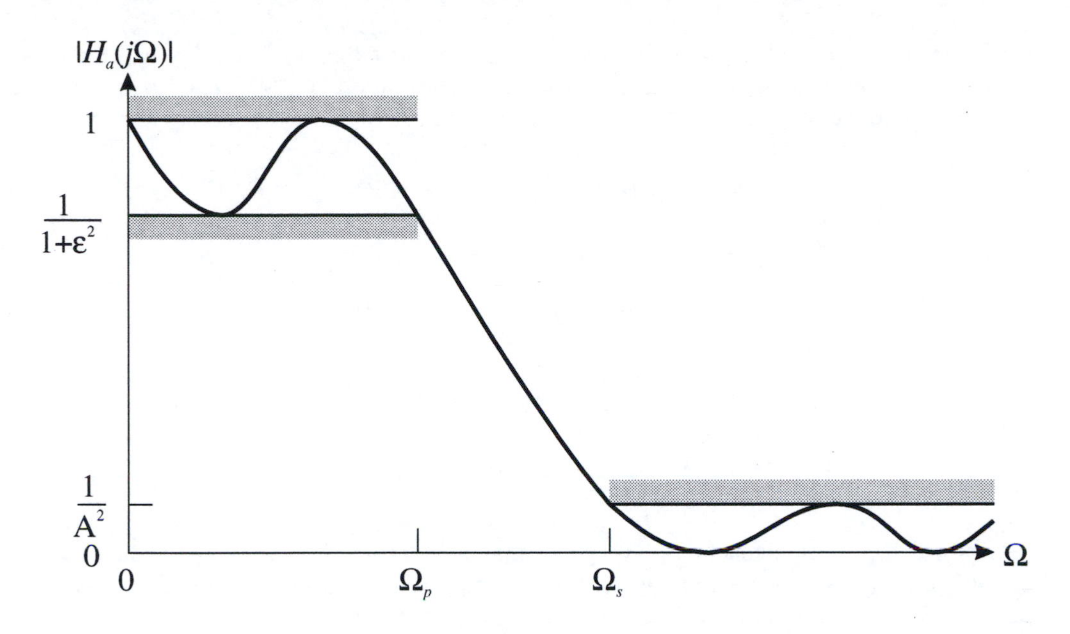

Figure 8.1: Analog lowpass filter specifications

- Know how to map cutoff frequencies of digital filters (of any type) into digital lowpass prototype filter frequencies, using the parameters of frequency-band transformation.
- Study and practice arbitrary frequency-selective (i.e., LP, HP, BP, BS) digital-filter designs, using the MATLAB functions developed in the text as well as those provided by MATLAB.

8.1 Design of Analog Lowpass Filters

To design analog filters, one has to first study analog-filter specifications and the properties of the magnitude-squared response used in specifying analog filters. This will lead into the characteristics and design of three widely used analog *prototype* filters — namely, *Butterworth*, *Chebyshev*, and *elliptic* filters. Study the following review topics, the related MATLAB functions, and the many examples given in the text to solve the problems given at the end of the chapter.

Study Topics: Characteristics and Parameters of Butterworth, Chebyshev, and Elliptic Filters

1. In many applications, the analog lowpass filter specifications are given on the magnitude-square response, $|H_a(j\Omega)|^2$, as shown in Figure 8.1. Here,

$$\begin{array}{ccccll} \dfrac{1}{1+\epsilon^2} & \leq & |H_a(\Omega)|^2 & \leq & 1, & |\Omega| \leq \Omega_{\mathrm{p}} \\ 0 & \leq & |H_a(\Omega)|^2 & \leq & \dfrac{1}{A^2}, & \Omega_s \leq |\Omega| \end{array} \tag{8.1}$$

where ϵ is a passband *ripple parameter*, Ω_p is the passband cutoff frequency in rad/sec, A is a stopband *attenuation parameter*, and Ω_s is the stopband cutoff in rad/sec. The parameters ϵ and A are related to the passband ripple R_p and stopband attenuation A_s by

$$R_p = -10\log_{10}\frac{1}{1+\epsilon^2} \Longrightarrow \epsilon = \sqrt{10^{R_p/10}-1} \tag{8.2}$$

and

$$A_s = -10\log_{10}\frac{1}{A^2} \Longrightarrow A = 10^{A_s/20} \tag{8.3}$$

2. The specification for the magnitude-squared response, $|H_a(\Omega)|^2$, can be expressed as the s-domain function

$$H_a(s)H_a(-s) = |H_a(\Omega)|^2\Big|_{\Omega=s/j} \tag{8.4}$$

from which the system function $H_a(s)$ of the analog filter can be extracted. To obtain a *causal* and *stable* filter $H_a(s)$, we assign all left-half poles and zeros of $H_a(s)H_a(-s)$ to $H_a(s)$ as shown in Figure 8.2. The resulting filter is called a *minimum-phase* filter.

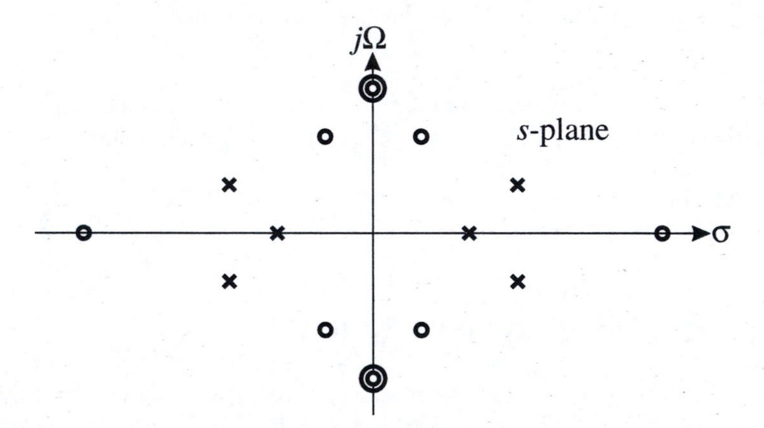

Figure 8.2: Typical pole–zero pattern of $H_a(s)H_a(-s)$

3. The Butterworth filters have monotone behavior both in the passband and in the stopband. The magnitude-squared response of an Nth-order Butterworth lowpass filter is given by

$$|H_a(\Omega)|^2 = \frac{1}{1+\left(\dfrac{\Omega}{\Omega_c}\right)^{2N}} \tag{8.5}$$

where N is the order of the filter and Ω_c is the cutoff frequency in rad/sec. In the s-domain, the response is given by

$$H_a(s)H_a(-s) = |H_a(\Omega)|^2\Big|_{\Omega=s/j} = \frac{(j\Omega)^{2N}}{s^{2N}+(j\Omega_c)^{2N}} \tag{8.6}$$

The poles of $H_a(s)H_a(-s)$ from (8.6) are given by

$$p_k = (-1)^{\frac{1}{2N}}(j\Omega) = \Omega_c e^{j\frac{\pi}{2N}(2k+N+1)}, \; k = 0, 1, \ldots, 2N-1 \tag{8.7}$$

Thus, the $2N$ poles of $H_a(s)H_a(-s)$ are equally distributed on a circle of radius Ω_c with angular spacing π/N radians and are symmetrically located with respect to the $j\Omega$ axis. Also, no pole ever falls on the imaginary axis. A stable and causal filter $H_a(s)$ can now be specified by selecting poles in the left half-plane, and $H_a(s)$ can be written in the form

$$H_a(s) = \frac{\Omega_c^N}{\prod\limits_{\text{LHP poles}} (s - p_k)} \tag{8.8}$$

The MATLAB function `[z,p,k]=buttap(N)` designs a *normalized* (i.e., $\Omega_c = 1$) Butterworth analog prototype filter of order N; the text function `[b,a]=U_buttap(N,Omegac)` designs the unnormalized Butterworth analog prototype filter.

4. Given the design specifications Ω_p, R_p, Ω_s, and A_s, the order N of the Butterworth filter is given by

$$N = \left\lceil \frac{\log_{10}\left[\left(10^{R_p/10} - 1\right)\left(10^{A_s/10} - 1\right)\right]}{2\log_{10}\left(\Omega_p/\Omega_s\right)} \right\rceil \tag{8.9}$$

where $\lceil\cdot\rceil$ is the `ceil` operation. The cutoff frequency, Ω_c, is chosen either as

$$\Omega_c = \frac{\Omega_p}{\sqrt[2N]{\left(10^{R_p/10} - 1\right)}} \tag{8.10}$$

which satisfies specifications exactly at Ω_p, or as

$$\Omega_c = \frac{\Omega_s}{\sqrt[2N]{\left(10^{A_s/10} - 1\right)}} \tag{8.11}$$

which satisfies specifications exactly at Ω_s. The function `[b,a]=afd_butt(Wp,Ws,Rp,As)` designs a Butterworth lowpass filter when given its specifications.

5. The Chebyshev-I filters exhibit equiripple behavior in the passband, but monotone behavior in the stopband. The magnitude-squared response of a Chebyshev-I filter is given by

$$|H_a(\Omega)|^2 = \frac{1}{1 + \epsilon^2 \mathrm{T}_N^2\left(\frac{\Omega}{\Omega_c}\right)} \tag{8.12}$$

where N is the order of the filter, ϵ is the passband ripple factor, and $\mathrm{T}_N(x)$ is the Nth order Chebyshev polynomial given by

$$\mathrm{T}_N(x) = \begin{cases} \cos\left(N\cos^{-1}(x)\right) & 0 \le x \le 1 \\ \cosh\left(N\cosh^{-1}(x)\right) & 1 < x < \infty \end{cases} \quad \text{where } x = \frac{\Omega}{\Omega_c} \tag{8.13}$$

To determine a causal and stable $H_a(s)$, we first obtain the roots of the polynomial

$$1 + \epsilon^2 \mathrm{T}_N^2 \left(\frac{s}{j\Omega_c} \right) \tag{8.14}$$

If $p_k = \sigma_k + j\Omega_k$, $k = 0, \ldots, N-1$ are the (left half-plane) roots of the above polynomial, then

$$\begin{aligned} \sigma_k &= (a\Omega_c)\cos\left[\frac{\pi}{2} + \frac{(2k+1)\pi}{2N}\right] \\ \Omega_k &= (b\Omega_c)\sin\left[\frac{\pi}{2} + \frac{(2k+1)\pi}{2N}\right] \end{aligned} \quad k = 0, \ldots N-1 \tag{8.15}$$

where

$$a = \frac{1}{2}\left(\sqrt[N]{\alpha} - \sqrt[N]{1/\alpha} \right), \; b = \frac{1}{2}\left(\sqrt[N]{\alpha} + \sqrt[N]{1/\alpha} \right), \text{ and } \alpha = \frac{1}{\epsilon} + \sqrt{1 + \frac{1}{\epsilon^2}} \tag{8.16}$$

Now the system function is given by

$$H_a(s) = \frac{K}{\prod_k (s - p_k)} \tag{8.17}$$

where K is a normalizing factor chosen to make

$$H_a(0) = \begin{cases} 1, & N \text{ odd} \\ \frac{1}{\sqrt{1+\epsilon^2}}, & N \text{ even} \end{cases} \tag{8.18}$$

The MATLAB function `[z,p,k]=cheb1ap(N,Rp)` designs a *normalized* (i.e., $\Omega_c = 1$) Chebyshev-I analog prototype filter of order N; the text function `[b,a]=U_chb1ap(N,Rp,Omegac)` designs the unnormalized Chebyshev-I analog prototype filter.

6. Given the design specifications Ω_p, Ω_s, R_p, and A_s, the order N of the Chebyshev-I filter is given by

$$N = \left\lceil \frac{\log_{10}\left[g + \sqrt{g^2 - 1}\right]}{\log_{10}\left[\Omega_r + \sqrt{\Omega_r^2 - 1}\right]} \right\rceil \tag{8.19}$$

where

$$g = \sqrt{\left(A^2 - 1\right)/\epsilon^2} \tag{8.20}$$

and

$$\Omega_c = \Omega_p \text{ and } \Omega_r = \frac{\Omega_s}{\Omega_p} \tag{8.21}$$

From Equations (8.2) and (8.3), the remaining parameters are given by

$$\epsilon = \sqrt{10^{0.1R_p} - 1} \text{ and } A = 10^{A_s/20}$$

The function `[b,a]=afd_chb1(Wp,Ws,Rp,As)` designs a Chebyshev-I lowpass filter when given its specifications.

7. Chebyshev-II filters exhibit equiripple behavior in the stopband but monotone behavior in the passband and are related to the Chebyshev-I filters through a simple transformation. If we replace the term $\epsilon^2 \mathrm{T}_N^2(\Omega/\Omega_c)$ in (8.12) by its reciprocal and also the argument $x = \Omega/\Omega_c$ by its reciprocal, we obtain the magnitude-squared response of Chebyshev-II filters as

$$|H_a(\Omega)|^2 = \frac{1}{1 + \left[\epsilon^2 \mathrm{T}_N^2(\Omega_c/\Omega)\right]^{-1}} \tag{8.22}$$

The MATLAB function `[z,p,k]=cheb2ap(N,As)` designs a *normalized* (i.e., $\Omega_c = 1$) Chebyshev-II analog prototype filter of order N; the text function `[b,a]=U_chb2ap(N,As,Omegac)` designs the unnormalized Chebyshev-I analog prototype filter. The design equations for the Chebyshev-II prototype are similar to those of the Chebyshev-I, except that $\Omega_c = \Omega_s$. The function `[b,a]=afd_chb2(Wp,Ws,Rp,As)` designs a Chebyshev-II lowpass filter when given its specifications.

8. Elliptic filters exhibit equiripple behavior in both the passband and the stopband. The magnitude-squared response of elliptic filters is given by

$$|H_a(\Omega)|^2 = \frac{1}{1 + \epsilon^2 \mathrm{U}_N^2\left(\dfrac{\Omega}{\Omega_c}\right)} \tag{8.23}$$

where N is the order, ϵ is the passband ripple (which is related to R_p), and $\mathrm{U}_N(\cdot)$ is the Nth order Jacobian elliptic function. The order N of the filter is given by

$$N = \frac{K(k)K\left(\sqrt{1-k_1^2}\right)}{K(k_1)\,K\left(\sqrt{1-k^2}\right)} \tag{8.24}$$

where $k = \dfrac{\Omega_p}{\Omega_s}$, $k_1 = \dfrac{\epsilon}{\sqrt{A^2-1}}$, and

$$K(x) = \int_0^{\pi/2} \frac{d\theta}{\sqrt{1 - x^2 \sin^2\theta}} \tag{8.25}$$

is the complete elliptic integral of the first kind. MATLAB provides the function `ellipke` to compute the integral numerically. The MATLAB function `[z,p,k]=ellipap(N,Rp,As)` designs a *normalized* elliptic analog prototype filter of order N; the text function `[b,a]=u_elipap(N,Rp,As,Omegac)` designs the unnormalized elliptic analog prototype filter. Using the filter-order computation formula given in (8.24), the function `[b,a]=afd_elip(Wp,Ws,Rp,As)` designs an elliptic lowpass filter when given its specifications.

9. Elliptic filters provide optimal performance in the magnitude-squared response, but have highly nonlinear phase response in the passband (which is undesirable in many applications). The Butterworth filters generally require a higher order N (more poles) to achieve the same stopband attenuation, but they exhibit a fairly linear phase response in their passband. The Chebyshev filters have phase characteristics that lie somewhere in between. Therefore, in practical applications, the choice of a prototype filter depends both on the filter order (which influences processing speed and implementation complexity) and on the phase characteristics (which control the distortion).

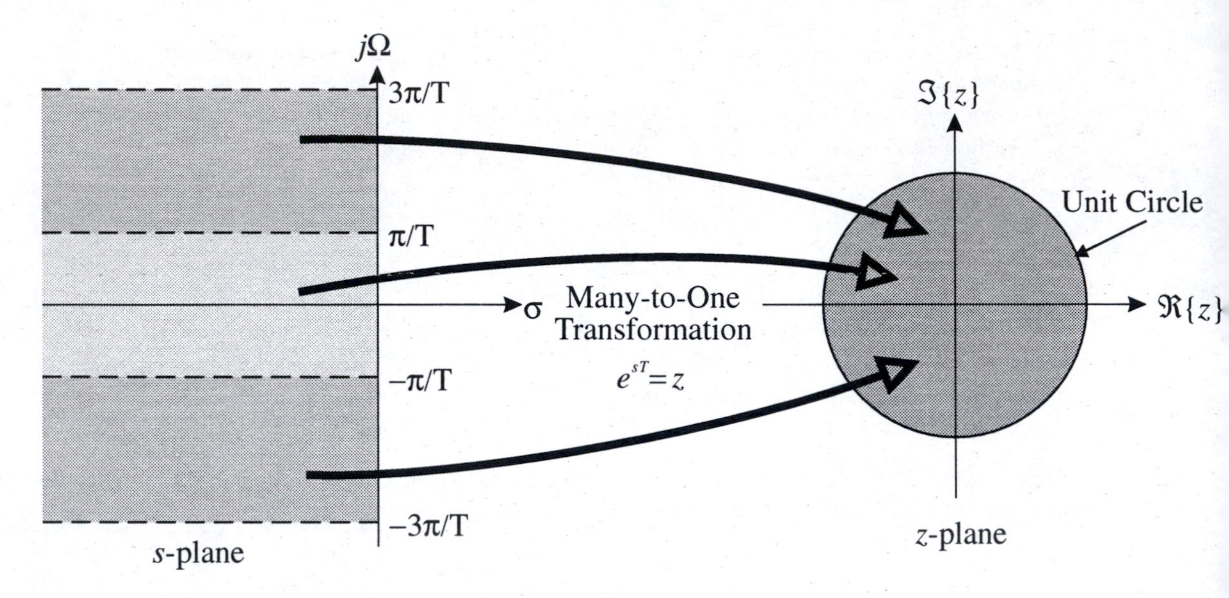

Figure 8.3: Complex-plane mapping in impulse-invariance transformation

8.2 Analog-to-Digital Filter Transformations

The next step in designing IIR digital filters is to transform analog filters into digital filters. These transformations are complex-valued mappings that are derived by preserving different aspects of analog and digital filters. The impulse-invariance mapping preserves the shape of the impulse response. The best mapping is a *bilinear* transformation, which preserves the system function representation from the analog to the digital domain. These topics are covered in the following review topics. Also, study the several MATLAB functions and examples that are given in the text to solve the practice problems given at the end of the chapter.

Study Topics: Filter Transformations

1. In the impulse invariance-transformation, the shape of the analog filter impulse response $h_a(t)$ is "captured" in the digital filter by the sampling operation

$$h(n) = h_a(t = nT) \tag{8.26}$$

in which T is a parameter that plays a minor role in the procedure. The system functions $H(z)$ and $H_a(s)$ are related through the frequency-domain aliasing formula (Chapter 4):

$$H(z)|_{z=e^{st}} = \frac{1}{T}\sum_{k=-\infty}^{\infty} H_a(s - j\frac{2\pi}{T}k) \tag{8.27}$$

2. The digital and analog frequencies are related by $\omega = \Omega T$, and the corresponding complex frequencies are related by $z = e^{sT}$. The complex-plane transformation under this mapping is shown in Figure 8.3.

3. The complex-plane mapping $z = e^{sT}$ maps the left-half s-plane onto the inside of the unit circle in the z-plane. Thus, this mapping is a stable one, but it is not a unique one. It is a *many-to-one* mapping that is also illustrated by (8.27). Therefore, this mapping suffers from the aliasing problem and can be used only for mapping sufficiently narrowband lowpass filters.

4. The design procedure using impulse-invariance mapping is as follows: Given the digital lowpass filter specifications ω_p, ω_s, R_p, and A_s, we first compute the analog band-edge frequencies, using

$$\Omega_p = \frac{\omega_p}{T} \text{ and } \Omega_s = \frac{\omega_s}{T} \tag{8.28}$$

in which any value of T is chosen ($T = 1$ is preferred). Then we design an analog lowpass filter $H_a(s)$, using any one of the prototype filters to satisfy the specifications Ω_p, Ω_s, R_p, and A_s. Finally, we map the poles $\{p_k\}$ of $H_a(s)$ into the corresponding poles $\left\{e^{p_k T}\right\}$ of the digital filter $H(z)$; that is,

$$H_a(s) = \frac{C(s)}{D(s)} = \sum_{k=1}^{N} \frac{R_k}{s - p_k} \longrightarrow H(z) = \sum_{k=1}^{N} \frac{R_k}{1 - e^{p_k T} z^{-1}} = \frac{B(z)}{A(z)} \tag{8.29}$$

The MATLAB function `[b,a] = imp_invr(c,d,T)` transforms the numerator and denominator polynomials of $H_a(s)$ into those of $H(z)$.

5. In the bilinear transformation, the complex frequencies are related by

$$s = \frac{2}{T}\frac{1 - z^{-1}}{1 + z^{-1}} \Longrightarrow z = \frac{1 + sT/2}{1 - sT/2} \tag{8.30}$$

The complex-plane mapping is shown in Figure 8.4. The real frequencies are related by

$$\omega = 2\tan^{-1}\left(\frac{\Omega T}{2}\right) \quad \text{or} \quad \Omega = \frac{2}{T}\tan\left(\frac{\omega}{2}\right) \tag{8.31}$$

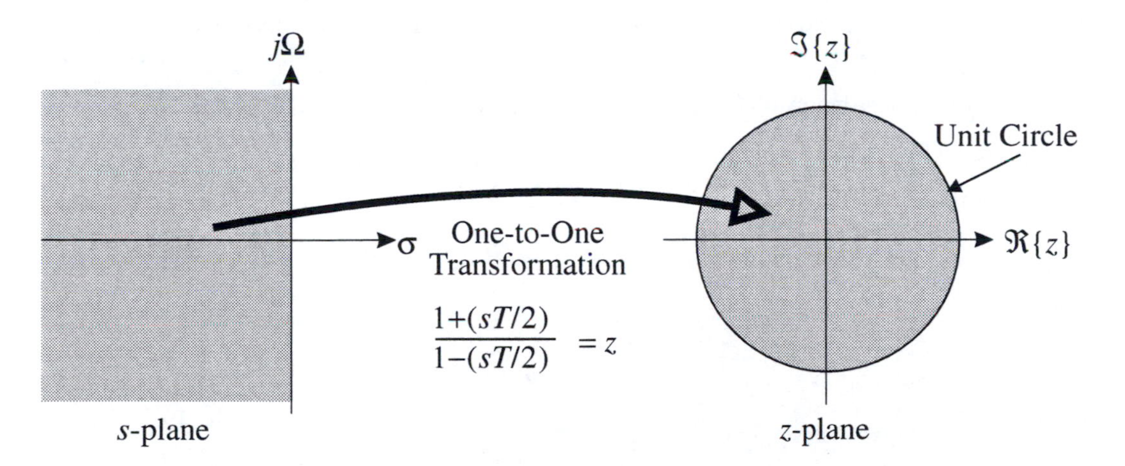

Figure 8.4: Complex-plane mapping in bilinear transformation

6. In this mapping, the entire left-half s-plane is mapped onto the inside of the unit circle in the z-plane in a unique fashion. Thus, this mapping is the best one, because it is a stable mapping, there is no aliasing problem, and there is no restriction on the type of filter that can be transformed.

7. The design procedure using the bilinear transformation is as follows: Given the digital lowpass filter specifications ω_p, ω_s, R_p, and A_s, we first determine the analog band-edge frequencies, using

$$\Omega_p = \frac{2}{T}\tan\left(\frac{\omega_p}{2}\right), \quad \Omega_s = \frac{2}{T}\tan\left(\frac{\omega_s}{2}\right) \tag{8.32}$$

in which any value of T is chosen ($T = 2$ is preferred). Then we design an analog lowpass filter $H_a(s)$, using any one of the prototype filters to satisfy the specifications Ω_p, Ω_s, R_p, and A_s. Finally, we set

$$H(z) = H_a\left(\frac{2}{T}\frac{1-z^{-1}}{1+z^{-1}}\right)$$

and simplify to obtain $H(z)$ as a rational function in z^{-1}. The MATLAB function `[b,a] = bilinear(c,d,T)` transforms the numerator and denominator polynomials of $H_a(s)$ into those of $H(z)$.

8.3 Frequency-Band Transformations

The final step in the IIR digital-filter design procedure is to convert a lowpass digital filter into any other type of frequency-selective multiband digital filter (including a lowpass filter). It is accomplished by transforming the frequency axis (or band) of a lowpass filter so that it behaves as another frequency-selective filter. These transformations on the complex variable z are very similar to bilinear transformations, and the design equations are algebraic. In this section, we present the use of the z-plane mapping and the design procedure to obtain a highpass digital filter. Several MATLAB functions that perform the transformation are then described. Study the following review topics, the related MATLAB functions, and the many examples given in the text to solve the practice problems given at the end of the chapter.

Study Topics: Frequency-Band Transformations for Digital Filters

1. Let $H_{\text{LP}}(Z)$ be the given prototype *lowpass* digital filter, and let $H(z)$ be another frequency-selective digital filter. Then a mapping of the form

$$Z^{-1} = G(z^{-1})$$

transforms

$$H(z) = H_{\text{LP}}(Z)|_{Z^{-1}=G(z^{-1})}$$

so that $H(z)$ is the desired digital filter.

2. Such a mapping $G(\cdot)$ must be a rational function in z^{-1}, so that $H(z)$ is implementable. Furthermore, the unit circle of the Z-plane must map onto the unit circle of the z-plane, and, for stable filters, the

inside of the unit circle of the Z-plane must map onto the inside of the unit circle of the z-plane. The mapping that satisfies these requirements is of the form

$$Z^{-1} = G\left(z^{-1}\right) = \pm \prod_{k=1}^{K} \frac{z^{-1} - \alpha_k}{1 - \alpha_k z^{-1}} \tag{8.33}$$

where $|\alpha_k| < 1$ and K are parameters that are to be determined for the given specifications. The most widely used transformations are given in Table 8.1. The text function `[bz,az] = zmapping(bZ,aZ,Nz,Dz)` transforms the numerator and denominator polynomials of $H_{\text{LP}}(Z)$ into those of $H(z)$ by using the numerator and denominator polynomials of the rational mapping $G(z^{-1})$.

3. The procedure for designing a highpass digital filter is as follows: Given the specifications ω_s, ω_p, R_p, and A_s of the highpass filter, we first obtain the specifications on an equivalent lowpass digital filter. We arbitrarily set a cutoff frequency ω_p' of the lowpass filter (e.g., $\omega_p' = 0.2\pi$), and, using the design relation from Table 8.1, we construct

$$\alpha = -\frac{\cos[(\omega_p + \omega_p')/2]}{\cos[(\omega_p - \omega_p')/2]} \tag{8.34}$$

With this value of α and ω_s, we compute the other cutoff frequency ω_s'. Now a lowpass digital filter $H_{\text{LP}}(Z)$ can be designed to satisfy the specifications ω_p', ω_s', R_p, and A_s. The desired highpass filter $H(z)$ can now be obtained as

$$H(z) = H_{\text{LP}}\left(Z = -\frac{z^{-1} + \alpha}{1 + \alpha z^{-1}}\right)$$

The text function `[b,a] = cheb1hpf(wp,ws,Rp,As)` implements this design procedure.

4. The highpass filter design procedure illustrated can easily be extended to other frequency-selective filters by using the transformation functions in Table 8.1. These are implemented in the MATLAB functions `butter`, `cheby1`, `cheby2`, and `ellip`.

8.4 Summary

MATLAB Functions

The following MATLAB functions are used in this chapter:

Analog-Filter Design

- `[C,B,A]=sdir2cas(b,a)`: Direct-form to cascade-form conversion in the s-plane
- `[db,mag,pha,w] = freqs_m(b,a,wmax)`: Modified version of `freqs` function
- `[b,a]=u_buttap(N,Omegac)`: Unnormalized Butterworth lowpass filter prototype
- `[b,a]=afd_butt(Wp,Ws,Rp,As)`: Butterworth lowpass filter design

Type of Transformation	Transformation	Parameters
Lowpass	$z^{-1} \longrightarrow \dfrac{z^{-1}-\alpha}{1-\alpha z^{-1}}$	$\omega_c' =$ cutoff frequency of new filter $\alpha = \dfrac{\sin\left[\left(\omega_c - \omega_c'\right)/2\right]}{\sin\left[\left(\omega_c + \omega_c'\right)/2\right]}$
Highpass	$z^{-1} \longrightarrow -\dfrac{z^{-1}+\alpha}{1+\alpha z^{-1}}$	$\omega_c' =$ cutoff frequency of new filter $\alpha = -\dfrac{\cos\left[\left(\omega_c + \omega_c'\right)/2\right]}{\cos\left[\left(\omega_c - \omega_c'\right)/2\right]}$
Bandpass	$z^{-1} \longrightarrow -\dfrac{z^{-2}-\alpha_1 z^{-1}+\alpha_2}{\alpha_2 z^{-2}-\alpha_1 z^{-1}+1}$	$\omega_\ell =$ lower cutoff frequency $\omega_u =$ upper cutoff frequency $\alpha_1 = -2\beta K/(K+1)$ $\alpha_2 = (K-1)/(K+1)$ $\beta = \dfrac{\cos\left[\left(\omega_u + \omega_\ell\right)/2\right]}{\cos\left[\left(\omega_u - \omega_\ell\right)/2\right]}$ $K = \cot\dfrac{\omega_u - \omega_\ell}{2}\tan\dfrac{\omega_c}{2}$
Bandstop	$z^{-1} \longrightarrow \dfrac{z^{-2}-\alpha_1 z^{-1}+\alpha_2}{\alpha_2 z^{-2}-\alpha_1 z^{-1}+1}$	$\omega_\ell =$ lower cutoff frequency $\omega_u =$ upper cutoff frequency $\alpha_1 = -2\beta/(K+1)$ $\alpha_2 = (K-1)/(K+1)$ $\beta = \dfrac{\cos\left[\left(\omega_u + \omega_\ell\right)/2\right]}{\cos\left[\left(\omega_u - \omega_\ell\right)/2\right]}$ $K = \tan\dfrac{\omega_u - \omega_\ell}{2}\tan\dfrac{\omega_c}{2}$

Table 8.1: Frequency transformation for digital filters (Prototype lowpass filter has cutoff frequency ω_c.)

- `[b,a]=u_chb1ap(N,Rp,Omegac)`: Unnormalized Chebyshev-I lowpass filter prototype
- `[b,a]=afd_chb1(Wp,Ws,Rp,As)`: Chebyshev-I lowpass filter design
- `[b,a]=u_chb2ap(N,As,Omegac)`: Unnormalized Chebyshev-II lowpass filter prototype
- `[b,a]=afd_chb2(Wp,Ws,Rp,As)`: Chebyshev-II lowpass filter design
- `[b,a]=u_elipap(N,Rp,As,Omegac)`: Unnormalized elliptic lowpass filter prototype
- `[b,a]=afd_elip(Wp,Ws,Rp,As)`: Elliptic lowpass filter design

Analog-to-Digital Filter Transformation

- `[b,a]=imp_invr(c,d,T)`: Impulse-invariance transformation
- `[b,a]=bilinear(c,d,Fs)`: Bilinear transformation

Frequency-Band Transformation

- `[bz,az]=zmapping(bZ,aZ,Nz,Dz)`: Frequency-band mapping
- `[b,a]=chb1hpf(wp,ws,Rp,As)`: Highpass filter design via Chebyshev-I prototype
- `[N,wn]=buttord(wp,ws,Rp,As)`: Order and cutoff frequency computation for Butterworth
- `[b,a]=butter(N,wn,type)`: Digital-filter design via Butterworth prototype
- `[N,wn]=chebyord(wp,ws,Rp,As)`: Order and cutoff frequency computation for Chebyshev
- `[b,a]=Cheby1(N,Rp,wn,type)`: Digital-filter design via Chebyshev-I prototype
- `[b,a]=Cheby2(N,As,wn,type)`: Digital-filter design via Chebyshev-II prototype
- `[N,wn]=ellipord(wp,ws,Rp,As)`: Order and cutoff frequency computation for elliptic
- `[b,a]=ellip(N,Rp,As,wn,type)`: Digital-filter design via elliptic prototype

8.5 Problems

P8.1 Design an analog Butterworth lowpass filter that has a 1 dB or lower ripple at 30 rad/sec and at least 30 dB of attenuation at 40 rad/sec. Determine the system function in a cascade form. Plot the magnitude response, the log-magnitude response in dB, the phase response, and the impulse response of the filter.

P8.2 Design a lowpass analog elliptic filter having all of the following characteristics:

- an acceptable passband ripple of 1 dB,
- passband cutoff frequency of 10 rad/sec
- stopband attenuation of 40 dB or greater beyond 15 rad/sec

Determine the system function in a rational-function form. Plot the magnitude response, the log-magnitude response in dB, the phase response, and the impulse response of the filter.

P8.3 A signal $x_a(t)$ contains two frequencies, 100 Hz and 130 Hz. We want to suppress the 130-Hz component to 50-dB attenuation while passing the 100-Hz component with less than 2-dB attenuation. Design a minimum-order Chebyshev-I analog filter to perform this filtering operation. Plot the log-magnitude response, and verify the design.

P8.4 Design an analog Chebyshev-II lowpass filter that has a 0.5 dB or better ripple at 250 Hz and at least 45 dB of attenuation at 300 Hz. Plot the magnitude response, the log-magnitude response in dB, the phase response, and the impulse response of the filter.

P8.5 Write a MATLAB function to design analog lowpass filters. The format of this function should be as follows:

```
function [b,a] = afd(type,Fp,Fs,Rp,As)
%
% function [b,a] = afd(type,Fp,Fs,Rp,As)
%   Designs analog lowpass filters
% type = 'butter' or 'cheby1' or 'cheby2' or 'ellip'
%   Fp = passband cutoff in Hz
%   Fs = stopband cutoff in Hz
%   Rp = passband ripple in dB
%   As = stopband attenuation in dB
```

Use the `afd_butt`, `afd_chb1`, `afd_chb2`, and `afd_elip` functions developed in this chapter. Check your function on specifications given in Problems P8.1 through P8.4.

P8.6 Design a lowpass digital filter to be used in a system, as shown,

$$x_a(t) \longrightarrow \boxed{\text{A/D}} \longrightarrow \boxed{H(z)} \longrightarrow \boxed{\text{D/A}} \longrightarrow y_a(t)$$

to satisfy the following requirements:

- sampling rate, 8000 sam/sec
- passband edge, 1500 Hz with ripple of 3 dB or less
- stopband edge, 2000 Hz with attenuation by 40 dB or more
- equiripple passband, but monotone stopband
- impulse-invariance method

(a) Choose $T = 1$ in the impulse-invariance method and construct the system function $H_a(s)$ in cascade form. Plot the log-magnitude response in dB and the impulse response $h(n)$.

(b) Choose $T = 1/8000$ in the impulse-invariance method and determine the system function $H_a(s)$ in cascade form. Plot the log-magnitude response in dB and the impulse response $h(n)$. Compare this design with that in part (a); and comment on the effect of T on the impulse-invariance design.

P8.7 Design a Butterworth digital lowpass filter to satisfy the following specifications:

$$\begin{array}{lll} \text{passband edge:} & 0.4\pi, & R_p = 0.5 \text{ dB} \\ \text{stopband edge:} & 0.6\pi, & As = 50 \text{ dB} \end{array}$$

Use the impulse-invariance method with $T = 2$. Construct the system function in the rational form, and plot the log-magnitude response in dB. Plot the impulse response $h(n)$ and the impulse response $h_a(t)$ of the analog prototype, and compare their shapes.

P8.8 Write a MATLAB function to design digital lowpass filters that use the impulse-invariance transformation. The format of this function should be as follows:

```
function [b,a] = dlpfd_ii(type,wp,ws,Rp,As,T)
%
% function [b,a] = dlpfd_ii(type,wp,ws,Rp,As,T)
%   Designs digital lowpass filters using impulse invariance
% type = 'butter' or 'cheby1'
%   wp = passband cutoff in Hz
%   ws = stopband cutoff in Hz
%   Rp = passband ripple in dB
%   As = stopband attenuation in dB
%    T = sampling interval
```

Use the afd function developed in Problem P8.5. Check your function on the specifications given in Problems P8.6 and P8.7.

P8.9 Consider the design of the lowpass Butterworth filter of Problem P8.7.

(a) Use the bilinear transformation technique outlined in this chapter and the bilinear function. Plot the log-magnitude response in dB. Compare the impulse responses of the analog prototype and the digital filter.

(b) Use the butter function, and compare this design with the one from part (a).

P8.10 Design the analog Chebyshev-I filter having the specification of Problem P8.6, using the bilinear transformation method. Compare the two designs.

P8.11 Design a digital lowpass filter, using an elliptic prototype, to satisfy the following requirements:

$$\begin{aligned} &\text{passband edge:} \quad 0.4\pi, \quad R_p = 1 \text{ dB} \\ &\text{stopband edge:} \quad 0.6\pi, \quad A_s = 60 \text{ dB} \end{aligned}$$

Use both the bilinear and the ellip functions, and compare your designs.

P8.12 Develop a MATLAB function to design a highpass digital filter by using the bilinear transformation. The format of this function should be as follows:

```
function [b,a] = dhpfd_bl(type,wp,ws,Rp,As)
% IIR Highpass filter design using bilinear transformation
% [b,a] = dhpfd_bl(type,wp,ws,Rp,As)
% type = 'butter' or 'cheby1' or 'cheby2' or 'ellip'
%    b = Numerator polynomial of the highpass filter
%    a = Denominator polynomial of the highpass filter
%   wp = Passband frequency in radians
%   ws = Stopband frequency in radians (wp < ws)
%   Rp = Passband ripple in dB
%   As = Stopband attenuation in dB
```

A highpass filter is to be designed to satisfy the following specifications:

$$\begin{array}{ll} \text{stopband edge: } 0.4\pi, & A_s = 60 \text{ dB} \\ \text{passband edge: } 0.6\pi, & R_p = 0.5 \text{ dB} \end{array}$$

(a) Use the `dhpfd_bl` function and the elliptic prototype to design this filter. Plot the log-magnitude response in dB of the designed filter.

(b) Use the `ellip` function for design, and plot the log-magnitude response in dB. Compare these two designs.

P8.13 Design a bandpass digital filter by using the `Cheby2` function. The specifications are as follows:

$$\begin{array}{lll} \text{lower stopband edge: } & 0.3\pi & \multirow{2}{*}{$A_s = 50 \text{ dB}$} \\ \text{upper stopband edge: } & 0.6\pi & \\ \text{lower passband edge: } & 0.4\pi & \multirow{2}{*}{$R_p = 0.5 \text{ dB}$} \\ \text{upper passband edge: } & 0.5\pi & \end{array}$$

Plot the impulse response and the log-magnitude response in dB of the designed filter.

Part II

Solutions to Problems

Appendix A

Discrete-Time Signals and Systems

P2.1 (a) $x_1(n) = \sum_{m=0}^{10} (m+1)\left[\delta(n-2m) - \delta(n-2m-1)\right], 0 \le n \le 25.$

```
clear; close all;
Hf_1 = figure('Units','normalized','position',[0.1,0.1,0.8,0.8],'color',[0,0,0]);
set(Hf_1,'NumberTitle','off','Name','P2.1ac');
%
% x1(n) = sum_{m=0}^{10} (m+1)*[delta(n-2*m)-delta(n-2*m-1)]
n1 = [0:25]; x1 = zeros(1,length(n1));
for m = 0:10
    x1 = x1 + (m+1)*(impseq(2*m,0,25) - impseq(2*m+1,0,25));
end
subplot(2,1,1); stem(n1,x1);
axis([min(n1)-1,max(n1)+1,min(x1)-2,max(x1)+2]);
xlabel('n'); ylabel('x1(n)'); title('Sequence x1(n)');
ntick = [n1(1):n1(length(n1))];
set(gca,'XTickMode','manual','XTick',ntick,'FontSize',10)
```

The plot of $x_1(n)$ is shown in Figure A.1.

(b) $x_2(n) = n^2\left[u(n+5) - u(n-6)\right] + 10\delta(n) + 20(0.5)^n\left[u(n-4) - u(n-10)\right].$

```
clear; close all;
Hf_1 = figure('Units','normalized','position',[0.1,0.1,0.8,0.8],'color',[0,0,0]);
set(Hf_1,'NumberTitle','off','Name','P2.1be');
%
% (b) x2(n) = (n^2)*[u(n+5)-u(n-6)]+10*delta(n)+20*(0.5)^n*[u(n-4)-u(n-10)]
n2 = -5:10; % Overall support of x2(n)
x2 = (n2.^2).*(stepseq(-5,-5,10)-stepseq(6,-5,10))+10*impseq(0,-5,10)+...
      20*((0.5).^n2).*(stepseq(4,-5,10)-stepseq(10,-5,10));
subplot(2,1,1); stem(n2,x2);
axis([min(n2)-1,max(n2)+1,min(x2)-2,max(x2)+2]);
xlabel('n'); ylabel('x1(n)'); title('Sequence x2(n)');
ntick = [n2(1):n2(length(n2))];
set(gca,'XTickMode','manual','XTick',ntick,'FontSize',10)
```

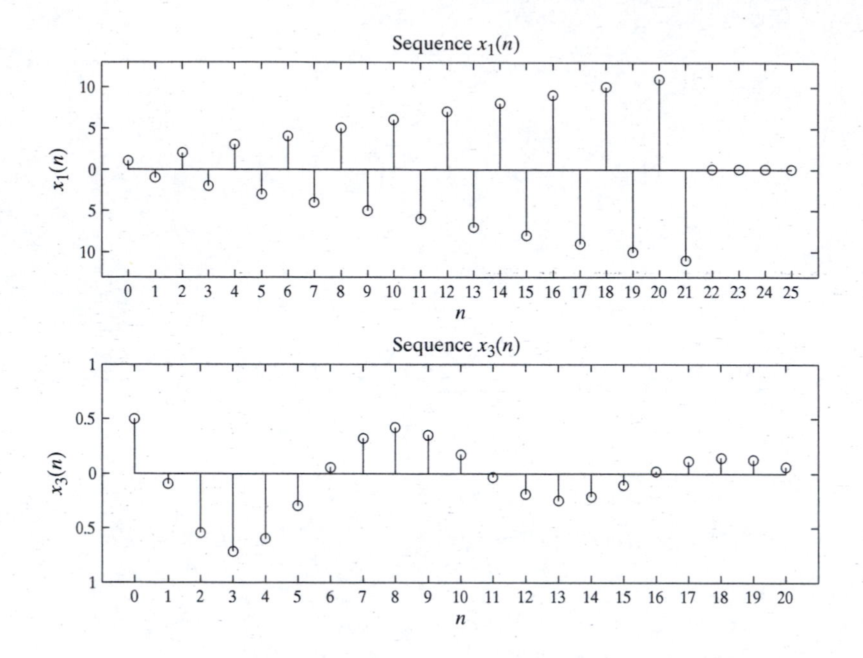

Figure A.1: Problem P2.1 sequence plots

The plot of $x_2(n)$ is shown in Figure A.3.

(c) $x_3(n) = (0.9)^n \cos(0.2\pi n + \pi/3), \quad 0 \le n \le 20.$

```
% x3(n) = (0.9)^n*cos(0.2*pi*n+pi/3); 0<=n<=20
n3 = [0:20];
x3 = ((0.9).^n3).*cos(0.2*pi*n3+pi/3);
subplot(2,1,2); stem(n3,x3);
axis([min(n3)-1,max(n3)+1,-1,1]);
xlabel('n'); ylabel('x3(n)'); title('Sequence x3(n)');
ntick = [n3(1):n3(length(n3))];
set(gca,'XTickMode','manual','XTick',ntick,'FontSize',10)
```

The plot of $x_3(n)$ is shown in Figure A.1.

(d) $x_4(n) = 10\cos\left(0.0008\pi n^2\right) + w(n)$, $0 \le n \le 100$, where $w(n)$ is a random sequence distributed uniformly between $[-1, 1]$.

```
clear; close all;
% (d) x4(n) = 10*cos(0.0008*pi*n.^2)+w(n); 0 <= n <= 100; w(n)~uniform[-1,1]
w = 2*(rand(1,101)-0.5);
n4 = [0:100]; x4 = 10*cos(0.0008*pi*n4.^2)+w;
subplot(2,1,2); stem(n4,x4); axis([min(n4)-1,max(n4)+1,min(x4)-2,max(x4)+2]);
xlabel('n'); ylabel('x4(n)'); title('Sequence x4(n)');
ntick = [n4(1):10:n4(length(n4))];
set(gca,'XTickMode','manual','XTick',ntick,'FontSize',10)
```

The plot of $x_4(n)$ is shown in Figure A.2, from which we observe that it is a noisy sinusoid with increasing frequency (or a noisy *chirp* signal).

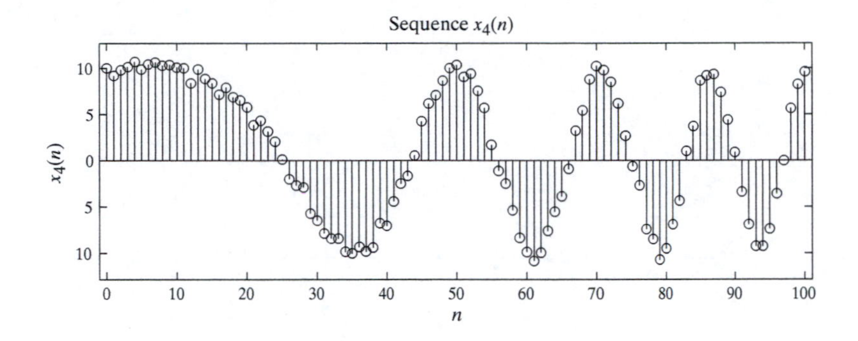

Figure A.2: Plot of the sequence $x_4(n)$ in Problem P2.1(d).

(e) $\tilde{x}_5(n) = \left\{\ldots, 1, 2, 3, \underset{\uparrow}{2}, 1, 2, 3, 2, 1, \ldots\right\}$; plot five periods.

```
% (e) x5(n) = {...,1,2,3,2,1,2,3,2,1,...}periodic. 5 periods
n5 = [-8:11]; x5 = [2,1,2,3];
x5 = x5'*ones(1,5); x5 = (x5(:))';
subplot(2,1,2); stem(n5,x5);
axis([min(n5)-1,max(n5)+1,0,4]);
xlabel('n'); ylabel('x5(n)'); title('Sequence x5(n)');
ntick = [n5(1):n5(length(n5))];
set(gca,'XTickMode','manual','XTick',ntick,'FontSize',10)
```

The plot of $x_5(n)$ is shown in Figure A.3.

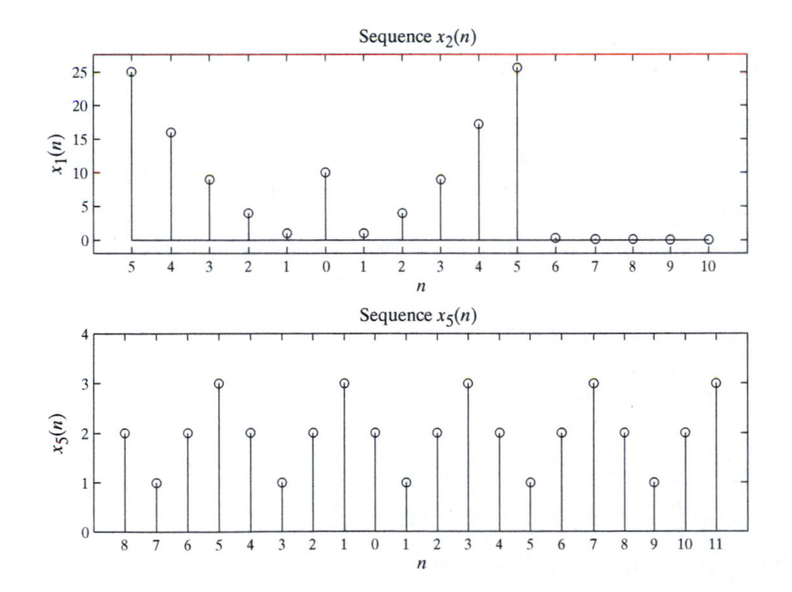

Figure A.3: Problem P2.1 sequence plots

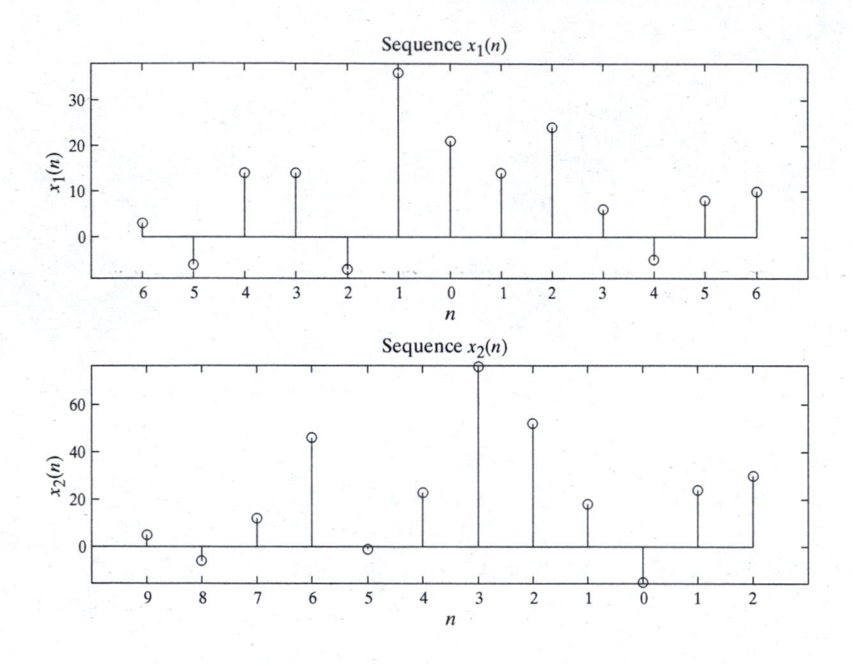

Figure A.4: Problem P2.2 sequence plots

P2.2 The sequence $x(n) = \{1, -2, 4, 6, -5, 8, 10\}$ is given.

(a) $x_1(n) = 3x(n+2) + x(n-4) - 2x(n)$.

```
clear; close all;
Hf_1 = figure('Units','normalized','position',[0.1,0.1,0.8,0.8],'color',[0,0,0]);
set(Hf_1,'NumberTitle','off','Name','P2.2ab');
n = [-4:2]; x = [1,-2,4,6,-5,8,10]; % given seq x(n)
%
% (a) x1(n) = 3*x(n+2) + x(n-4) - 2*x(n)
[x11,n11] = sigshift(3*x,n,-2);        % shift by -2 and scale by 3
[x12,n12] = sigshift(x,n,4);           % shift x(n) by 4
[x13,n13] = sigadd(x11,n11,x12,n12);   % add two sequences at time
[x1,n1] = sigadd(x13,n13,2*x,n);       % add two sequences
subplot(2,1,1); stem(n1,x1);
axis([min(n1)-1,max(n1)+1,min(x1)-2,max(x1)+2]);
xlabel('n'); ylabel('x1(n)'); title('Sequence x1(n)');
ntick = [n1(1):1:n1(length(n1))];
set(gca,'XTickMode','manual','XTick',ntick,'FontSize',10);
```

The plot of $x_1(n)$ is shown in Figure A.4.

(b) $x_2(n) = 5x(5+n) + 4x(n+4) + 3x(n)$.

```
% (b) x2(n) = 5*x(5+n) + 4*x(n+4) +3*x(n)
[x21,n21] = sigshift(5*x,n,-5);
[x22,n22] = sigshift(4*x,n,-4);
[x23,n23] = sigadd(x21,n21,x22,n22);
```

```
[x2,n2] = sigadd(x23,n23,3*x,n);
subplot(2,1,2); stem(n2,x2);
axis([min(n2)-1,max(n2)+1,min(x2)-0.5,max(x2)+0.5]);
xlabel('n'); ylabel('x2(n)'); title('Sequence x2(n)');
ntick = [n2(1):1:n2(length(n2))];
set(gca,'XTickMode','manual','XTick',ntick,'FontSize',10)
```

The plot of $x_2(n)$ is shown in Figure A.4.

(c) $x_3(n) = x\,(n+4)\,x(n-1) + x(2-n)x(n)$.

```
clear; close all;
Hf_1 = figure('Units','normalized','position',[0.1,0.1,0.8,0.8],'color',[0,0,0]);
set(Hf_1,'NumberTitle','off','Name','P2.2cd');
n = [-4:2]; x = [1,-2,4,6,-5,8,10]; % given seq x(n)
%
% (c) x3(n) = x(n+4)*x(n-1) + x(2-n)*x(n)
[x31,n31] = sigshift(x,n,-4);          % shift x(n) by -4
[x32,n32] = sigshift(x,n,1);           % shift x(n) by 1
[x33,n33] = sigmult(x31,n31,x32,n32); % multiply two sequences.
[x34,n34] = sigfold(x,n);              % fold x(n)
[x34,n34] = sigshift(x34,n34,2);       % shift x(-n) by 2
[x34,n34] = sigmult(x34,n34,x,n);      % shift x(-n) by 2
  [x3,n3] = sigadd(x33,n33,x34,n34);   % add two sequences
subplot(2,1,1); stem(n3,x3);
axis([min(n3)-1,max(n3)+1,min(x3)-2,max(x3)+2]);
xlabel('n'); ylabel('x3(n)'); title('Sequence x3(n)');
ntick = [n3(1):1:n3(length(n3))];
set(gca,'XTickMode','manual','XTick',ntick,'FontSize',10);
```

The plot of $x_3(n)$ is shown in Figure A.5.

(d) $x_4(n) = 2e^{0.5n}x(n) + \cos(0.1\pi n)\,x\,(n+2)\,,\ -10 \leq n \leq 10.$

```
% (d) x4(n) = 2*exp(0.5*n)*x(n)+cos(0.1*pi*n)*x(n+2); -10 <= n <= 10
n4 = [-10:10]; x41 = 2*exp(0.5*n4); x412 = cos(0.1*pi*n4);
[x42,n42] = sigmult(x41,n4,x,n);
[x43,n43] = sigshift(x,n,-2);
[x44,n44] = sigmult(x412,n4,x43,n43);
  [x4,n4] = sigadd(x42,n42,x44,n44);
subplot(2,1,2); stem(n4,x4);
axis([min(n4)-1,max(n4)+1,min(x4)-0.5,max(x4)+0.5]);
xlabel('n'); ylabel('x4(n)'); title('Sequence x4(n)');
ntick = [n4(1):1:n4(length(n4))];
set(gca,'XTickMode','manual','XTick',ntick,'FontSize',10)
```

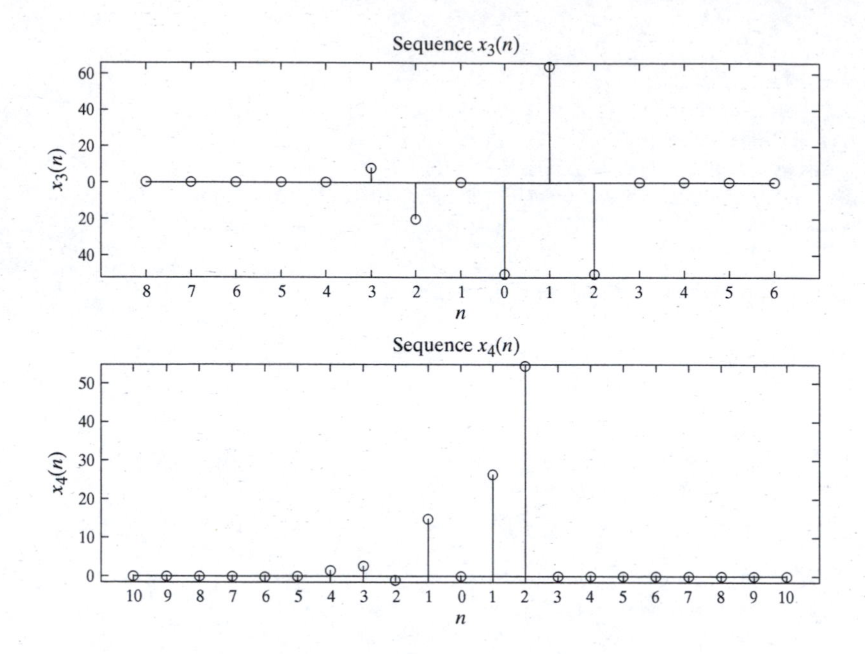

Figure A.5: Problem P2.2 sequence plots

The plot of $x_4(n)$ is shown in Figure A.5.

(e) $x_5(n) = \sum_{k=1}^{5} nx\,(n-k)$ where $x(n) = \{1, -2, 4, 6, -5, 8, 10\}$.

```
clear; close all;

n = [-4:2]; x = [1,-2,4,6,-5,8,10]; % given seq x(n)
% (e) x5(n) = sum_{k=1}^{5}n*x(n-k);
[x51,n51] = sigshift(x,n,1); [x52,n52] = sigshift(x,n,2);
[x5,n5] = sigadd(x51,n51,x52,n52);
[x53,n53] = sigshift(x,n,3); [x5,n5] = sigadd(x5,n5,x53,n53);
[x54,n54] = sigshift(x,n,4); [x5,n5] = sigadd(x5,n5,x54,n54);
[x55,n55] = sigshift(x,n,5); [x5,n5] = sigadd(x5,n5,x55,n55);
[x5,n5] = sigmult(x5,n5,n5,n5);
subplot(2,1,2); stem(n5,x5); axis([min(n5)-1,max(n5)+1,min(x5)-2,max(x5)+2]);
xlabel('n'); ylabel('x5(n)'); title('Sequence x5(n)');
ntick = [n5(1):1:n5(length(n5))];
set(gca,'XTickMode','manual','XTick',ntick,'FontSize',10);
```

The plot of $x_5(n)$ is shown in Figure A.6.

P2.3 A sequence $x(n)$ is periodic if $x\,(n+N) = x(n)$ for all n. Consider a complex exponential sequence $e^{j\omega_0 n} = e^{j2\pi f_0 n}$.

(a) Analytical proof: The preceding sequence is periodic if

$$e^{j2\pi f_0(n+N)} = e^{j2\pi f_0 n}$$

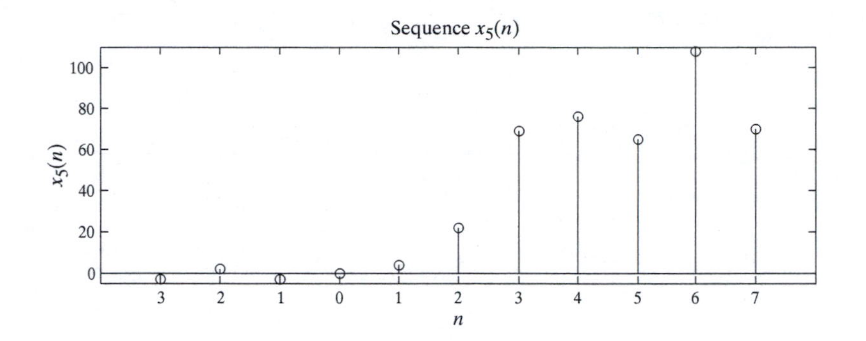

Figure A.6: Plot of the sequence $x_5(n)$ in Problem P2.2(e)

or if

$$e^{j2\pi f_0 N} = 1 \Rightarrow f_0 N = K \text{ (an integer)}$$

which proves the desired result.

(b) $x_1(n) = \cos(0.3\pi n), \quad -20 \leq n \leq 20.$

```
% (b) x1(n) = cos(0.3*pi*n)
x1 = cos(0.3*pi*n);
subplot(2,1,1); stem(n,x1);
axis([min(n)-1,max(n)+1,-1.1,1.1]);
ylabel('x1(n)'); title('Sequence cos(0.3*pi*n)');
ntick = [n(1):5:n(length(n))];
set(gca,'XTickMode','manual','XTick',ntick,'FontSize',10);
```

Because $f_0 = 0.3/2 = 3/20$, the sequence is periodic. From the plot in Figure A.7, we see that, in one period of 20 samples, $x_1(n)$ exhibits three cycles. This is true whenever K and N are relatively prime.

(c) $x_2(n) = \cos(0.3n), \quad -20 \leq n \leq 20.$

```
% (b) x2(n) = cos(0.3*n)
x2 = cos(0.3*n);
subplot(2,1,2); stem(n,x2);
axis([min(n)-1,max(n)+1,-1.1,1.1]);
ylabel('x2(n)'); title('Sequence cos(0.3*n)');
ntick = [n(1):5:n(length(n))];
set(gca,'XTickMode','manual','XTick',ntick,'FontSize',10);
```

In this case, f_0 is not a rational number; hence, the sequence $x_2(n)$ is not periodic. This can be clearly seen from the plot of $x_2(n)$ in Figure A.7.

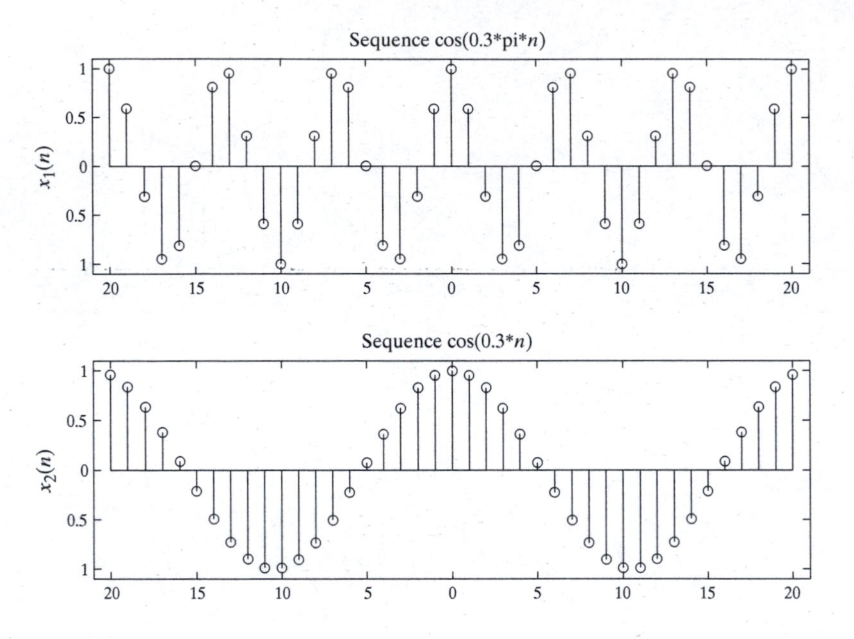

Figure A.7: Problem P2.3 sequence plots

P2.4 Even–odd decomposition of complex-valued sequences.

(a) MATLAB function evenodd:

```
function [xe, xo, m] = evenodd(x,n)
% Complex-valued signal decomposition into even and odd parts
% ---------------------------------------------------------------
% [xe, xo, m] = evenodd(x,n)
%
[xc,nc] = sigfold(conj(x),n);
[xe,m] = sigadd(0.5*x,n,0.5*xc,nc);
[xo,m] = sigadd(0.5*x,n,-0.5*xc,nc);
```

(b) Even–odd decomposition of $x(n) = 10e^{-j(0.4\pi n)}, 0 \le n \le 10$.

```
n = 0:10; x = 10*exp(-0.4*pi*n);
[xe,xo,neo] = evenodd(x,n);
Re_xe = real(xe); Im_xe = imag(xe);
Re_xo = real(xo); Im_xo = imag(xo);
% Plots of the sequences

subplot(2,2,1); stem(neo,Re_xe);
ylabel('Re{xe(n)}'); title('Real part of Even Seq.');
subplot(2,2,3); stem(neo,Im_xe);
xlabel('n'); ylabel('Im{xe(n)}'); title('Imag part of Even Seq.');
subplot(2,2,2); stem(neo,Re_xo);
ylabel('Re{xo(n)}'); title('Real part of Odd Seq.');
subplot(2,2,4); stem(neo,Im_xo);
xlabel('n'); ylabel('Im{xo(n)}'); title('Imag part of Odd Seq.');
```

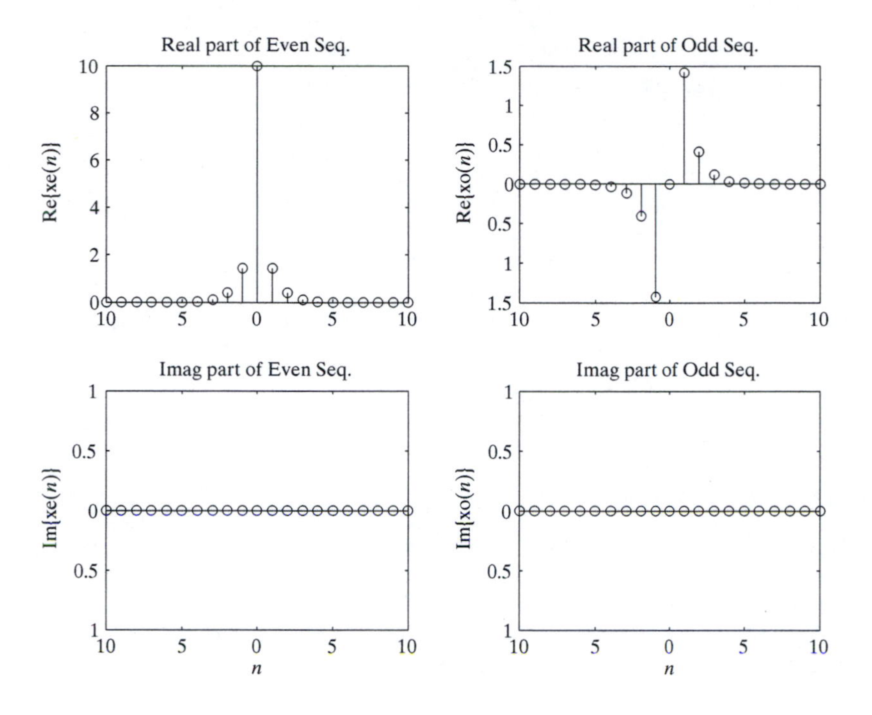

Figure A.8: Plots in Problem P2.4

The MATLAB verification plots are shown in Figure A.8.

P2.5 Three Systems are:

$$\begin{aligned} T_1\,[x(n)] &: \quad y(n) = 2^{x(n)} \\ T_2\,[x(n)] &: \quad y(n) = 3\,x(n) + 4 \\ T_3\,[x(n)] &: \quad y(n) = x(n) + 2\,x(n-1) - x(n-2) \end{aligned}$$

(a) System-1 $T_1[\cdot]$: Let $y_1(n)$ be the output due to the input $x_1(n)$ and let $y_2(n)$ be the output due to the input $x_2(n)$. Then

$$y_1(n) = 2^{x_1(n)}; \quad y_2(n) = 2^{x_2(n)}$$

Let $x_1(n) + x_2(n)$ be the input to the system. Then the output is

$$2^{x_1(n)+x_2(n)} = 2^{x_1(n)} 2^{x_2(n)} = y_1(n)\,y_2(n) \neq y_1(n) + y_2(n)$$

Hence, System-1 is not linear.

System-2 $T_2[\cdot]$: Let $y_1(n)$ be the output due to the input $x_1(n)$ and let $y_2(n)$ be the output due to the input $x_2(n)$. Then

$$y_1(n) = 3\,x_1(n) + 4; \quad y_2(n) = 3\,x_2(n) + 4$$

Let $x_1(n) + x_2(n)$ be the input to the system. Then the output is

$$3\,[x_1(n) + x_2(n)] + 4 = 3\,x_1(n) + 3x_2(n) + 4 = y_1(n) + y_2(n) - 4 \neq y_1(n) + y_2(n)$$

Hence, System-2 is not linear.

System-3 $T_3[\cdot]$: Let $y_1(n)$ be the output due to the input $x_1(n)$ and let $y_2(n)$ be the output due to the input $x_2(n)$. Then

$$y_1(n) = x_1(n) + 2\,x_1(n-1) - x_1(n-2); \quad y_2(n) = x_2(n) + 2\,x_2(n-1) - x_2(n-2)$$

Let $x_1(n) + x_2(n)$ be the input to the system. Then the output is

$$\begin{aligned}[x_1(n) + x_2(n)] + 2\,[x_1(n) + x_2(n)] - [x_1(n) + x_2(n)] &= [x_1(n) + 2\,x_1(n-1) - x_1(n-2)] \\ &\quad + x_2(n) + 2\,x_2(n-1) - x_2(n-2) \\ &= y_1(n) + y_2(n)\end{aligned}$$

Hence, System-3 is linear.

(b) MATLAB verification:

```
clc;

% System-1: y(n) = 2^{x(n)}
n = 0:100; Ln = length(n); x1 = rand(Ln,1); x2 = randn(Ln,1); y1 =
2.^x1; y2 = 2.^x2; y = 2.^(x1+x2); diff = sum(abs(y - (y1+y2)));
if (diff < 1e-5)
    disp(' *** System-1 is Linear *** ');
else
    disp(' *** System-1 is NonLinear *** ');
end

 *** System-1 is NonLinear ***

% System-2: y(n) = 3*x(n)+4
n = 0:100; Ln = length(n); x1 = rand(Ln,1); x2 = randn(Ln,1); y1 =
3*x1+4; y2 = 3*x2+4; y = 3*(x1+x2)+4; diff = sum(abs(y -
(y1+y2))); if (diff < 1e-5)
    disp(' *** System-2 is Linear *** ');
else
    disp(' *** System-2 is NonLinear *** ');
end

 *** System-2 is NonLinear ***

% System-3: y(n) = x(n) - 2*x(n-1) + x(n-2)
n = 0:100; Ln = length(n); x1 = rand(Ln,1); x2 = randn(Ln,1);
[x11,nx11] = sigshift(x1,n,1); [x12,nx12] = sigshift(x1,n,2);
[x21,nx21] = sigshift(x2,n,1); [x22,nx22] = sigshift(x2,n,2);
[y11,ny11] = sigadd(x1,n,-2*x11,nx11); [y1,ny1] =
sigadd(y11,ny11,x12,nx12); [y21,ny21] = sigadd(x2,n,-2*x21,nx21);
[y2,ny2] = sigadd(y21,ny21,x22,nx22); x12 = x1 + x2; [x121,nx121]
= sigshift(x12,n,1); [x122,nx122] = sigshift(x12,n,2); [y,ny] =
sigadd(x12,n,-2*x121,nx121); [y,ny] = sigadd(y,ny,x122,nx122);
diff = sum(abs(y - (y1+y2))); if (diff < 1e-5)
```

```
    disp(' *** System-3 is Linear *** ');
else
    disp(' *** System-3 is NonLinear *** ');
end

 *** System-3 is Linear ***
```

P2.6 Three Systems are:

$$T_1[x(n)] \quad : \quad y(n) = \sum_{k=0}^{n} x(k)$$
$$T_2[x(n)] \quad : \quad y(n) = \sum_{k=n-10}^{n+10} x(k)$$
$$T_3[x(n)] \quad : \quad y(n) = x(-n)$$

(a) System-1 $T_1[\cdot]$: Let $x(n-l)$ be the input [which is an l-sample shifted version of the input $x(n)$] to the system. Then the output is

$$\sum_{k=0}^{n} x(k-l) = \sum_{k=-l}^{n-l} x(k) \neq \sum_{k=0}^{n-l} x(k) = y(n-l)$$

Hence, System-1 is not time-invariant.

System-2 $T_2[\cdot]$: Let $x(n-l)$ be the input [which is an l-sample shifted version of the input $x(n)$] to the system. Then the output is

$$\sum_{k=n-10}^{n+10} x(k-l) = \sum_{k=n-l-10}^{n-l+10} x(k) = y(n-l)$$

Hence, System-2 is time-invariant.

System-3 $T_3[\cdot]$: Let $x(n-l)$ be the input [which is an l-sample shifted version of the input $x(n)$] to the system. Then the output is

$$x(-n-l) \neq x(-n+l) = y(n-l)$$

Hence, System-3 is not time-invariant.

(b) MATLAB verification:

```
clc;

% System-1: y(n) = sum_{k=0}^{k=n}x(k)
n = 0:100; Ln = length(n); x = randn(Ln,1); y = cumsum(x); xx =
[0;x]; yy = cumsum(xx); diff = sum(abs(y - yy(1:length(y)))); if
(diff < 1e-5)
    disp(' *** System-1 is Time-Invariant *** ');
else
    disp(' *** System-1 is Time-Varying *** ');
end
```

```
*** System-1 is Time-Varying ***

% System-2: y(n) = sum_{k=n-10}^{k=n+10}x(k)
n = 0:100; Ln = length(n); x = randn(Ln,1); T =
toeplitz([1,zeros(1,90)],[ones(1,11),zeros(1,90)]); y = T*x; xx =
[0;x]; T = toeplitz([1,zeros(1,91)],[ones(1,11),zeros(1,91)]); yy
= T*xx; yy = yy(2:end); diff = sum(abs(y - yy(1:length(y)))); if
(diff < 1e-5)
    disp(' *** System-2 is Time-Invariant *** ');
else
    disp(' *** System-2 is Time-Varying *** ');
end

*** System-2 is Time-Invariant ***

% System-3: y(n) = x(-n)
n = 0:100; Ln = length(n); x = randn(Ln,1); [y,ny] = sigfold(x,n);
[x1,nx1] = sigshift(x,n,1); [y1,ny1] = sigfold(x1,nx1); [yy,nyy] =
sigshift(y,ny,1); [diff,ndiff] = sigadd(yy,nyy,-y1,ny1); diff =
sum(abs(diff)); if (diff < 1e-5)
    disp(' *** System-3 is Time-Invariant *** ');
else
    disp(' *** System-3 is Time-Varying *** ');
end

*** System-3 is Time-Varying ***
```

P2.7 Three Systems in Problem P2.5 are:

$$\begin{aligned} T_1\left[x(n)\right] &: \quad y(n) = 2^{x(n)} \\ T_2\left[x(n)\right] &: \quad y(n) = 3\,x(n) + 4 \\ T_3\left[x(n)\right] &: \quad y(n) = x(n) + 2\,x(n-1) - x(n-2) \end{aligned}$$

(a) **BIBO stability**

System-1 $T_1[\cdot]$: Let $x(n)$ be bounded by a number M i.e., $|x(n)| \le M < \infty$. Consider

$$|y(n)| = \left|2^{x(n)}\right| \le 2^{|x(n)|} \le 2^M < \infty$$

Hence, System-1 is BIBO stable.

System-2 $T_2[\cdot]$: Let $x(n)$ be bounded by a number M i.e., $|x(n)| \le M < \infty$. Consider

$$|y(n)| = |3\,x(n) + 4| \le 3\,|x(n)| + 4 \le 3M + 4 < \infty$$

Hence, System-2 is BIBO stable.

System-3 $T_3[\cdot]$: Let $x(n)$ be bounded by a number M i.e., $|x(n)| \le M < \infty$. Consider

$$\begin{aligned} |y(n)| &= |\,x(n) + 2\,x(n-1) - x(n-2)| \\ &\le |x(n)| + 2\,|x(n-1)| + |x(n-2)| \\ &\le M + 2M + M = 4M < \infty \end{aligned}$$

Hence, System-3 is BIBO stable.

(b) **Causality**

System-1 $T_1[\cdot]$: Since the output $y(n) = 2^{x(n)}$ depends on the present value of the input $x(n)$, the System-1 is causal.

System-2 $T_2[\cdot]$: Since the output $y(n) = 3\,x(n) + 4$ depends on the present value of the input $x(n)$, the System-2 is causal.

System-3 $T_3[\cdot]$: Since the output $y(n) = x(n) + 2\,x(n-1) - x(n-2)$ depends on the present and past two values of the input $x(n)$, the System-3 is causal.

Three Systems in Problem P2.6 are:

$$
\begin{aligned}
T_1\,[x(n)] &: \quad y(n) = \sum_{k=0}^{n} x(k) \\
T_2\,[x(n)] &: \quad y(n) = \sum_{k=n-10}^{n+10} x(k) \\
T_3\,[x(n)] &: \quad y(n) = x(-n)
\end{aligned}
$$

(a) **Stability**

System-1 $T_1[\cdot]$: Let $x(n)$ be bounded by a number M i.e., $|x(n)| \le M < \infty$. Consider

$$
\begin{aligned}
|y(n)| &= \left|\sum_{k=0}^{n} x(k)\right| \le \sum_{k=0}^{n} |x(k)| \le \sum_{k=0}^{n} M \\
&= (n+1)\,M \underset{n\to\infty}{\longrightarrow} \infty
\end{aligned}
$$

Hence, System-1 is not BIBO stable.

System-2 $T_2[\cdot]$: Let $x(n)$ be bounded by a number M i.e., $|x(n)| \le M < \infty$. Consider

$$
\begin{aligned}
|y(n)| &= \left|\sum_{k=n-10}^{n+10} x(k)\right| \le \sum_{k=n-10}^{n+10} |x(k)| \le \sum_{k=n-10}^{n+10} M \\
&= 21M < \infty
\end{aligned}
$$

Hence, System-2 is BIBO stable.

System-3 $T_3[\cdot]$: Let $x(n)$ be bounded by a number M i.e., $|x(n)| \le M < \infty$. Consider

$$
|y(n)| = |\,x(-n)| \le M < \infty
$$

Hence, System-3 is BIBO stable.

(b) **Causality**

System-1 $T_1[\cdot]$: Let $n = -1$. Then to determine $y(-1) = \sum_{k=0}^{-1} x(k) = x(-1) + x(0)$, we need $x(0)$, which is the future value when $n = -1$. Since the output $y(n)$ depends on the future values of the input $x(n)$ for $n < 0$ the System-1 is noncausal.

System-2 $T_2[\cdot]$: Here again the output $y(n) = x(n-10) + \cdots + x(n+10)$ depends on the future values of the input. Hence, the System-2 is noncausal.

System-3 $T_3[\cdot]$: Let $n = -1$. Then $y(-1) = x(1)$. Thus we need the future value of the input when $n = -1$. Since the output $y(n)$ depends on the future values of the input $x(n)$ for $n < 0$ the System-3 is noncausal.

P2.8 Properties of linear convolution.

$$\begin{aligned} x_1(n) * x_2(n) &= x_2(n) * x_1(n) &&: \text{Commutation} \\ [x_1(n) * x_2(n)] * x_3(n) &= x_1(n) * [x_2(n) * x_3(n)] &&: \text{Association} \\ x_1(n) * [x_2(n) + x_3(n)] &= x_1(n) * x_2(n) + x_1(n) * x_3(n) &&: \text{Distribution} \\ x(n) * \delta(n - n_0) &= x(n - n_0) &&: \text{Identity} \end{aligned}$$

(a) Commutation:

$$\begin{aligned} x_1(n) * x_2(n) &= \sum_{k=-\infty}^{\infty} x_1(k) x_2(\underbrace{n-k}_{=m}) = \sum_{m=-\infty}^{\infty} x_1(n-m) x_2(m) \\ &= \sum_{m=-\infty}^{\infty} x_2(m) x_1(n-m) = x_2(n) * x_1(n) \end{aligned}$$

Association:

$$\begin{aligned} [x_1(n) * x_2(n)] * x_3(n) &= \left[\sum_{k=-\infty}^{\infty} x_1(k) x_2(n-k) \right] * x_3(n) \\ &= \sum_{m=-\infty}^{\infty} \sum_{k=-\infty}^{\infty} x_1(k) x_2(m-k) x_3(n-m) \\ &= \sum_{k=-\infty}^{\infty} x_1(k) \left[\sum_{m=-\infty}^{\infty} x_2(\underbrace{m-k}_{=\ell}) x_3(n-m) \right] \\ &= \sum_{k=-\infty}^{\infty} x_1(k) \left[\sum_{m=-\infty}^{\infty} x_2(\ell) x_3(n-k-\ell) \right] \\ &= \sum_{k=-\infty}^{\infty} x_1(k) [x_2(n-k) * x_3(n-k)] = x_1(n) * [x_2(n) * x_3(n)] \end{aligned}$$

Distribution:

$$\begin{aligned} x_1(n) * [x_2(n) + x_3(n)] &= \sum_{k=-\infty}^{\infty} x_1(k) [x_2(n-k) + x_3(n-k)] \\ &= \sum_{k=-\infty}^{\infty} x_1(k) x_2(n-k) + \sum_{k=-\infty}^{\infty} x_1(k) x_3(n-k) \\ &= x_1(n) * x_2(n) + x_1(n) * x_3(n) \end{aligned}$$

Identity:

$$x(n) * \delta(n - n_0) = \sum_{k=-\infty}^{\infty} x(k) \delta(n - n_0 - k) = x(n - n_0)$$

because $\delta(n - n_0 - k) = 1$ for $k = n - n_0$ and is zero elsewhere.

(b) Verification by using MATLAB:

```
n1 = -10:20; x1 = n1;
n2 = 0:30; x2 = cos(0.1*pi*n2);
n3 = -5:10; x3 = (1.2).^n3;

% Commutative Property
[y1,ny1] = conv_m(x1,n1,x2,n2);
[y2,ny2] = conv_m(x2,n2,x1,n1);
ydiff = max(abs(y1-y2))
ydiff =
  4.2633e-014
ndiff = max(abs(ny1-ny2))
ndiff =
     0

% Associative Property
[y1,ny1] = conv_m(x1,n1,x2,n2);
[y1,ny1] = conv_m(y1,ny1,x3,n3);
[y2,ny2] = conv_m(x2,n2,x3,n3);
[y2,ny2] = conv_m(x1,n1,y2,ny2);
ydiff = max(abs(y1-y2))
ydiff =
  6.8212e-013
ndiff = max(abs(ny1-ny2))
ndiff =
     0

% Distributive Property
[y1,ny1] = sigadd(x2,n2,x3,n3);
[y1,ny1] = conv_m(x1,n1,y1,ny1);
[y2,ny2] = conv_m(x1,n1,x2,n2);
[y3,ny3] = conv_m(x1,n1,x3,n3);
[y2,ny2] = sigadd(y2,ny2,y3,ny3);
ydiff = max(abs(y1-y2))
ydiff =
  1.7053e-013
ndiff = max(abs(ny1-ny2))
ndiff =
     0

% Identity Property
n0 = fix(100*(rand(1,1)-0.5));
[dl,ndl] = impseq(n0,n0,n0);
[y1,ny1] = conv_m(x1,n1,dl,ndl);
[y2,ny2] = sigshift(x1,n1,n0);
ydiff = max(abs(y1-y2))
ydiff =
     0
ndiff = max(abs(ny1-ny2))
```

```
ndiff =
     0
```

P2.9 Linear convolution as a matrix-vector multiplication. Consider the sequences

$$x(n) = \{1, 2, 3, 4\} \text{ and } h(n) = \{3, 2, 1\}$$

(a) The linear convolution of the above two sequences is

$$y(n) = \{3, 8, 14, 20, 11, 4\}$$

(b) The vector representation of the above operation is

$$\underbrace{\begin{bmatrix} 3 \\ 8 \\ 14 \\ 20 \\ 11 \\ 4 \end{bmatrix}}_{\mathbf{y}} = \underbrace{\begin{bmatrix} 3 & 0 & 0 & 0 \\ 2 & 3 & 0 & 0 \\ 1 & 2 & 3 & 0 \\ 0 & 1 & 2 & 3 \\ 0 & 0 & 1 & 2 \\ 0 & 0 & 0 & 1 \end{bmatrix}}_{\mathbf{H}} \underbrace{\begin{bmatrix} 1 \\ 2 \\ 3 \\ 4 \end{bmatrix}}_{\mathbf{x}}$$

(c) Note that the matrix **H** has an interesting structure. Each diagonal of **H** contains the same number. Such a matrix is called a Toeplitz matrix. It is characterized by the following property

$$[\mathbf{H}]_{i,j} = [\mathbf{H}]_{i-j}$$

which is similar to the definition of time-invariance.

(d) Note carefully that the first column of **H** contains the impulse-response vector $h(n)$ followed by number of zeros equal to the number of $x(n)$ values minus one. The first row contains the first element of $h(n)$ followed by the same number of zeros as in the first column. Using this information and the above property, we can generate the whole Toeplitz matrix.

P2.10 (a) The MATLAB function conv_tp:

```
function [y,H]=conv_tp(h,x)
% Linear Convolution using Toeplitz Matrix
% ----------------------------------------
% [y,H] = conv_tp(h,x)
% y = output sequence in column vector form
% H = Toeplitz matrix corresponding to sequence h so that y = Hx
% h = Impulse response sequence in column vector form
% x = input sequence in column vector form
%
Nx = length(x); Nh = length(h);
hc = [h; zeros(Nx-1, 1)];
hr = [h(1),zeros(1,Nx-1)];
H = toeplitz(hc,hr);
y = H*x;
```

(b) MATLAB verification:

```
x = [1,2,3,4]'; h = [3,2,1]';
[y,H] = conv_tp(h,x); y = y', H
y =
     3     8    14    20    11     4
H =
     3     0     0     0
     2     3     0     0
     1     2     3     0
     0     1     2     3
     0     0     1     2
     0     0     0     1
```

P2.11 Let $x(n) = (0.8)^n u(n)$.

(a) Convolution $y(n) = x(n) * x(n)$:

$$\begin{aligned} y(n) &= \sum_{k=-\infty}^{\infty} x(k)x(n-k) = \sum_{k=0}^{\infty} (0.8)^k (0.8)^{n-k} u(n-k) \\ &= \left[\sum_{k=0}^{n} (0.8)^k (0.8)^n (0.8)^{-k}\right] u(n) = (0.8)^n \left[\sum_{k=0}^{n} (8/8)^k\right] u(n) \\ &= (0.8)^n (n+1) u(n) = (n+1)(0.8)^n u(n) \end{aligned}$$

```
clear; close all;
Hf_1 = figure('Units','normalized','position',[0.1,0.1,0.8,0.8],'color',[0,0,0]);
set(Hf_1,'NumberTitle','off','Name','P2.15');
% (a) analytical solution: y(n) = (n+1)*(0.8)^(n+1)*u(n)
na = [0:50]; ya = (na+1).*(0.8).^(na);
subplot(2,1,1); stem(na,ya); axis([-1,51,-1,3]);
xlabel('n'); ylabel('ya(n)'); title('Analytical computation');
```

(b) To use MATLAB's `filter` function, we have to represent one of the $x(n)$ sequences by the coefficients of an equivalent difference equation. The MATLAB solution by using the `filter` function is

```
% (b) use of the filter function
nb = [0:50]; x = (0.8).^nb;
yb = filter(1,[1, -0.8],x);
subplot(2,1,2); stem(nb,yb); axis([-1,51,-1,3])
xlabel('n'); ylabel('yb(n)'); title('Filter output');
%
error = max(abs(ya-yb))
error =
  4.4409e-016
```

The analytical solution to the convolution in (a) is the exact answer. In the `filter` function approach of (b), the infinite-duration sequence $x(n)$ is represented exactly by coefficients of an equivalent filter. Therefore, the filter solution should be exact except that it is evaluated up to the length of the input sequence. The plots of this solution are shown in Figure A.9.

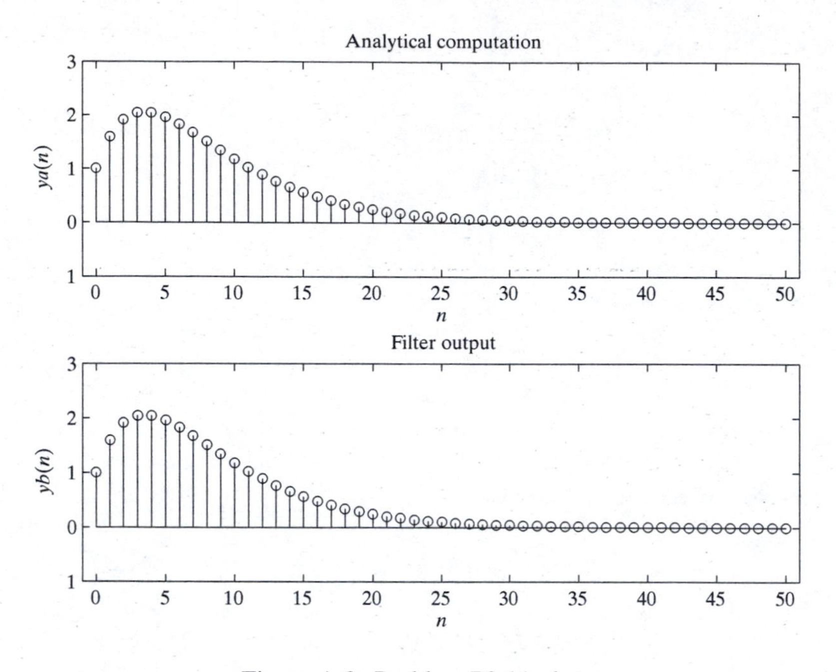

Figure A.9: Problem P2.11 plots

P2.12 An LSI system is described by

$$y(n) - 0.5y(n-1) + 0.25y(n-2) = x(n) + 2x(n-1) + x(n-3)$$

(a) The characteristic polynomial is given by

$$D(z) = z^2 - 0.5z + 0.25 = \left(z - 0.5e^{j\pi/3}\right)\left(z - 0.5e^{-j\pi/3}\right)$$

The roots of this polynomial have absolute values that are less than 1. Hence, for causal operation, the system is stable.

(b) MATLAB script:

```
close all; set(0,'defaultaxesfontsize',8);
b = [1,2,0,1]; a = [1,-0.5,0.25];
[Delta,n] = impseq(0,0,100);
h = filter(b,a,Delta);
subplot(2,1,1); stem(n,h,'filled'); axis([-1,101,-1,3]);
ylabel('h(n)','fontsize',10);
title('Impulse Response','fontsize',10);
```

From the plot, the stability of the system is obvious. The plot of the impulse response is shown in Figure A.10.

(c) MATLAB script:

```
n = -10:100;
x = 5*ones(size(n)) + 3*cos(0.2*pi*n) + 4*sin(0.6*pi*n);
y = filter(b,a,x);
n = n(11:end); y = y(11:end);
subplot(2,1,2); stem(n,y,'filled'); axis([-1,101,0,50]);
xlabel('n','fontsize',10), ylabel('y(n)','fontsize',10);
title('Output Response','fontsize',10);
```

The plot of the output response $y(n)$ is shown in Figure A.10.

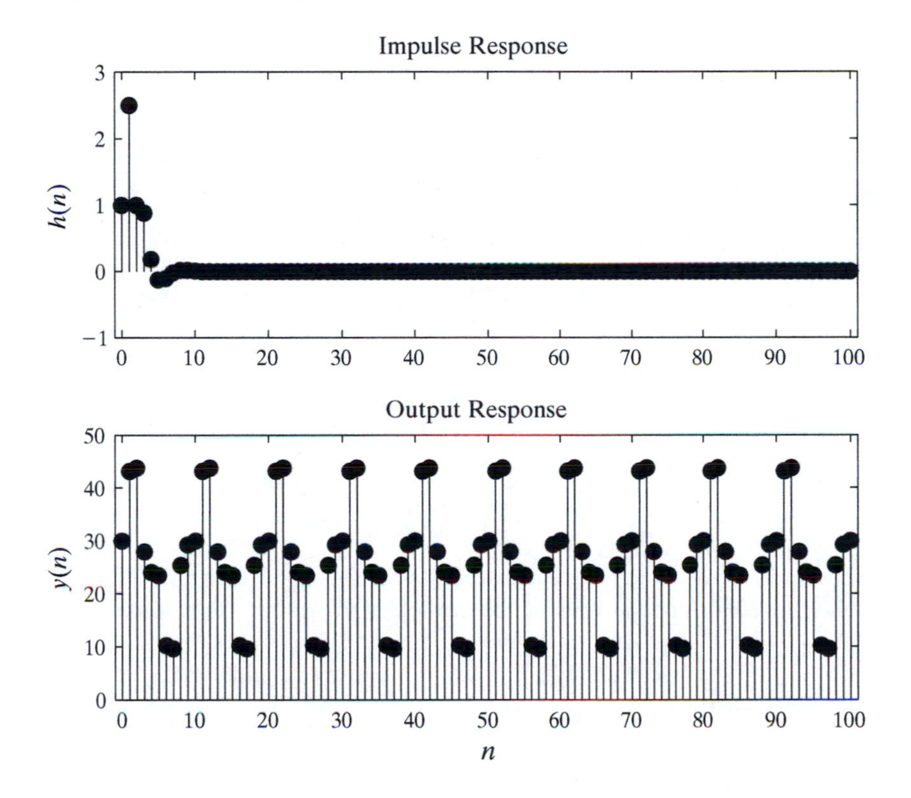

Figure A.10: Problem P2.12 plots

Appendix B

The z-transform

P3.1 Computation of z-transforms by using definition.

(a) The given sequence is $x(n) = (4/3)^n \, u(1-n)$. Hence, the z-transform is

$$\begin{aligned} X(z) &= \sum_{-\infty}^{1} \left(\frac{4}{3}\right)^n z^{-n} = \sum_{-\infty}^{1} \left(\frac{4}{3z}\right)^n \\ &= \sum_{-1}^{\infty} \left(\frac{3z}{4}\right)^n = \frac{4}{3z} \sum_{0}^{\infty} \left(\frac{3z}{4}\right)^n \\ &= \frac{4}{3z} \frac{1}{1-(3z/4)} = \frac{-16/9}{z\,(z-4/3)}, \quad |z| < \frac{4}{3} \end{aligned}$$

MATLAB verification: The sequence is a left-sided sequence, so the difference equation (and hence the `filter` function) should be run backward in time. This means that $X(z)$ should be a rational function in z. Furthermore, because $x(n)$ begins at $n = 1$ and continues (backward in time) to $-\infty$, we will advance $x(n)$ by one sample and simulate $zX(z)$ for verification purposes, where

$$zX(z) = \frac{4/3}{1-(3/4)\,z}$$

from the first term on the right-hand side of $X(z)$.

```
%(c) x(n) = (4/3)^n*u(1-n)
b = [4/3]; a = [1, -3/4];              % Difference equation
delta = [1, zeros(1,7)];                % Input sequence

% filter solution
x = filter(b,a,delta)
x =
  Columns 1 through 7
    1.3333    1.0000    0.7500    0.5625    0.4219    0.3164    0.2373
  Column 8
    0.1780
```

```
% simulation of x(n)
n = [1:-1:-6]; x = (4/3).^n
x =
  Columns 1 through 7
    1.3333    1.0000    0.7500    0.5625    0.4219    0.3164    0.2373
  Column 8
    0.1780
```

(b) The given sequence $x(n) = 2^{-|n|} + 3^{-|n|}$ can be rearranged as

$$x(n) = 2^{-n}u(n) - \left[-2^{n}u(-n-1)\right] + 3^{-n}u(n) - \left[-3^{n}u(-n-1)\right]$$

i. The z-transform is

$$\begin{aligned} X(z) &= \underbrace{\frac{1}{1-2^{-1}z^{-1}}}_{|z|>2^{-1}} - \underbrace{\frac{1}{1-2z^{-1}}}_{|z|<2} + \underbrace{\frac{1}{1-3^{-1}z^{-1}}}_{|z|>3^{-1}} - \underbrace{\frac{1}{1-3z^{-1}}}_{|z|<3} \\ &= \underbrace{\frac{-2+5z^{-1}}{1-5z^{-1}+6z^{-2}}}_{|z|<2} + \underbrace{\frac{2-(5/6)z^{-1}}{1-(5/6)z^{-1}+(1/6)z^{-2}}}_{|z|>2^{-1}} \end{aligned} \quad \text{(B.1)}$$

which, after simplification, becomes

$$X(z) = \frac{-4.1667z^{-1} + 11.6667z^{-2} - 4.1667z^{-3} + 0.5z^{-4}}{1 - 5.8333z^{-1} + 10.3333z^{-2} - 5.8333z^{-3} + 1.0000z^{-4}}, \; 0.5 < |z| < 2$$

ii. MATLAB verification: The sequence is two-sided, so it is not possible to derive the difference equation in both directions by using the `filter` function. However, we can verify the step in (B.1) by generating positive-time and negative-time sequences, as shown in the following:

```
%(d) x(n) = (2)^(-|n|)+(3)^(-|n|)
R = [1;-1;1;-1];                    % residues
p = [1/2;2;1/3;3];                  % poles
[b,a] = residuez(R,p,[])            % Difference equation coefficients
b =
         0   -4.1667   11.6667   -4.1667
a =
    1.0000   -5.8333   10.3333   -5.8333    1.0000

% Forward difference equation
Rf = [1;1]; pf = [1/2;1/3];
[bf,af] = residuez(Rf,pf,[])
bf =
    2.0000   -0.8333
af =
    1.0000   -0.8333    0.1667
[delta,nf]= impseq(0,0,30);
xf = filter(bf,af,delta);
```

```
% Backward difference equation
Rb = [-1;-1]; pb = [2;3];
[bb,ab] = residuez(Rb,pb,[])
bb =
    -2      5
ab =
     1     -5      6
[delta,nb]= impseq(0,0,29);
xb = filter(fliplr(bb),fliplr(ab),delta);

% Total solution
x1 = [fliplr(xb),xf];

% simulation of x(n)
n = [-fliplr(nb+1),nf];
x2 = 2.^(-abs(n)) + 3.^(-abs(n));

% difference
diff = max(abs(x1-x2))
diff =
  1.1102e-016
```

P3.2 Computation of z-transform by using properties and table.

(a) The given sequence $x(n)$ can be rearranged as

$$\begin{aligned} x(n) &= \left(\frac{1}{3}\right)^n u(n-2) + (0.9)^{n-3}\, u(n) \\ &= \left(\frac{1}{3}\right)^2 \left(\frac{1}{3}\right)^{n-2} u(n-2) + (0.9)^{-3}\,(0.9)^n\, u(n) \\ &= \frac{1}{9}\left(\frac{1}{3}\right)^{n-2} u(n-2) + \frac{1000}{729}\,(0.9)^n\, u(n) \end{aligned}$$

The z-transform is

$$\begin{aligned} X(z) &= \frac{1}{9} z^{-2} \mathcal{Z}\left[\left(\frac{1}{3}\right)^n u(n)\right] + \frac{1000}{729} \mathcal{Z}\left[(0.9)^n\, u(n)\right] \\ &= \frac{1}{9} z^{-2} \left(\frac{1}{1-\frac{1}{3}z^{-1}}\right) + \frac{1000}{729}\left(\frac{1}{1-0.9z^{-1}}\right) \end{aligned}$$

which, after simplification, becomes

$$X(z) = \frac{\frac{1000}{729} - \frac{1000}{2187}z^{-1} + \frac{1}{9}z^{-2} - 0.1z^{-3}}{1 - \frac{37}{30}z^{-1} + 0.3z^{-2}}$$

MATLAB verification:

```
% Sequence:
% x(n) = (1/3)^n*u(n-2) + (0.9)^(n-3)*u(n)
%
% Analytical Expression of X(z)
% X(z) = ((1000/729) - (1000/2187)*z^(-1) + (1/9)*z^(-2) -0.1*z^(-3)
% X(z) = -----------------------------------------------
%                1 - (37/30)*z^(-1) + 0.3*z^(-2)

% Matlab verification
b = [1000/729, -1000/2187, 1/9, -0.1]; a = [1, -37/30, 0.3];
delta = impseq(0,0,7); format long
xb1 = filter(b,a,delta)
xb1 =
  Columns 1 through 4
   1.37174211248285   1.23456790123457   1.22222222222222   1.03703703703704
  Columns 5 through 8
   0.91234567901235   0.81411522633745   0.73037174211248   0.65655724737083
%
% check
n = 0:7;
xb2 = ((1/3).^n).*stepseq(2,0,7) + ((0.9).^(n-3)).*stepseq(0,0,7)
xb2 =
  Columns 1 through 4
   1.37174211248285   1.23456790123457   1.22222222222222   1.03703703703704
  Columns 5 through 8
   0.91234567901235   0.81411522633745   0.73037174211248   0.65655724737083
%
error = abs(max(xb1-xb2)), format short;
error =
    4.440892098500626e-016
```

(b) The given sequence $x(n)$ can be rearranged as

$$
\begin{aligned}
x(n) &= \left(\frac{1}{2}\right)^n \cos\left(\frac{\pi n}{4} - 45^\circ\right) u(n-1) \\
&= \tfrac{1}{2}\left(\frac{1}{2}\right)^{n-1} \cos\left(\frac{\pi n}{4} - \frac{\pi}{4}\right) u(n-1) \\
&= \tfrac{1}{2}\left(\frac{1}{2}\right)^{n-1} \cos\left\{\frac{\pi}{4}(n-1)\right\} u(n-1)
\end{aligned}
$$

The z-transform is

$$
\begin{aligned}
X(z) &= \frac{1}{2} z^{-1} \mathcal{Z}\left[\left(\frac{1}{2}\right)^n \cos\left(\frac{\pi}{4} n\right) u(n)\right] \\
&= \frac{1}{2} z^{-1} \left(\frac{1 - \frac{1}{2} z^{-1} \cos(\pi/4)}{1 - z^{-1}\cos(\pi/4) + \frac{1}{4} z^{-2}}\right)
\end{aligned}
$$

which, after simplification, becomes

$$X(z) = \frac{0.5z^{-1} - \frac{1}{4\sqrt{2}}z^{-2}}{1 - \frac{1}{\sqrt{2}}z^{-1} + 0.25z^{-2}}, \ |z| > 0.5$$

MATLAB verification:

```
% Sequence:
% x(n) = (1/2)^n*cos(pi*n/4-pi/4)*u(n-1)
%
% Analytical Expression of X(z)
%          0.5*z^(-1) - 1/(4*sq(2))*z^(-2)
% X(z) = --------------------------------
%        1 - 1/sq(2)*z^(-1) + 0.25*z^(-2)

% Matlab verification
b = [0, 0.5, -1/(4*sqrt(2))]; a = [1, -1/sqrt(2), 0.25];
delta = impseq(0,0,7); format long
xb1 = filter(b,a,delta)
xb1 =
  Columns 1 through 4
                  0   0.50000000000000   0.17677669529664   0.00000000000000
  Columns 5 through 8
  -0.04419417382416  -0.03125000000000  -0.01104854345604   0.00000000000000
%
% check
n = 0:7;
xb2 = ((1/2).^n).*cos(pi*n/4-pi/4).*stepseq(1,0,7)
xb2 =
  Columns 1 through 4
                  0   0.50000000000000   0.17677669529664   0.00000000000000
  Columns 5 through 8
  -0.04419417382416  -0.03125000000000  -0.01104854345604   0.00000000000000
%
error = abs(max(xb1-xb2)), format short;
error =
    6.938893903907228e-018
```

P3.3 Computation of z-transforms by using properties.

(a) The z-transform of $x(n)$ is $X(z) = \left(1 + 2z^{-1}\right)$, $|z| \neq 0$. Consider

$$x_2(n) = \left(1 + n + n^2\right)x(n) = x(n) + n[x(n)] + n[nx(n)]$$

Then

$$\begin{aligned} X_2(z) &= X(z) + \left\{-z\frac{d}{dz}X(z)\right\} + \left[-z\frac{d}{dz}\left\{-z\frac{d}{dz}X(z)\right\}\right] \\ &= X(z) - z\frac{d}{dz}X(z) + z\frac{d}{dz}\left\{z\frac{d}{dz}X(z)\right\} \\ &= \left(1 + 2z^{-1}\right) - z\left\{-2z^{-2}\right\} + 2z^{-1} = 1 + 6z^{-1}, \ z \neq 0. \end{aligned}$$

(b) The z-transform of a sequence $x(n)$ is $X(z) = \left(1 + 2z^{-1}\right)$, $|z| \neq 0$. A new sequence is $x_3(n) = \left(\frac{1}{2}\right)^n x(n-2)$. Using the time-shift property, we have

$$\mathcal{Z}[x(n-2)] = z^{-2}X(z) = z^{-2} + 2z^{-3}$$

with no change in ROC. Now, using the frequency-shift property, we obtain

$$\begin{aligned} \mathcal{Z}\left[\left(\frac{1}{2}\right)^n x(n-2)\right] &= X_3(z) = \left[z^{-2} + 2z^{-3}\right]\Big|_{z \to z/0.5} \\ &= (2z)^{-2} + 2(2z)^{-3} = 0.25z^{-2} + 0.25z^{-3} \end{aligned}$$

with ROC scaled by$(1/2)$. Because the old ROC is $|z| > 0$, the new ROC is also $|z| > 0$.

P3.4 The inverse z-transform of $X(z)$ is $x(n) = 2^{-n}u(n)$. Then

$$\begin{aligned} \mathcal{Z}^{-1}\left[zX\left(z^{-1}\right)\right] &= x(-(n+1)) = x(-n-1) \\ &= 2^{-(-n-1)}u(-n-1) = 2^{n+1}u(-n-1) \end{aligned}$$

P3.5 Inverse z-transform using PFE.

(a) The z-transform $X(z)$ of a sequence is given as

$$X(z) = \frac{1 - z^{-1} - 4z^{-2} + 4z^{-3}}{1 - \frac{11}{4}z^{-1} + \frac{13}{8}z^{-2} - 0.25z^{-3}}, \text{ absolutely summable sequence}$$

The partial fractions are computed by using the `residuez` function.

```
b = [1,-1,-4,4]; a = [1,-11/4,13/8,-1/4];
[R,p,k] = residuez(b,a)
R =
    0.0000
  -10.0000
   27.0000
p =
    2.0000
    0.5000
    0.2500
k =
   -16
```

Therefore,

$$X(z) = -16 + \frac{0}{1 - 2z^{-1}} - \frac{10}{1 - 0.5z^{-1}} + \frac{27}{1 - 0.25z^{-1}}, \; 0.5 < |z| < 2$$

Hence, from the $\mathcal{Z}$-transform table,

$$x(n) = -16\delta(n) - 10(0.5)^n u(n) + 27(0.25)^n u(n)$$

MATLAB verification:

```
[delta,n] = impseq(0,0,7);
xb1 = filter(b,a,delta)
xb1 =
  Columns 1 through 4
   1.00000000000000   1.75000000000000  -0.81250000000000  -0.82812500000000
  Columns 5 through 8
  -0.51953125000000  -0.28613281250000  -0.14965820312500  -0.07647705078125
xb2 = -16*delta - 10*(0.5).^n + 27*(0.25).^n
xb2 =
  Columns 1 through 4
   1.00000000000000   1.75000000000000  -0.81250000000000  -0.82812500000000
  Columns 5 through 8
  -0.51953125000000  -0.28613281250000  -0.14965820312500  -0.07647705078125
error = abs(max(xb1-xb2))
error =
     0
```

(b) The z-transform $X(z)$ of a sequence is given as

$$X(z) = \frac{z}{z^3 + 2z^2 + 1.25z + 0.25}, \qquad |z| > 1$$

The partial fractions are computed by using the `residuez` function.

```
b = [0,0,1]; a = [1,2,1.25,0.25];
[R,p,k] = residuez(b,a), echo on;
R =
    4.0000
    0.0000 - 0.0000i
   -4.0000 + 0.0000i
p =
   -1.0000
   -0.5000 + 0.0000i
   -0.5000 - 0.0000i
k =
     []
```

Therefore,

$$X(z) = \frac{4}{1+z^{-1}} + \frac{0}{1+0.5z^{-1}} + \frac{-4}{\left(1+0.5z^{-1}\right)^2}$$

Hence, from the $\mathcal{Z}$-transform table:

$$\begin{aligned} x(n) &= 4(-1)^n u(n) + (-4)\left(-\tfrac{1}{0.5}\right)(n+1)(-0.5)^{n+1} u(n+1) \\ &= 4(-1)^n u(n) + 8(n+1)(-0.5)^{n+1} u(n) \end{aligned}$$

MATLAB verification:

```
[delta,n] = impseq(0,0,7);
xd1 = filter(b,a,delta)
xd1 =
         0         0    1.0000   -2.0000    2.7500   -3.2500    3.5625   -3.7500
xd2 = 4*(-1).^n + (8*(n+1)).*((-0.5).^(n+1))
xd2 =
         0         0    1.0000   -2.0000    2.7500   -3.2500    3.5625   -3.7500
error = abs(max(xd1-xd2))
error =
     0
```

P3.6 The z-transform $X(z)$ is given as

$$X(z) = \frac{2 + 3z^{-1}}{1 - z^{-1} + 0.81z^{-2}}, \quad |z| > 0.9$$

(a) Sequence $x(n)$ is in a form that contains no complex numbers. Compare the denominator $1 - z^{-1} + 0.81z^{-2}$ with the denominator $1 - 2az^{-1}\cos\omega_0 + a^2z^{-2}$ of the sin(cos) transform pairs.

$$\begin{aligned} 1 - z^{-1} + 0.81z^{-2} &= 1 - 2az^{-1}\cos\omega_0 + a^2z^{-2} \\ &\Rightarrow \; a^2 = 0.81, \quad \cos\omega_0 = \frac{1}{2a} \end{aligned}$$

or

$$a = 0.9, \; \cos\omega_0 = \frac{1}{1.8} \Rightarrow \omega_0 = 0.3125\pi, \text{ and } \sin\omega_0 = 0.8315$$

Now $X(z)$ can be put into the form

$$\begin{aligned} X(z) &= \frac{2 - z^{-1} + 4z^{-1}}{1 - 2\,(0.9)\,z^{-1}\frac{1}{1.8} + (0.9)^2\,z^{-2}}, \quad |z| > 0.9 \\ &= \frac{2\left(1 - 0.9z^{-1}\frac{1}{1.8}\right)}{1 - 2\,(0.9)\,z^{-1}\frac{1}{1.8} + (0.9)^2\,z^{-2}} + 5.3452\frac{0.9\,(0.8315)\,z^{-1}}{1 - 2\,(0.9)\,z^{-1}\frac{1}{1.8} + (0.9)^2\,z^{-2}} \end{aligned}$$

Finally, after table lookup,

$$x(n) = \left[2\,(0.9)^n\cos(0.3125\pi) + (5.3452)\,(0.9)^n\sin(0.3125\pi)\right]u(n)$$

For another approach using PFE and residues, examine the following MATLAB script.

```
clear all
% (a) Use of the residue and transform table
b=[2, 3]; a=[1,-1,0.81];
[R,p,k] = residuez(b,a);
R_real = (real(R))
R_real =
     1
```

```
      1
R_imag = (imag(R))
R_imag =
   -2.6726
    2.6726
p_magn = (abs(p))
p_magn =
    0.9000
    0.9000
p_angl = (angle(p))/pi
p_angl =
    0.3125
   -0.3125
[delta,n] = impseq(0,0,19);
xa = (p_magn(1).^n).*(2*R_real(1)*cos(p_angl(1)*pi*n)...
                                   -2*R_imag(1)*sin(p_angl(1)*pi*n));
%
% Print Response
fprintf(1,'\n Hence the sequence x(n) is \n')

 Hence the sequence x(n) is
fprintf(1,'\n\tx(n) = (%1.1f)^n * (%1.0f*cos(%1.4f*pi*n)...
                   - (%2.4f)*sin(%1.4f*pi*n)\n\n',...
    p_magn(1),2*R_real(1),p_angl(1),2*R_imag(1),p_angl(1));

    x(n) = (0.9)^n * (2*cos(0.3125*pi*n) - (-5.3452)*sin(0.3125*pi*n)
```

(b) MATLAB verification:

```
xb = filter(b,a,delta);
error = abs(max(xa-xb))
error =
  1.5543e-015
```

P3.7 System representations.

(a) The impulse response is $h(n) = 2\,(0.5)^n\,u(n)$. The system-function representation is

$$H(z) = Z\,[h(n)] = \frac{2}{1 - 0.5z^{-1}},\ |z| > 0.5$$

The difference-equation representation is

$$\frac{Y(z)}{X(z)} = \frac{2}{1 - 0.5z^{-1}} \Rightarrow Y(z) - 0.5z^{-1}Y(z) = 2X(z)$$

or

$$y(n) = 2x(n) + 0.5y\,(n-1)$$

The pole–zero description is given by a zero at $z = 0$ and a pole at $z = 0.5$. Finally, to compute output $y(n)$ when $x(n) = (1/4)^n u(n)$, we use the z-transform approach (because the ROCs overlap):

$$\begin{aligned} Y(z) &= H(z)X(z) = \frac{2}{1-0.5z^{-1}} \times \frac{1}{1-0.25z^{-1}} \\ &= \frac{4}{1-0.5z^{-1}} - \frac{2}{1-0.25z^{-1}}, \ |z| > 0.5 \end{aligned}$$

Hence,

$$y(n) = 4\,(0.5)^n\, u(n) - 2\,(0.25)^n\, u(n)$$

(b) Impulse response of an LTI system:

$$h(n) = n\,[u(n) - u\,(n-10)] = \left[\underset{\uparrow}{0}, 1, 2, \ldots, 9\right]$$

i) System-function representation:

$$H(z) = z^{-1} + 2z^{-2} + \ldots + 9z^{-9} = \sum_{k=1}^{9} kz^{-k}$$

ii) Difference-equation representation:

$$y(n) = \sum_{k=1}^{9} kx\,(n-k)$$

iii) Pole–zero plot — MATLAB script:

```
clear, close all;
hb = [0:9]; ha = [1,0]; zplane(hb,ha);
```

The pole–zero plot is shown in Figure B.1.

iv) The output $y(n)$ when the input is $x\,(n) = (1/4)^n\, u(n)$:

$$\begin{aligned} Y(z) &= H(z)X(z) = \left(z^{-1} + 2z^{-2} + \ldots + 9z^{-9}\right)\frac{1}{1-0.25z^{-1}}, \ |z| > 0.25 \\ &= \frac{z^{-1} + 2z^{-2} + \ldots + 9z^{-9}}{1-0.25z^{-1}} \\ &= \frac{3029220}{1-0.25z^{-1}} - 3029220 - 757304z^{-1} - 189324z^{-2} - 47328z^{-3} \\ &\quad -11828z^{-4} - 2952z^{-5} - 732z^{-6} - 176z^{-7} - 36z^{-8} \end{aligned}$$

where the PFE was performed by using MATLAB.

```
clear, close all;
hb = [0:9]; ha = [1,0]; xb = [1]; xa = [1,-0.25];
yb = hb; ya = xa;
[R,p,k] = residuez(yb,ya)
```

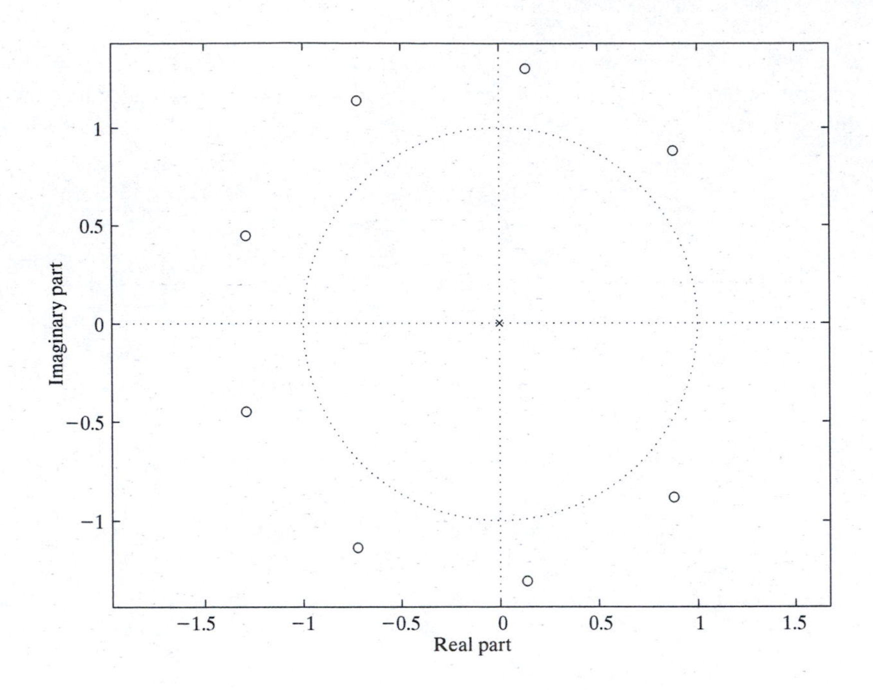

Figure B.1: Pole–zero plot in Problem P3.7(b)

```
R =
      3029220
p =
     0.2500
k =
   Columns 1 through 6
     -3029220      -757304      -189324       -47328       -11828        -2952
   Columns 7 through 9
         -732         -176          -36
```

Hence,

$$y(n) = 3029220\,(0.25)^n\,u(n) - 3029220\delta(n) - 757304\delta\,(n-1) - \cdots \\ -176\delta\,(n-7) - 36\delta\,(n-8)$$

P3.8 A stable system has the following pole–zero locations:

$$z_1 = j, \quad z_2 = -j, \quad p_1 = -\frac{1}{2} + j\frac{1}{2}, \quad p_2 = -\frac{1}{2} - j\frac{1}{2}$$

It is also known that the frequency-response function $H\left(e^{j\omega}\right)$ evaluated at $\omega = 0$ is equal to 0.8:

$$H\left(e^{j0}\right) = 0.8$$

(a) System function $H(z)$ and region of convergence:

$$H(z) = K\frac{(z-j)\,(z+j)}{(z+0.5-j0.5)\,(z+0.5+j0.5)} = K\frac{z^2+1}{z^2+z+0.5}, \; |z| > \frac{1}{\sqrt{2}}$$

Now @ $z = e^{j0} = 1$, $H(1) = 0.8$ is given, hence

$$0.8 = K\frac{1+1}{1+1+0.5} = K\frac{2}{2.5} \Rightarrow K = 1$$

or

$$\boxed{H(z) = \frac{z^2+1}{z^2+z+0.5}, \ |z| > \frac{1}{\sqrt{2}}}$$

(b) Difference-equation representation:

$$H(z) = \frac{z^2+1}{z^2+z+0.5} = \frac{1+z^{-2}}{1+z^{-1}+0.5z^{-2}} = \frac{Y(z)}{X(z)}$$

After cross-multiplying and inverse transforming,

$$\boxed{y(n) + y(n-1) + 0.5y(n-2) = x(n) + x(n-1)}$$

(c) Steady-state response $y_{ss}(n)$ if the input is

$$x(n) = \frac{1}{\sqrt{2}} \sin\left(\frac{\pi n}{2}\right) u(n)$$

From the z-transform table,

$$X(z) = \left(\frac{1}{\sqrt{2}}\right) \frac{z}{z^2+1}, \ |z| > 1$$

Hence,

$$\begin{aligned} Y(z) = H(z)X(z) &= \frac{z^2+1}{z^2+z+0.5}\left(\frac{1}{\sqrt{2}}\right)\frac{z}{z^2+1} \\ &= \left(\frac{1}{\sqrt{2}}\right)\frac{z}{z^2+z+0.5}, \ |z| > \frac{1}{\sqrt{2}} \end{aligned}$$

Thus, the poles of $Y(z)$ are the poles of $H(z)$, which are inside the unit circle. Therefore, there is *no* steady-state response: $y_{ss}(n) = 0$.

(d) Transient response $y_{tr}(n)$: Because $y_{ss}(n) = 0$, the total response $y(n) = y_{tr}(n)$. From the $Y(z)$ expression from part (c),

$$Y(z) = \left(\frac{1}{\sqrt{2}}\right)\frac{z}{z^2+z+0.5} = \sqrt{2}\frac{\frac{1}{\sqrt{2}}\frac{1}{\sqrt{2}}z}{1+z^{-1}+\left(\frac{1}{\sqrt{2}}\right)^2 z^{-2}}, \ |z| > \frac{1}{\sqrt{2}}$$

Hence, by table lookup, we have

$$y(n) = y_{tr}(n) = \sqrt{2}\left(\frac{1}{\sqrt{2}}\right)^n \sin(0.75\pi n)\, u(n)$$

P3.9 A digital filter is described by the difference equation

$$y(n) = x(n) + x(n-1) + 0.9y(n-1) - 0.81y(n-2)$$

(a) Magnitude and phase of the frequency response: MATLAB script

```
clear; close all;
Hf_1 = figure('Units','normalized','position',[0.1,0.1,0.8,0.8],'color',[0,0,0]);
set(Hf_1,'NumberTitle','off','Name','P3.9a');
% (a) Magnitude and Phase Plots
b = [1, 1]; a = [1,-0.9,0.81];
w = [0:1:500]*pi/500; H = freqz(b,a,w);
magH = abs(H); phaH = angle(H)*180/pi;
subplot(2,1,1); plot(w/pi,magH); axis([0,1,0,12]); grid
title('Magnitude Response'); xlabel('frequency in pi units'); ylabel('|H|');
subplot(2,1,2); plot(w/pi,phaH); axis([0,1,-180,180]); grid
title('Phase Response'); xlabel('frequency in pi units'); ylabel('Degrees');
%
w = [pi/3,pi]; H = freqz(b,a,w); magH = abs(H); phaH = angle(H)*180/pi;
fprintf(1,'\n At w = pi/3 the magnitude is %1.4f and the phase is %3.4f...
                                 degrees \n', magH(1), phaH(1));

 At w = pi/3 the magnitude is 10.5215 and the phase is -58.2595 degrees
fprintf(1,'\n At w =  pi  the magnitude is %1.4f and the phase is %3.4f...
                                 degrees \n', magH(2), phaH(2));

 At w =  pi  the magnitude is 0.0000 and the phase is -90.0000 degrees
```

The frequency-response plots are shown in Figure B.2.

(b) Steady-state response: MATLAB script

```
n = [0:200]; x = sin(pi*n/3) + 5*cos(pi*n);
y = filter(b,a,x);
n = n(101:201); x = x(101:201); y = y(101:201);  % Steady-state section
Hf_2 = figure('Units','normalized','position',[0.1,0.1,0.8,0.8],'color',[0,0,0]);
set(Hf_2,'NumberTitle','off','Name','P4.16b');
subplot(2,1,1); plot(n,x); title('Input sequence'); axis([100,200,-6,6])
xlabel('n'); ylabel('x(n)');
subplot(2,1,2); plot(n,y); title('Output sequence'); axis([100,200,-12,12])
xlabel('n'); ylabel('y(n)');
```

The steady-state response is shown in Figure B.3. It shows that the $\omega = \pi$ frquency is suppressed and that the only component in the output is due to the $\omega = \pi/3$ frequency, which is amplified by about 10.

P3.10 Difference-equation solution using one-sided z-transform approach:

$$y(n) = 0.5y(n-1) + 0.25y(n-2) + x(n), \; n \geq 0$$
$$x(n) = (0.8)^n u(n)$$
$$y(-1) = 1, \; y(-2) = 2$$

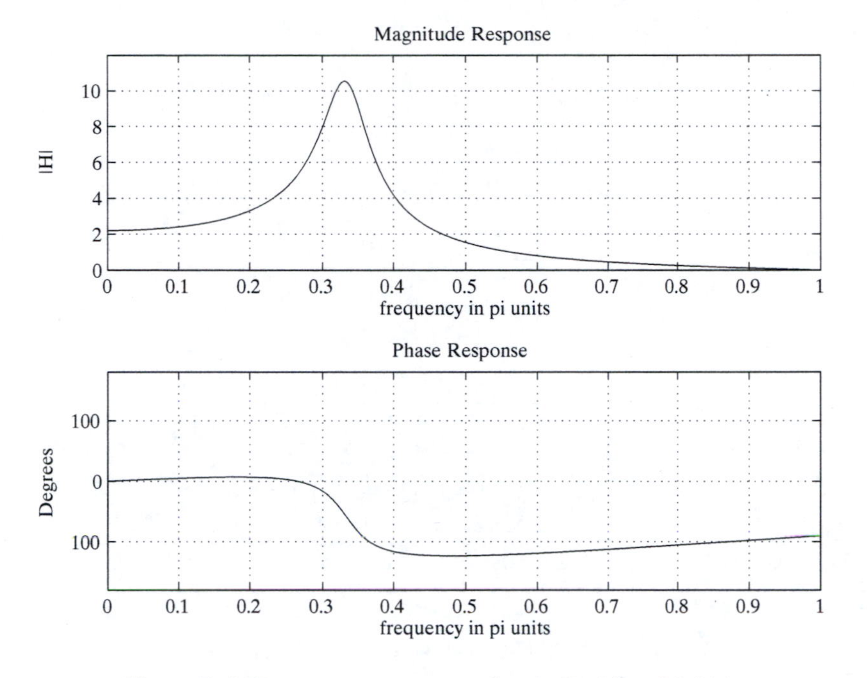

Figure B.2: Frequency-response plots in Problem P3.9(a)

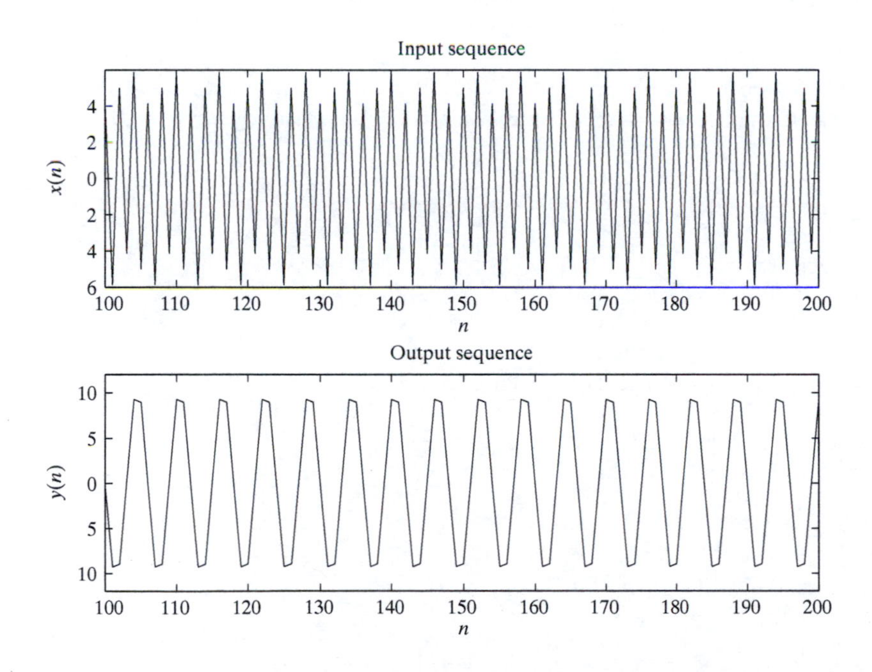

Figure B.3: Steady-state response in Problem P3.9(b)

Taking one-sided z-transform yeilds:

$$Y^{+}(z) = 0.5\left[z^{-1}Y^{+}(z) + y(-1)\right] + 0.25\left[z^{-2}Y^{+}(z) + y(-2) + z^{-1}y(-1)\right] + X^{+}(z)$$

or

$$\begin{aligned} Y^{+}(z)\left(1 - 0.5z^{-1} - 0.25z^{-2}\right) &= 0.5 + 0.5 + 0.25z^{-1} + \frac{1}{1 - 0.8z^{-1}} \\ &= \underbrace{1 + 0.25z^{-1}}_{\text{Equi. I.C. Input}} + \frac{1}{1 - 0.8z^{-1}} = \frac{1 - 0.55z^{-1} - 0.2z^{-2} + 1}{1 - 0.8z^{-1}} \end{aligned}$$

or

$$\begin{aligned} Y^{+}(z) &= \frac{2 - 0.55z^{-1} - 0.2z^{-2}}{\left(1 - 0.8z^{-1}\right)\left(1 - 0.5z^{-1} - 0.25z^{-2}\right)} \\ &= \frac{1 - 0.55z^{-1} - 0.2z^{-2}}{1 - 1.3z^{-1} + 0.15z^{-2} + 0.2z^{-3}} \\ &= \frac{65.8702}{1 - 0.8090z^{-1}} + \frac{-64}{1 - 0.8z^{-1}} + \frac{0.1298}{1 + 0.3090z^{-1}} \end{aligned}$$

Hence,

$$y(n) = 65.8702\,(0.8090)^{n} - 64\,(0.8)^{n} + 0.1298\,(-0.3090)^{n}, \quad n \geq 0$$

MATLAB verification:

```
b = [1]; a = [1, -0.5, -0.25];          % Difference equation
yic = [1,2];                            % Initial conditions
n = [0:20]; x = (0.8).^n;               % Input sequence

% Numerical Solution
V = filtic(b,a,yic), echo on;           % equivalent initial condition input
V =
    1.0000    0.2500
y1 = filter(b,a,x,V);                   % Output sequence

% Analytical solution
b1 = conv(V,[1,-0.8])+[1,0,0]           % Num of Y(z)
b1 =
    2.0000   -0.5500   -0.2000
a1 = conv([1,-0.8],[1,-0.5,-0.25])      % Denom of Y(z)
a1 =
    1.0000   -1.3000    0.1500    0.2000
[R,p,k] = residuez(b1,a1), echo on;     % PFE
R =
   65.8702
  -64.0000
    0.1298
p =
    0.8090
    0.8000
   -0.3090
```

```
k =
     []
y2 = zeros(1,21); L = length(R);       %
for l = 1:L                            % Assemble
    y2 = y2 + R(l)*(p(l).^n);          %         Output
end                                    %                Sequence
%
error = abs(max(y1-y2))                % Difference
error =
  1.1169e-013
```

P3.11 A causal, linear, time-invariant system is given by

$$y(n) = y(n-1) + y(n-2) + x(n-1)$$

(a) System Function: Taking z-transform of both sides, we obtain

$$Y(z) = z^{-1}Y(z) + z^{-2}Y(z) + z^{-1}X(z) \Rightarrow H(z) \stackrel{\text{def}}{=} \frac{Y(z)}{X(z)} = \frac{z^{-1}}{1 - z^{-1} - z^{-2}}, |z| > r_0$$

where r_0 is the magnitude of the largest pole (because the system is causal).

(b) Pole–zero plot and the ROC:

$$H(z) = \frac{z}{z^2 - z - 1} = \frac{z}{\left(z - \dfrac{1+\sqrt{5}}{2}\right)\left(z - \dfrac{1-\sqrt{5}}{2}\right)}$$

The zero is @ $z = 0$, and poles are @ $z = \frac{1\pm\sqrt{5}}{2}$. Hence, the ROC is $|z| > \frac{1+\sqrt{5}}{2} = 1.618$.

(c) Impulse response: Using PFE,

$$H(z) = \frac{z}{\left(z - \frac{1+\sqrt{5}}{2}\right)\left(z - \frac{1-\sqrt{5}}{2}\right)} = \frac{1}{\sqrt{5}}\frac{z}{\left(z - \frac{1+\sqrt{5}}{2}\right)} - \frac{1}{\sqrt{5}}\frac{z}{\left(z - \frac{1-\sqrt{5}}{2}\right)}, |z| > \frac{1+\sqrt{5}}{2}$$

Hence,

$$h(n) = \frac{1}{\sqrt{5}}\left(\frac{1+\sqrt{5}}{2}\right)^n u(n) - \frac{1}{\sqrt{5}}\left(\frac{1-\sqrt{5}}{2}\right)^n u(n)$$

(d) Clearly, the system is not stable: $h(n) \nearrow \infty$ as $n \nearrow \infty$. For a stable unit-sample response, the ROC should be

$$\frac{1-\sqrt{5}}{2} < |z| < \frac{1+\sqrt{5}}{2}$$

Then

$$h(n) = -\frac{1}{\sqrt{5}}\left(\frac{1+\sqrt{5}}{2}\right)^n u(-n-1) - \frac{1}{\sqrt{5}}\left(\frac{1-\sqrt{5}}{2}\right)^n u(n)$$

P3.12 The difference equation is

$$y(n) = \frac{1}{4} y(n-1) + x(n) + 3x(n-1), \; n \geq 0; \quad y(-1) = 2$$

with the input $x(n) = e^{j\pi n/4} u(n)$. Taking a one-sided z-transform of the difference equation, we obtain

$$Y^{+}(z) = \frac{1}{4}\left[z^{-1}Y^{+}(z) + y(-1)\right] + \frac{1}{1 - e^{j\pi/4}z^{-1}} + 3\frac{z^{-1}}{1 - e^{j\pi/4}z^{-1}}.$$

Substituting the initial condition and rearranging, we obtain

$$Y^{+}(z)\left[1 - \frac{1}{4}z^{-1}\right] = \frac{1}{2} + \frac{1 + 3z^{-1}}{1 - e^{j\pi/4}z^{-1}}. \tag{B.2}$$

The second term on the right-hand side provides the zero-state response of the system. Thus,

$$\begin{aligned} Y_{zs}^{+}(z) &= \frac{1 + 3z^{-1}}{\left(1 - \frac{1}{4}z^{-1}\right)\left(1 - e^{j\pi/4}z^{-1}\right)} \\ &= \frac{4.4822e^{-j0.8084}}{1 - \frac{1}{4}z^{-1}} + \frac{3.8599e^{j2.1447}}{1 - e^{j\pi/4}z^{-1}} \end{aligned}$$

Hence, the zero-state response is

$$y_{zs}(n) = \left[4.4822e^{-j0.8084}\left(\frac{1}{4}\right)^{n} + 3.8599e^{j2.1447}e^{j\pi n/4}\right]u(n)$$

The steady-state part of the total response is due to simple poles on the unit circle. The pole on the unit circle at $z = e^{j\pi/4}$ is due to the input sequence. From (B.2), the total response is

$$\begin{aligned} Y^{+}(z) &= \left(\frac{1}{1 - \frac{1}{4}z^{-1}}\right)\left(\frac{1}{2} + \frac{1 + 3z^{-1}}{1 - e^{j\pi/4}z^{-1}}\right) \\ &= \frac{3/2 + (2.6464 - j0.3536)\,z^{-1}}{\left(1 - \frac{1}{4}z^{-1}\right)\left(1 - e^{j\pi/4}z^{-1}\right)} \\ &= \frac{4.4822e^{-j0.8084}}{1 - \frac{1}{4}z^{-1}} + \frac{3.6129e^{j2.0282}}{1 - e^{j\pi/4}z^{-1}}. \end{aligned}$$

The steady-state response is given by the second term on the right-hand side. Thus,

$$y_{ss}(n) = 3.6129e^{j2.0282}e^{j\pi n/4}u(n) = 3.6129e^{j(\pi n/4 - 2.0282)}u(n)$$

Appendix C

Discrete-Time Fourier Transform

P4.1 A MATLAB function to compute DTFT:

```
function [X] = dtft(x,n,w)
% Computes Discrete-time Fourier Transform
% [X] = dtft(x,n,w)
%
% X = DTFT values computed at w frequencies
% x = finite duration sequence over n (row vector)
% n = sample position row vector
% W = frequency row vector

X = x*exp(-j*n'*w);
```

P4.2 Numerical computation of DTFT.

(a) $x(n) = \left\{\underset{\uparrow}{4}, 3, 2, 1, 2, 3, 4\right\}$

```
clear; close all;
Hf_1 = figure('Units','normalized','position',[0.1,0.1,0.8,0.8],'color',[0,0,0]);
set(Hf_1,'NumberTitle','off','Name','P4.2a');
%
n = 0:6; x = [4,3,2,1,2,3,4];
w = [0:1:500]*pi/500;
X = dtft(x,n,w); magX = abs(X); phaX = angle(X);
%
% Magnitude Response Plot
subplot(2,1,1); plot(w/pi,magX);grid;
xlabel('frequency in pi units'); ylabel('|X|');
title('Magnitude Response');
wtick = [0:0.2:1]; magtick = [0;10;20];
set(gca,'XTickMode','manual','XTick',wtick)
set(gca,'YTickMode','manual','YTick',magtick)
%
% Phase response plot
```

```
subplot(2,1,2); plot(w/pi,phaX*180/pi);grid;
xlabel('frequency in pi units'); ylabel('Degrees');
title('Phase Response'); axis([0,1,-180,180])
wtick = [0:0.2:1]; phatick = [-180;0;180];
set(gca,'XTickMode','manual','XTick',wtick)
set(gca,'YTickMode','manual','YTick',phatick)
```

The plots are shown in Figure C.1. The angle plot for this signal is a linear function of ω.

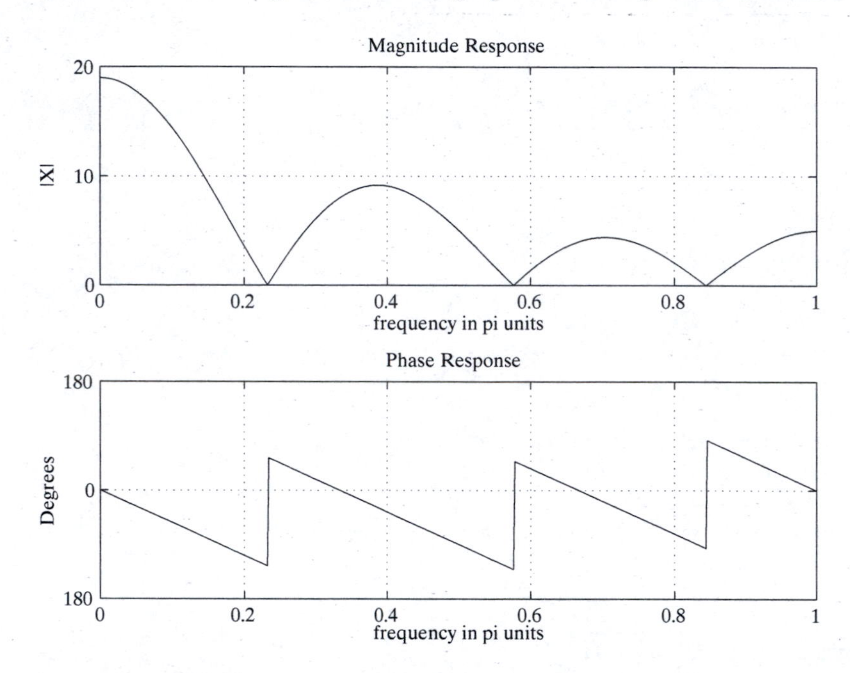

Figure C.1: Frequency-response plots in Problem P4.2(a)

(b) $x(n) = \left\{ \underset{\uparrow}{4}, 3, 2, 1, 1, 2, 3, 4 \right\}$

```
clear; close all;
Hf_1 = figure('Units','normalized','position',[0.1,0.1,0.8,0.8],'color',[0,0,0]);
set(Hf_1,'NumberTitle','off','Name','P4.2b');
%
% (d) DTFT of x(n) = {4,3,2,1,1,2,3,4}
n = 0:7; x = [4,3,2,1,1,2,3,4];
w = [0:1:500]*pi/500;
X = dtft(x,n,w); magX = abs(X); phaX = angle(X);
%
% Magnitude Response Plot
subplot(2,1,1); plot(w/pi,magX);grid;
xlabel('frequency in pi units'); ylabel('|X|');
title('Magnitude Response');
wtick = [0:0.2:1]; magtick = [0;10;20];
set(gca,'XTickMode','manual','XTick',wtick)
```

```
set(gca,'YTickMode','manual','YTick',magtick)
%
% Phase response plot
subplot(2,1,2); plot(w/pi,phaX*180/pi);grid;
xlabel('frequency in pi units'); ylabel('Degrees');
title('Phase Response'); axis([0,1,-180,180])
wtick = [0:0.2:1]; phatick = [-180;0;180];
set(gca,'XTickMode','manual','XTick',wtick)
set(gca,'YTickMode','manual','YTick',phatick)
```

The plots are shown in Figure C.2. The angle plot for this signal is a linear function of ω.

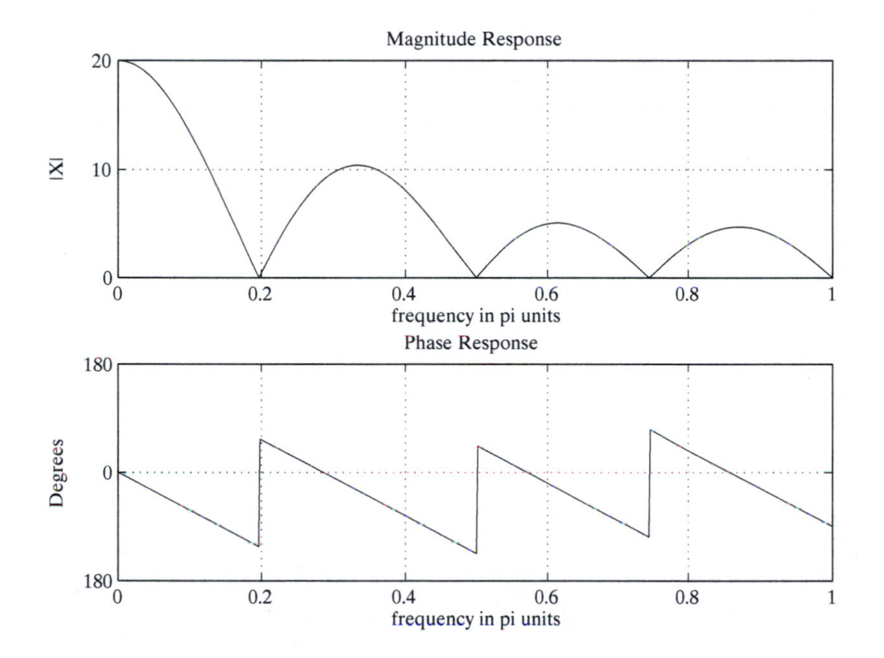

Figure C.2: Frequency-response plots in Problem P4.2(b)

P4.3 Analytical computation of DTFT.

(a) $x(n) = 3\,(0.9)^n\,u(n)$. The DTFT is given by

$$
\begin{aligned}
X(\omega) &= 3\sum_{n=0}^{\infty}(0.9)^n\,e^{-j\omega n} = 3\sum_{n=0}^{\infty}\left(0.9e^{-j\omega}\right)^n \\
&= \frac{3}{1-0.9e^{-j\omega}}
\end{aligned}
$$

MATLAB script for magnitude and angle plot:

```
% Problem~P4.3 : Magnitude and Angle Plot of DTFT
clear; close all;
Hf_1 = figure('Units','normalized','position',[0.1,0.1,0.8,0.8],'color',[0,0,0]);
set(Hf_1,'NumberTitle','off','Name','P4.3');
w = [0:200]*pi/200; Z = exp(-j*w); ZZ = Z.^2;

% (a) x(n) = 3*(0.9)^n*u(n)
X = 3*(1-0.9*Z).^(-1); X_mag = abs(X); X_pha = angle(X)*180/pi;
subplot(5,2,1); plot(w/pi,X_mag); axis([0,1,0,max(X_mag)]);
title('Magnitude Plots','fontweight','bold'); ylabel('a.');
set(gca,'YTickMode','manual','YTick',[0,max(X_mag)],'FontSize',10);
subplot(5,2,2); plot(w/pi,X_pha); axis([0,1,-180,180]);
title('Angle Plots','fontweight','bold')
set(gca,'YTickMode','manual','YTick',[-180,0,180],'FontSize',10);
```

The plots are given in Figure C.3.

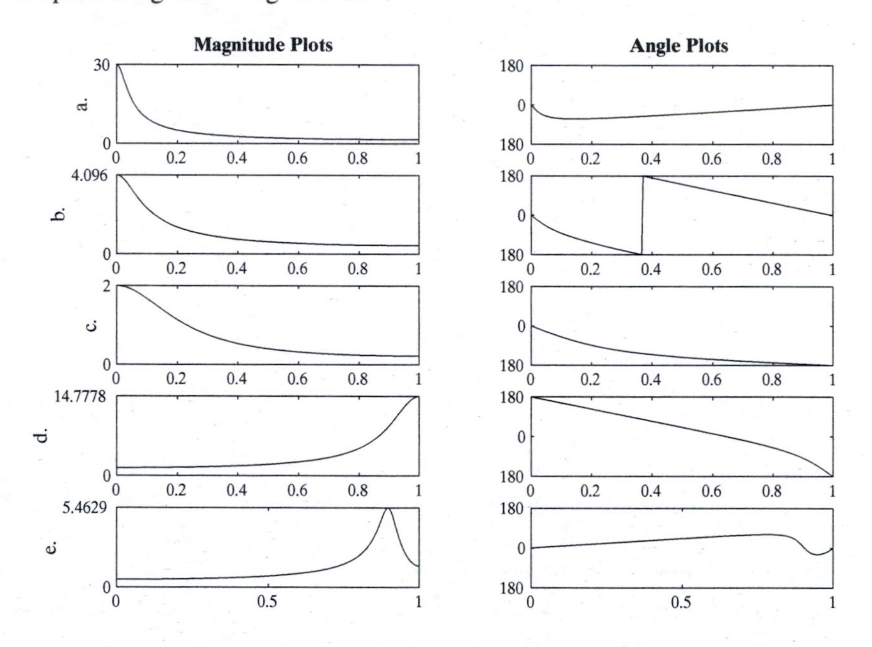

Figure C.3: Problem P4.3 DTFT plots

(b) $x(n) = 2\,(0.8)^{n+2}\,u\,(n-2)$. The sequence can be written as

$$\begin{aligned} x(n) &= 2\,(0.8)^{n-2+4}\,u\,(n-2) = 2\,(0.8)^{4}\,(0.8)^{n-2}\,u\,(n-2) \\ &= 0.8192\,(0.8)^{n-2}\,u\,(n-2) \end{aligned}$$

Now, using properties of DTFT, we obtain

$$X(\omega) = 0.8192\frac{e^{-j2\omega}}{1-0.8e^{-j\omega}}$$

MATLAB script for magnitude and angle plot:

```
% (b) x(n) = 2*(0.8)^(n+2)*u(n-2)
X = 0.8192*ZZ./(1-0.8*Z); X_mag = abs(X); X_pha = angle(X)*180/pi;
subplot(5,2,3); plot(w/pi,X_mag); axis([0,1,0,max(X_mag)]);  ylabel('b.');
set(gca,'YTickMode','manual','YTick',[0,max(X_mag)],'FontSize',10);
subplot(5,2,4); plot(w/pi,X_pha); axis([0,1,-180,180]);
set(gca,'YTickMode','manual','YTick',[-180,0,180],'FontSize',10);
```

The plots are given in Figure C.3.

(c) $x(n) = n\,(0.5)^u\,u(n)$. The DTFT of $(0.5)^n\,u(n)$ is given by $1/\left(1-0.5e^{-j\omega}\right)$, that is,

$$\sum_{n=0}^{\infty}(0.5)^n\,e^{-j\omega n} = \frac{1}{1-0.5e^{-j\omega}}$$

Differentiating both sides with respect to ω, we obtain

$$-j\sum_{n=0}^{\infty}n\,(0.5)^n\,e^{-j\omega n} = (-1)\,\frac{1}{\left(1-0.5e^{-j\omega}\right)^2}\left[-0.5e^{-j\omega}\,(-j)\right]$$

Hence,

$$\begin{aligned} X(\omega) &= \mathfrak{F}\left\{n\,(0.5)^n\,u\,(n)\right\} = \sum_{n=0}^{\infty}n\,(0.5)^n\,e^{-j\omega n} \\ &= \frac{0.5e^{-j\omega}}{\left(1-0.5e^{-j\omega}\right)^2} \end{aligned}$$

MATLAB script for magnitude and angle plot:

```
% (c) x(n) = n*(0.5)^n*u(n)
X = (0.5)*Z./((1-0.5*Z).^(2)); X_mag = abs(X); X_pha = angle(X)*180/pi;
subplot(5,2,5); plot(w/pi,X_mag); axis([0,1,0,max(X_mag)]);  ylabel('c.');
set(gca,'YTickMode','manual','YTick',[0,max(X_mag)],'FontSize',10);
subplot(5,2,6); plot(w/pi,X_pha); axis([0,1,-180,180]);
set(gca,'YTickMode','manual','YTick',[-180,0,180],'FontSize',10);
```

The plots are given in Figure C.3.

(d) $x(n) = (n+2)\,(-0.7)^{n-1}\,u\,(n-2)$. The sequence $x(n)$ can be arranged as

$$\begin{aligned} x(n) &= (n-2+4)\,(-0.7)^{n-2+1}\,u\,(n-2) \\ &= (-0.7)\,(n-2+4)\,(-0.7)^{n-2}\,u\,(n-2) \\ &= (-0.7)\,(n-2)\,(-0.7)^{n-2}\,u\,(n-2) + 4\,(-0.7)\,(-0.7)^{n-2}\,u\,(n-2) \\ &= (-0.7)\,(n-2)\,(-0.7)^{n-2}\,u\,(n-2) - 2.8\,(-0.7)^{n-2}\,u\,(n-2) \end{aligned}$$

Using properties and the result from part (c), we obtain,

$$\begin{aligned} X(\omega) &= (-0.7)\,e^{-j2\omega}\frac{(-0.7)\,e^{-j\omega}}{\left[1+0.7e^{-j\omega}\right]^2} - 2.8\frac{e^{-j\omega}}{1+0.7e^{-j\omega}} \\ &= \frac{0.49e^{-j3\omega}}{\left[1+0.7e^{-j\omega}\right]^2} - 2.8\frac{e^{-j\omega}}{1+0.7e^{-j\omega}} \end{aligned}$$

MATLAB script for magnitude and angle plot:

```
% (d) x(n) = (n+2)*(-0.7).^(n-1)*u(n-1)
X = (0.49)*(Z.*ZZ)./((1+0.7*Z).^2) - (2.8*ZZ)./(1+0.7*Z);
X_mag = abs(X); X_pha = angle(X)*180/pi;
subplot(5,2,7); plot(w/pi,X_mag); axis([0,1,0,max(X_mag)]); ylabel('d.');
set(gca,'YTickMode','manual','YTick',[0,max(X_mag)],'FontSize',10);
subplot(5,2,8); plot(w/pi,X_pha); axis([0,1,-180,180]);
set(gca,'YTickMode','manual','YTick',[-180,0,180],'FontSize',10);
```

The plots are given in Figure C.3.

(e) $x(n) = 5\,(-0.9)^n \cos(0.1\pi n)\,u(n)$. The sequence can be written as

$$\begin{aligned} x(n) &= 5\,(-0.9)^n \frac{e^{j0.1\pi n} + e^{-j0.1\pi n}}{2} u(n) \\ &= \frac{5}{2}\left(-0.9e^{j0.1\pi}\right)^n u(n) + \frac{5}{2}\left(-0.9e^{-j0.1\pi}\right)^n u(n) \end{aligned}$$

Hence, the DTFT is given by

$$\begin{aligned} X(\omega) &= \frac{5/2}{1+0.9e^{j0.1\pi}e^{-j\omega}} + \frac{5/2}{1+0.9e^{-j0.1\pi}e^{-j\omega}} \\ &= \frac{1+0.9\cos(0.1\pi)\,e^{-j\omega}}{1+1.8\cos(0.1\pi)\,e^{-j\omega}+0.81e^{-j2\omega}} \end{aligned}$$

MATLAB script for magnitude and angle plot:

```
% (e) x(n) = 5*(-0.9).^n*cos(0.1*pi*n)*u(n)
X = (1+0.9*cos(0.1*pi)*Z)./(1+1.8*cos(0.1*pi)*Z+(0.81)*ZZ);
X_mag = abs(X); X_pha = angle(X)*180/pi;
subplot(5,2,9); plot(w/pi,X_mag); axis([0,1,0,max(X_mag)]); ylabel('e.');
set(gca,'XTickMode','manual','XTick',[0,0.5,1],'FontSize',10);
set(gca,'YTickMode','manual','YTick',[0,max(X_mag)],'FontSize',10);
subplot(5,2,10); plot(w/pi,X_pha); axis([0,1,-180,180]);
set(gca,'XTickMode','manual','XTick',[0,0.5,1],'FontSize',10);
```

The plots are given in Figure C.3.

P4.4 This problem is solved by using MATLAB.

```
% Problem~P4.4 : DTFT of Rectangular pulse
clear; close all;
Hf_1 = figure('Units','normalized','position',[0.1,0.1,0.8,0.8],'color',[0,0,0]);
set(Hf_1,'NumberTitle','off','Name','P4.4');
w = [-100:100]*pi/100;

% x(n) = -N:N;
% (a) N = 5
N = 5; n = -N:N; x = ones(1,length(n)); X = dtft(x,n,w); X = real(X); X = X/max(X);
subplot(2,2,1); plot(w/pi,X); axis([-1,1,min(X),1]);
title('DTFT for N = 5','fontweight','bold'); ylabel('X');
set(gca,'XTickMode','manual','XTick',[-1,0,1],'FontSize',10);
set(gca,'YTickMode','manual','YTick',[min(X),0,1],'FontSize',10); grid;
```

```
% (b) N = 15
N = 15; n = -N:N; x = ones(1,length(n)); X = dtft(x,n,w); X = real(X); X = X/max(X);
subplot(2,2,2); plot(w/pi,X); axis([-1,1,min(X),1]);
title('DTFT for N = 15','fontweight','bold'); ylabel('X');
set(gca,'XTickMode','manual','XTick',[-1,0,1],'FontSize',10);
set(gca,'YTickMode','manual','YTick',[min(X),0,1],'FontSize',10); grid;

% (c) N = 25
N = 25; n = -N:N; x = ones(1,length(n)); X = dtft(x,n,w); X = real(X); X = X/max(X);
subplot(2,2,3); plot(w/pi,X); axis([-1,1,min(X),1]);
title('DTFT for N = 25','fontweight','bold'); ylabel('X');
xlabel('frequency in pi units');
set(gca,'XTickMode','manual','XTick',[-1,0,1],'FontSize',10);
set(gca,'YTickMode','manual','YTick',[min(X),0,1],'FontSize',10); grid;

% (d) N = 100
N = 100; n = -N:N; x = ones(1,length(n)); X = dtft(x,n,w); X = real(X); X = X/max(X);
subplot(2,2,4); plot(w/pi,X); axis([-1,1,min(X),1]);
title('DTFT for N = 100','fontweight','bold'); ylabel('X');
xlabel('frequency in pi units');
set(gca,'XTickMode','manual','XTick',[-1,0,1],'FontSize',10);
set(gca,'YTickMode','manual','YTick',[min(X),0,1],'FontSize',10); grid;
```

The plots are shown in Figure C.4. These plots show that the DTFT of a rectangular pulse is similar to a sinc function and that, as N increases, the function becomes narrower and narrower.

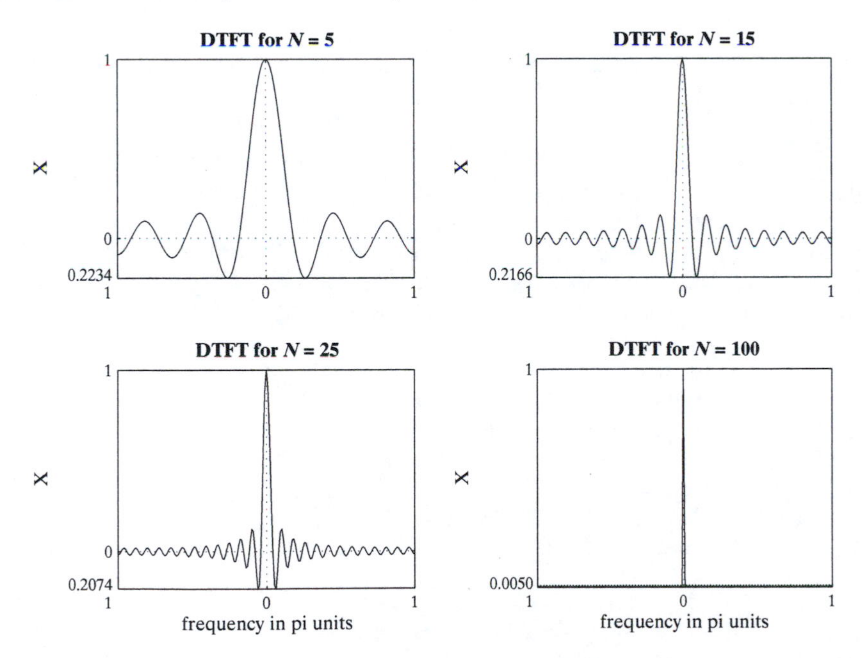

Figure C.4: Problem P4.4 DTFT plots

P4.5 DTFT of a symmetric triangular pulse.

$$\mathcal{T}_N(n) = \left[1 - \frac{|n|}{N}\right] \mathcal{R}_N(n)$$

This problem is solved by using MATLAB. It uses the function `dtft` described in P4.1:

```
function [X] = dtft(x,n,w)
% Computes Discrete-time Fourier Transform
% [X] = dtft(x,n,w)
%
% X = DTFT values computed at w frequencies
% x = finite duration sequence over n (row vector)
% n = sample position row vector
% W = frequency row vector
X = x*exp(-j*n'*w);
```

The MATLAB script for the problem is

```
clear; close all;

w = [-100:100]*pi/100;

% x(n) = (1-abs(n)/N)*R_N(n);
% (a) N = 5
N = 5; n = -N:N; x = 1-abs(n)/N; X = dtft(x,n,w); X = real(X); X = X/max(X);
subplot(2,2,1); plot(w/pi,X); axis([-1,1,0,1]);
title('DTFT for N = 5','fontweight','bold'); ylabel('X');
set(gca,'XTickMode','manual','XTick',[-1,0,1],'FontSize',10);
set(gca,'YTickMode','manual','YTick',[min(X),0,1],'FontSize',10); grid;

% (b) N = 15
N = 15; n = -N:N; x = 1-abs(n)/N; X = dtft(x,n,w); X = real(X); X = X/max(X);
subplot(2,2,2); plot(w/pi,X); axis([-1,1,0,1]);
title('DTFT for N = 15','fontweight','bold'); ylabel('X');
set(gca,'XTickMode','manual','XTick',[-1,0,1],'FontSize',10);
set(gca,'YTickMode','manual','YTick',[min(X),0,1],'FontSize',10); grid;

% (c) N = 25
N = 25; n = -N:N; x = 1-abs(n)/N; X = dtft(x,n,w); X = real(X); X = X/max(X);
subplot(2,2,3); plot(w/pi,X); axis([-1,1,0,1]);
title('DTFT for N = 25','fontweight','bold'); ylabel('X');
xlabel('frequency in pi units');
set(gca,'XTickMode','manual','XTick',[-1,0,1],'FontSize',10);
set(gca,'YTickMode','manual','YTick',[min(X),0,1],'FontSize',10); grid;

% (d) N = 100
N = 100; n = -N:N; x = 1-abs(n)/N; X = dtft(x,n,w); X = real(X); X = X/max(X);
subplot(2,2,4); plot(w/pi,X); axis([-1,1,0,1]);
title('DTFT for N = 100','fontweight','bold'); ylabel('X');
xlabel('frequency in pi units');
```

```
set(gca,'XTickMode','manual','XTick',[-1,0,1],'FontSize',10);
set(gca,'YTickMode','manual','YTick',[min(X),0,1],'FontSize',10); grid;
```

The DTFT plots are shown in Figure C.5. These plots show that the DTFT of a triangular pulse is similar to a $(\text{sinc})^2$ function and that, as N increases, the function becomes narrower and narrower.

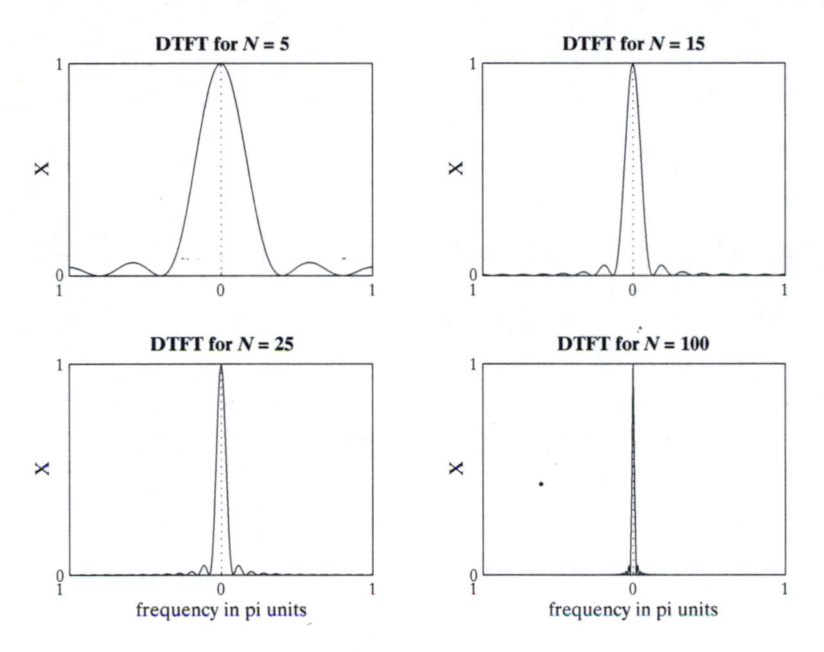

Figure C.5: DTFT plots in Problem P4.5

P4.6 Consider $x_e(n) = \frac{1}{2}\left[x(n) + x^*(-n)\right]$. Using $\mathcal{F}\left[x^*(-n)\right] = X^*(\omega)$, we obtain

$$\begin{aligned}\mathcal{F}\left[x_e(n)\right] &= \mathcal{F}\left[\frac{1}{2}\left\{x(n) + x^*(-n)\right\}\right] = \frac{1}{2}\left\{\mathcal{F}\left[x(n)\right] + \mathcal{F}\left[x^*(-n)\right]\right\} \\ &= \frac{1}{2}\left\{X(\omega) + X^*(\omega)\right\} \triangleq X_R(\omega).\end{aligned}$$

Similarly,

$$\begin{aligned}\mathcal{F}\left[x_o(n)\right] &= \mathcal{F}\left[\frac{1}{2}\left\{x(n) - x^*(-n)\right\}\right] = \frac{1}{2}\left\{\mathcal{F}\left[x(n)\right] - \mathcal{F}\left[x^*(-n)\right]\right\} \\ &= \frac{1}{2}\left\{X(\omega) - X^*(\omega)\right\} \triangleq jX_I(\omega).\end{aligned}$$

MATLAB verification by using $x(n) = e^{j0.1\pi n}\left[u(n) - u(n-20)\right]$:

```
clear; close all;
%
n = 0:20; x = exp(0.1*pi*n);
w = [-100:100]*pi/100; X = dtft(x,n,w);
```

```
XR = real(X); XI = imag(X);
[xe,xo,neo] = evenodd(x,n);
Xe = dtft(xe,neo,w); Xo = dtft(xo,neo,w);
diff_e = max(abs(XR-Xe))
diff_e =
  5.5511e-017
diff_o = max(abs(j*XI-Xo))
diff_o =
  6.9389e-017
```

P4.7 Gain and phase plots.

(a) A digital filter is described by the difference equation

$$y(n) = x(n) + 2x(n-1) + x(n-2) - 0.5y(n-1) + 0.25y(n-2)$$

The frequency response $H(\omega)$ of the filter, from substituting $x(n) = e^{j\omega n}$ and $y(n) = H(\omega)e^{j\omega n}$ in the difference equation and simplifying, is

$$H(\omega) = \frac{1 + 2e^{-j\omega} + e^{-j2\omega}}{1 + 0.5e^{-j\omega} - 0.25e^{-j2\omega}}$$

```
clear; close all;
Hf_1 = figure('Units','normalized','position',[0.1,0.1,0.8,0.8],'color',[0,0,0]);
set(Hf_1,'NumberTitle','off','Name','P4.7a');
%
% Analytical calculations of Frequency Response using diff eqn
b = [1,2,1]; a = [1,0.5,0.25];
w = [0:1:500]*pi/500; kb = 0:length(b)-1; ka = 0:length(a)-1;
H = (b*exp(-j*kb'*w)) ./ (a*exp(-j*ka'*w));
magH = abs(H); phaH = angle(H);
%
% Magnitude Responses Plot
subplot(2,1,1); plot(w/pi,magH);grid; axis([0,1,0,3]);
xlabel('frequency in pi units'); ylabel('|H|');
title('Magnitude Response');
wtick = [0:0.2:1]; magtick = [0:3];
set(gca,'XTickMode','manual','XTick',wtick)
set(gca,'YTickMode','manual','YTick',magtick)
%
% Phase response plot
subplot(2,1,2); plot(w/pi,phaH*180/pi);grid;
xlabel('frequency in pi units'); ylabel('Degrees');
title('Phase Response'); axis([0,1,-180,180])
wtick = [0:0.2:1]; phatick = [-180;0;180];
set(gca,'XTickMode','manual','XTick',wtick)
set(gca,'YTickMode','manual','YTick',phatick)
```

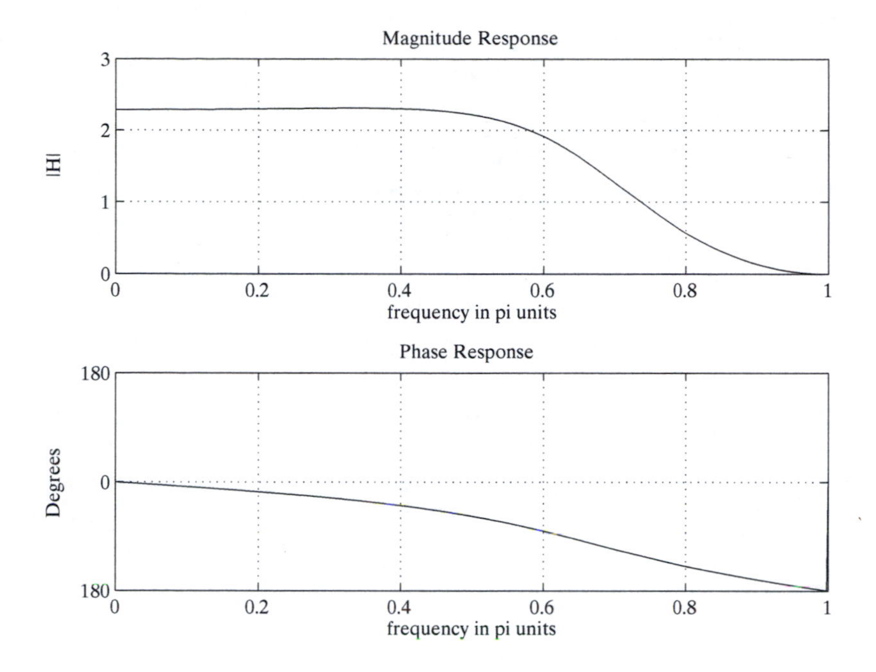

Figure C.6: Problem P4.7(a) plots

The magnitude and phase response plots are shown in Figure C.6.

(b) A digital filter is described by the difference equation

$$y(n) = 2x(n) + x(n-1) - 0.25y(n-1) + 0.25y(n-2)$$

The frequency response $H(\omega)$ of the filter, from substituting $x(n) = e^{j\omega n}$ and $y(n) = H(\omega)e^{j\omega n}$ in the difference equation and simplifying, is

$$H(\omega) = \frac{2 + e^{-j\omega}}{1 + 0.25e^{-j\omega} - 0.25e^{-j2\omega}}$$

```
clear; close all;
Hf_1 = figure('Units','normalized','position',[0.1,0.1,0.8,0.8],'color',[0,0,0]);
set(Hf_1,'NumberTitle','off','Name','P4.7b');
%
% Analytical calculations of Frequency Response using diff eqn
b = [2,1]; a = [1,0.25,-0.25];
w = [0:1:500]*pi/500; kb = 0:length(b)-1; ka = 0:length(a)-1;
H = (b*exp(-j*kb'*w)) ./ (a*exp(-j*ka'*w));
magH = abs(H); phaH = angle(H);
%
% Magnitude Responses Plot
subplot(2,1,1); plot(w/pi,magH);grid; axis([0,1,0,3]);
xlabel('frequency in pi units'); ylabel('|H|');
title('Magnitude Response');
wtick = [0:0.2:1]; magtick = [0:3];
```

```
set(gca,'XTickMode','manual','XTick',wtick)
set(gca,'YTickMode','manual','YTick',magtick)
%
% Phase response plot
subplot(2,1,2); plot(w/pi,phaH*180/pi);grid;
xlabel('frequency in pi units'); ylabel('Degrees');
title('Phase Response'); axis([0,1,-180,180])
wtick = [0:0.2:1]; phatick = [-180;0;180];
set(gca,'XTickMode','manual','XTick',wtick)
set(gca,'YTickMode','manual','YTick',phatick)
```

The magnitude and phase response plots are shown in Figure C.7.

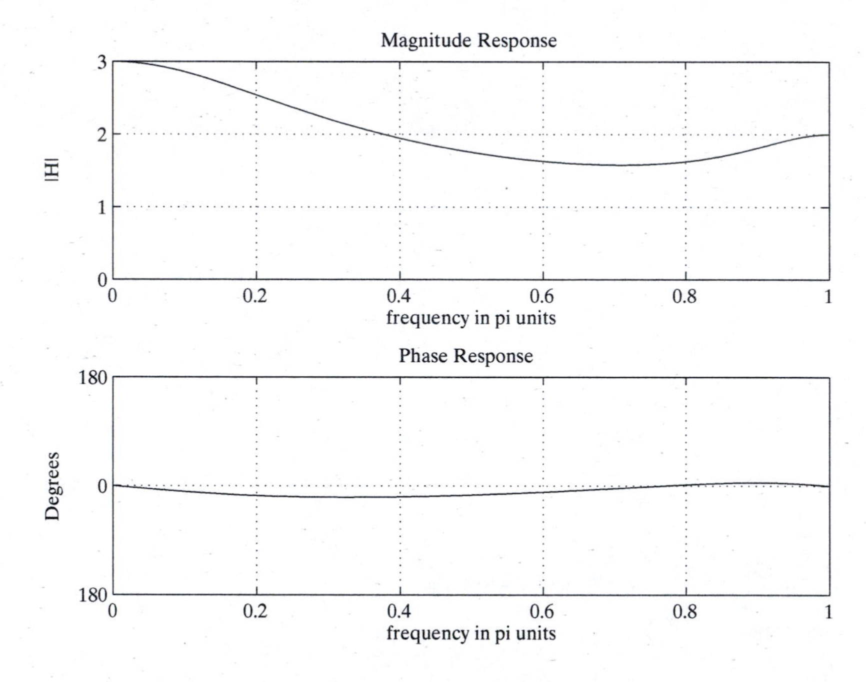

Figure C.7: Problem P4.7(b) plots

P4.8 Sampling frequency $F_s = 100$ sam/sec (or sampling interval $T_s = 0.01$ sec/sam) and impulse response $h(n) = (0.5)^n\, u(n)$.

(a) $x_a(t) = 3\cos(20\pi t)$. Hence, $x(n) = x_a(nT_s) = 3\cos(0.2\pi n)$. Therefore, the digital frequency is 0.2π rad/sam.

(b) The steady-state response when $x(n) = 3\cos(0.2\pi n)$: The frequency response is

$$H(\omega) = F[h(n)] = F\left[(0.5)^n\, u(n)\right] = \frac{1}{1 - 0.5e^{j\omega}}$$

At $\omega = 0.2\pi$, the response is

$$H(0.2\pi) = \frac{1}{1 - 0.5e^{j0.2\pi}} = 0.6969\left(\angle - 0.2063^c\right)$$

Hence,

$$y_{ss}(n) = 3\,(0.6969)\cos\left(0.2\pi n - 0.2363\right)$$

which, after D/A conversion, gives $y_{ss}(t)$ as

$$y_{ss,a}(t) = 2.0907\cos\left(20\pi t - 0.2363\right)$$

(c) The steady-state DC gain is obtained by setting $\omega = 0$, which is equivalent to $H(0) = 2$. Hence, $y_{ss}(n) = 3\,(2) = y_{ss,a}(t) = 6$.

(d) Aliased frequencies of F_0 for the given sampling rate F_s are $F_0 + kF_s$. Now, for $F_0 = 10$ Hz and $F_s = 100$, the aliased frequencies are $10 + 100k = \{110, 210, \ldots\}$. Therefore, two other $x_a(t)$'s are

$$3\cos\left(220\pi t\right) \text{ and } 3\cos\left(420\pi t\right)$$

(e) The prefilter should be a lowpass filter with its cutoff frequency at 50 Hz.

P4.9 An analog signal $x_a(t) = \cos(20\pi t), 0 \le t \le 1$, is sampled at $T_s = 0.01$-, 0.05-, and 0.1-sec intervals.

(a) Plots of $x(n)$ for each T_s. MATLAB script:

```
clear; close all;
%
t = 0:0.001:1; xa = cos(20*pi*t);
% (a) Plots of sequences

Ts = 0.01; N1 = round(1/Ts); n1 = 0:N1; x1 = cos(20*pi*n1*Ts);
subplot(3,1,1); plot(t,xa,n1*Ts,x1,'o'); axis([0,1,-1.1,1.1]);
ylabel('x1(n)'); title('Sampling of xa(t) using Ts=0.01');
set(gca,'xtickmode','manual','xtick',[0:1]);
Ts = 0.05; N2 = round(1/Ts); n2 = 0:N2; x2 = cos(20*pi*n2*Ts);
subplot(3,1,2); plot(t,xa,n2*Ts,x2,'o'); axis([0,1,-1.1,1.1]);
ylabel('x2(n)'); title('Sampling of xa(t) using Ts=0.05');
set(gca,'xtickmode','manual','xtick',[0:1]);
Ts = 00.1; N3 = round(1/Ts); n3 = 0:N3; x3 = cos(20*pi*n3*Ts);
subplot(3,1,3); plot(t,xa,n3*Ts,x3,'o'); axis([0,1,-1.1,1.1]);
ylabel('x3(n)'); title('Sampling of xa(t) using Ts=0.1');
set(gca,'xtickmode','manual','xtick',[0:1]);xlabel('t,sec');
```

The plots are shown in Figure C.8.

(b) Reconstruction from $x(n)$ by using the sinc interpolation. MATLAB script:

```
% (b) Reconstruction using sinc function

Ts = 0.01; Fs = 1/Ts;
xa1 = x1*sinc(Fs*(ones(length(n1),1)*t-(n1*Ts)'*ones(1,length(t))));
subplot(3,1,1);plot(t,xa1); axis([0,1,-1.1,1.1]);
ylabel('xa(t)'); title('Reconstruction of xa(t) when Ts=0.01');
set(gca,'xtickmode','manual','xtick',[0:1]);
```

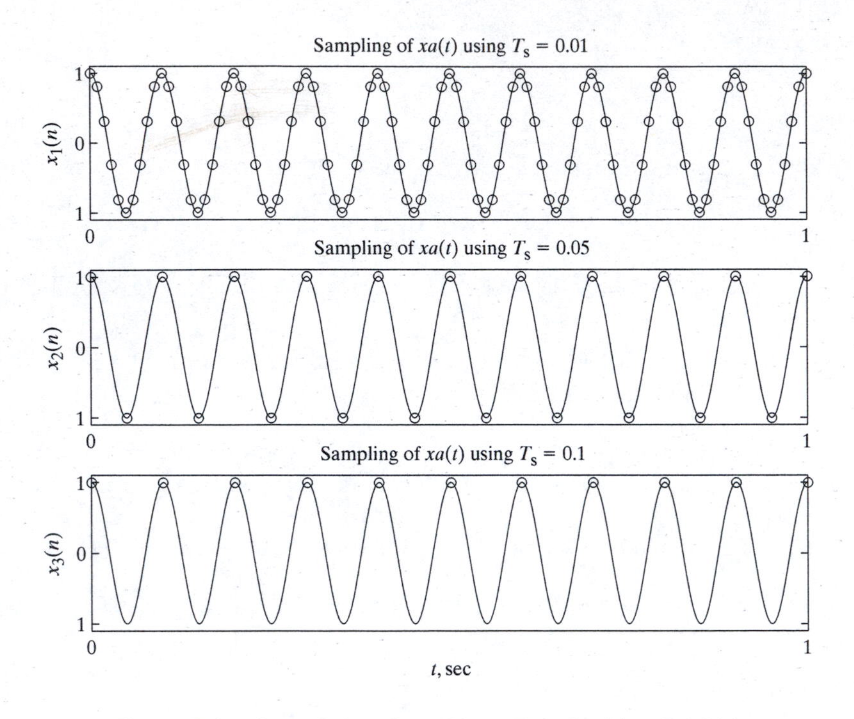

Figure C.8: Plots of $x(n)$ for various T_s in Problem P4.9(a)

```
Ts = 0.05; Fs = 1/Ts;
xa2 = x2*sinc(Fs*(ones(length(n2),1)*t-(n2*Ts)'*ones(1,length(t))));
subplot(3,1,2);plot(t,xa2); axis([0,1,-1.1,1.1]);
ylabel('xa(t)'); title('Reconstruction of xa(t) when Ts=0.05');
set(gca,'xtickmode','manual','xtick',[0:1]);
Ts = 0.1; Fs = 1/Ts;
xa3 = x3*sinc(Fs*(ones(length(n3),1)*t-(n3*Ts)'*ones(1,length(t))));
subplot(3,1,3);plot(t,xa3); axis([0,1,-1.1,1.1]);
ylabel('xa(t)'); title('Reconstruction of xa(t) when Ts=0.1');
set(gca,'xtickmode','manual','xtick',[0:1]);xlabel('t,sec');
```

The reconstruction is shown in Figure C.9.

(c) Reconstruction from $x(n)$ using the cubic spline interpolation. MATLAB script:

```
% (c) Reconstruction using cubic spline interpolation

Ts = 0.01; Fs = 1/Ts;
xa1 = spline(Ts*n1,x1,t);
subplot(3,1,1);plot(t,xa1); axis([0,1,-1.1,1.1]);
ylabel('xa(t)'); title('Reconstruction of xa(t) when Ts=0.01');
set(gca,'xtickmode','manual','xtick',[0:1]);
Ts = 0.05; Fs = 1/Ts;
xa2 = spline(Ts*n2,x2,t);
subplot(3,1,2);plot(t,xa2); axis([0,1,-1.1,1.1]);
ylabel('xa(t)'); title('Reconstruction of xa(t) when Ts=0.05');
```

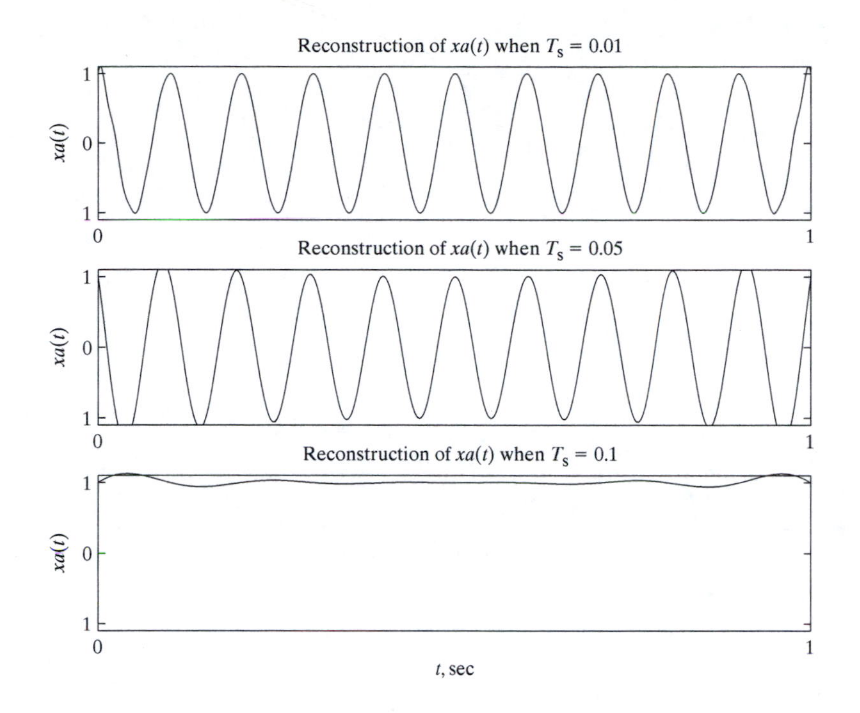

Figure C.9: The sinc interpolation in Problem P4.9(b)

```
set(gca,'xtickmode','manual','xtick',[0:1]);
Ts = 0.1; Fs = 1/Ts;
xa3 = spline(Ts*n3,x3,t);
subplot(3,1,3);plot(t,xa3); axis([0,1,-1.1,1.1]);
ylabel('xa(t)'); title('Reconstruction of xa(t) when Ts=0.1');
set(gca,'xtickmode','manual','xtick',[0:1]);xlabel('t,sec');
```

The reconstruction is shown in Figure C.10.

(d) Comments: From the plots in the figures, it is clear that reconstructions from samples at $T_s = 0.01$ and 0.05 depict the original frequency (excluding end effects), but that reconstructions for $T_s = 0.1$ show the original frequency aliased to zero. Furthermore, the cubic spline interpolation yields a better reconstruction than the sinc interpolation; that is, the sinc interpolation is more susceptible to boundary effects.

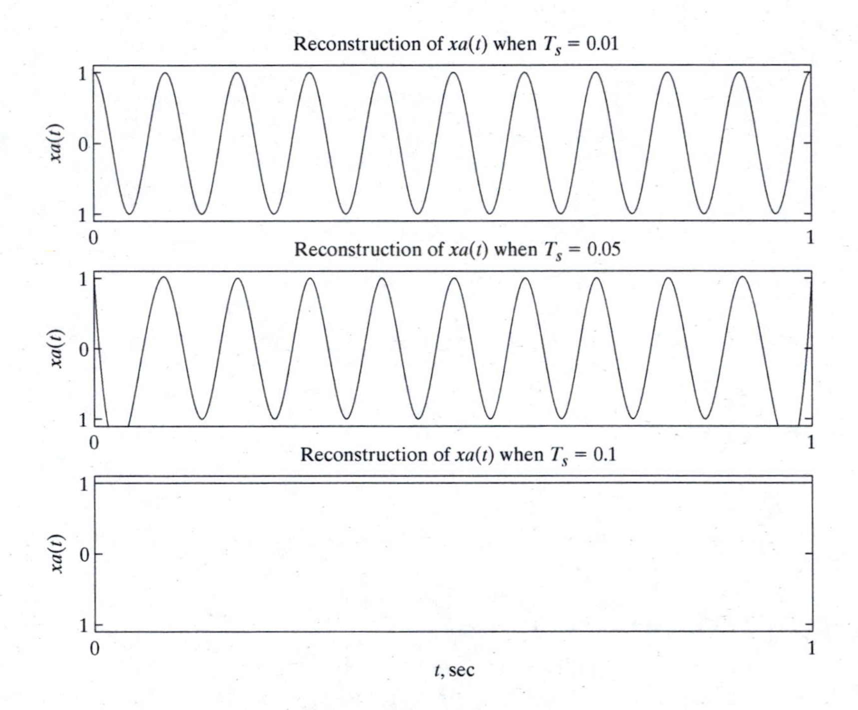

Figure C.10: The cubic spline interpolation in Problem P4.9(c)

Appendix D

The Discrete Fourier Transform

P5.1 (a) Periodic sequence: $\tilde{x}_1(n) = \left\{\ldots, \underset{\uparrow}{2}, 0, 2, 0, 2, 0, 2, 0, 2, 0, \ldots\right\}$. Now,

$$\tilde{X}_1(k) = \sum_{n=0}^{N-1} \tilde{x}_1(n) W_N^{nk}; \;\; N = 4; \;\; W_4 = e^{-j2\pi/4} = -j$$

Hence,

$$\begin{aligned}
\tilde{X}_1(0) &= 2(1) + 0(1) + 2(1) + 0(1) &= 4 \\
\tilde{X}_1(1) &= 2(1) + 0(-j) + 2(-1) + 0(j) &= 0 \\
\tilde{X}_1(2) &= 2(1) + 0(-1) + 2(1) + 0(-1) &= 4 \\
\tilde{X}_1(3) &= 2(1) + 0(j) + 2(-1) + 0(-j) &= 0
\end{aligned}$$

MATLAB verification:

```
xtilde1 = [2,0,2,0]; N = length(xtilde1);
[Xtilde1] = dft(xtilde1,N)
Xtilde1 =
   4.0000                 0 - 0.0000i   4.0000 + 0.0000i        0 - 0.0000i
```

(b) Periodic sequence: $\tilde{x}_2(n) = \left\{\ldots, \underset{\uparrow}{0}, 0, 1, 0, 0, 0, 0, 1, 0, 0, \ldots\right\}$. Now,

$$\tilde{X}_2(k) = \sum_{n=0}^{N-1} \tilde{x}_2(n) W_N^{nk}; \;\; N = 5; \;\; W_5 = e^{-j2\pi/5} = 0.3090 - j0.9511$$

Hence,

$$\begin{aligned}
\tilde{X}_2(0) &= 0 + 0 + 1(1) + 0 + 0 &= 1 \\
\tilde{X}_2(1) &= 0 + 0 + 1(0.3090 - j0.9511)^2 + 0 + 0 &= -0.8090 - j0.5878 \\
\tilde{X}_2(2) &= 0 + 0 + 1(0.3090 - j0.9511)^4 + 0 + 0 &= 0.3090 + j0.9511 \\
\tilde{X}_2(3) &= 0 + 0 + 1(0.3090 - j0.9511)^6 + 0 + 0 &= 0.3090 - j0.9511 \\
\tilde{X}_2(4) &= 0 + 0 + 1(0.3090 - j0.9511)^8 + 0 + 0 &= -0.8090 + j0.5878
\end{aligned}$$

MATLAB verification:

```
xtilde2 = [0,0,1,0,0]; N = length(xtilde2);
[Xtilde2] = dft(xtilde2,N)
Xtilde2 =
  Columns 1 through 4
   1.0000             -0.8090 - 0.5878i   0.3090 + 0.9511i   0.3090 - 0.9511i
  Column 5
  -0.8090 + 0.5878i
```

(c) Periodic sequence: $\tilde{x}_3(n) = \left\{\dots, \underset{\uparrow}{3}, -3, 3, -3, 3, -3, 3, -3, \dots\right\}$. Now,

$$\tilde{X}_3(k) = \sum_{n=0}^{N-1} \tilde{x}_3(n) W_N^{nk}; \; N = 4; \; W_4 = e^{-j2\pi/4} = -j$$

Hence,

$$\begin{aligned}
\tilde{X}_3(0) &= 3(1) - 3(1) + 3(1) - 3(1) &&= 0 \\
\tilde{X}_3(1) &= 3(1) - 3(-j) + 3(-1) - 3(j) &&= 0 \\
\tilde{X}_3(2) &= 3(1) - 3(-1) + 3(1) - 3(-1) &&= 12 \\
\tilde{X}_3(3) &= 3(1) - 3(j) + 3(-1) - 3(-j) &&= 0
\end{aligned}$$

MATLAB verification:

```
xtilde = [3,-3,3,-3]; N = length(xtilde);
[Xtilde] = dft(xtilde,N)
Xtilde =
        0             0.0000 - 0.0000i  12.0000 + 0.0000i   0.0000 - 0.0000i
```

(d) Periodic sequence: $\tilde{x}_4(n) = \left\{\dots, \underset{\uparrow}{j}, j, -j, -j, j, j, -j, -j, \dots\right\}$. Now,

$$\tilde{X}_4(k) = \sum_{n=0}^{N-1} \tilde{x}_4(n) W_N^{nk}; \; N = 4; \; W_4 = e^{-j2\pi/4} = -j$$

Hence,

$$\begin{aligned}
\tilde{X}_4(0) &= j(1) + j(1) - j(1) - j(1) &&= 0 \\
\tilde{X}_4(1) &= j(1) + j(-j) - j(-1) - j(j) &&= 2 + j2 \\
\tilde{X}_4(2) &= j(1) + j(-1) - j(1) - j(-1) &&= 0 \\
\tilde{X}_4(3) &= j(1) + j(j) - j(-1) - j(-j) &&= -2 + j2
\end{aligned}$$

MATLAB verification:

```
xtilde = [j,j,-j,-j]; N = length(xtilde4);
[Xtilde4] = dft(xtilde4,N)
Xtilde4 =
        0             2.0000 + 2.0000i        0             -2.0000 + 2.0000i
```

(e) Periodic sequence: $\tilde{x}_5(n) = \left\{\ldots, \underset{\uparrow}{1}, j, j, 1, 1, j, j, 1, j, j, \ldots\right\}$. Now,

$$\tilde{X}_5(k) = \sum_{n=0}^{N-1} \tilde{x}_5(n) W_N^{nk}; \; N = 4; \; W_4 = e^{-j2\pi/4} = -j$$

Hence,

$$\begin{aligned}
\tilde{X}_5(0) &= 1(1) + j(1) + j(1) + 1(1) &&= 2 + 2j \\
\tilde{X}_5(1) &= 1(1) + j(-j) + j(-1) + 1(j) &&= 2 \\
\tilde{X}_5(2) &= 1(1) + j(-1) + j(1) + 1(-1) &&= 0 \\
\tilde{X}_5(3) &= 1(1) + j(j) + j(-1) + 1(-j) &&= -2j
\end{aligned}$$

MATLAB verification:

```
xtilde5 = [1,j,j,1]; N5 = length(xtilde5);
[Xtilde5] = dft(xtilde5,N5)
Xtilde5 =
   2.0000 + 2.0000i   2.0000 + 0.0000i   0.0000 - 0.0000i   0.0000 - 2.0000i
```

P5.2 (a) Periodic DFS sequence: $\tilde{X}_1(k) = \{5, -2j, 3, 2j\}$, $N = 4$. Now,

$$\tilde{x}_1(n) = \frac{1}{N} \sum_{k=0}^{N-1} \tilde{X}_1(k) W_N^{-nk}; \; N = 4; \; W_4^{-1} = e^{j2\pi/4} = j$$

Hence,

$$\begin{aligned}
\tilde{x}_1(0) &= [5(1) - 2j(1) + 3(1) + 2j(1)]/4 &&= 2 \\
\tilde{x}_1(1) &= [5(1) - 2j(j) + 3(-1) + 2j(-j)]/4 &&= 1.5 \\
\tilde{x}_1(2) &= [5(1) - 2j(-1) + 3(1) + 2j(-1)]/4 &&= 2 \\
\tilde{x}_1(3) &= [5(1) - 2j(-j) + 3(-1) + 2j(j)]/4 &&= -0.5
\end{aligned}$$

MATLAB verification:

```
Xtilde1 = [5,-2*j,3,2*j]; N1 = length(Xtilde1);
[xtilde1] = real(idfs(Xtilde1,N1))
xtilde1 =
    2.0000    1.5000    2.0000   -0.5000
```

(b) Periodic DFS sequence: $\tilde{X}_2(k) = \{4, -5, 3, -5\}$. Now,

$$\tilde{x}_2(n) = \frac{1}{N} \sum_{k=0}^{N-1} \tilde{X}_2(k) W_N^{nk}; \; N = 4; \; W_4 = e^{-j2\pi/4} = -j$$

Hence,

$$\begin{aligned}
\tilde{x}_2(0) &= [4(1) - 5(1) + 3(1) - 5(1)]/4 &&= -0.75 \\
\tilde{x}_2(1) &= [4(1) - 5(-j) + 3(-1) - 5(j)]/4 &&= 0.25 \\
\tilde{x}_2(2) &= [4(1) - 5(-1) + 3(1) - 5(-1)]/4 &&= 4.25 \\
\tilde{x}_2(3) &= [4(1) - 5(j) + 3(-1) - 5(-j)]/4 &&= 0.25
\end{aligned}$$

MATLAB verification:

```
Xtilde2 = [4,-5,3,-5]; N = length(Xtilde2);
[xtilde2] = real(idfs(Xtilde2,N))
xtilde2 =
   -0.7500    0.2500    4.2500    0.2500
```

(c) Periodic DFS sequence: $\tilde{X}_3(k) = \{1, 2, 3, 4, 5\}$. Now,

$$\tilde{x}_3(n) = \frac{1}{N}\sum_{k=0}^{N-1} \tilde{X}_3(k)\, W_N^{nk};\; N = 5;\; W_5 = e^{-j2\pi/5} = 0.3090 - j0.9511$$

Hence,

$$\begin{aligned}
\tilde{x}_3(0) &= \left[1W_5^0 + 2W_5^0 + 3W_5^0 + 4W_5^0 + 5W_5^0\right]/5 &&= 3\\
\tilde{x}_3(1) &= \left[1W_5^0 + 2W_5^1 + 3W_5^2 + 4W_5^3 + 5W_5^4\right]/5 &&= -0.5000 - j0.6882\\
\tilde{x}_3(2) &= \left[1W_5^0 + 2W_5^2 + 3W_5^4 + 4W_5^6 + 5W_5^8\right]/5 &&= -0.5000 - j0.1625\\
\tilde{x}_3(3) &= \left[1W_5^0 + 2W_5^3 + 3W_5^6 + 4W_5^9 + 5W_5^{12}\right]/5 &&= -0.5000 + j0.1625\\
\tilde{x}_3(4) &= \left[1W_5^0 + 2W_5^4 + 3W_5^8 + 4W_5^{12} + 5W_5^{16}\right]/5 &&= -0.5000 + j0.6882
\end{aligned}$$

MATLAB verification:

```
Xtilde3 = [1,2,3,4,5]; N = length(Xtilde3);
[xtilde3] = idfs(Xtilde3,N)
xtilde3 =
  Columns 1 through 4
   3.0000             -0.5000 - 0.6882i  -0.5000 - 0.1625i  -0.5000 + 0.1625i
  Column 5
  -0.5000 + 0.6882i
```

(d) Periodic DFS sequence: $\tilde{X}(k) = \{0, 0, 2, 0\}$. Now,

$$\tilde{x}(n) = \frac{1}{N}\sum_{k=0}^{N-1} \tilde{X}(k)\, W_N^{nk};\; N = 4;\; W_4 = e^{-j2\pi/4} = -j$$

Hence,

$$\begin{aligned}
\tilde{x}(0) &= [0(1) + 0(1) + 2(1) + 0(1)]/4 &&= 0.5\\
\tilde{x}(1) &= [0(1) + 0(-j) + 2(-1) + 0(j)]/4 &&= -0.5\\
\tilde{x}(2) &= [0(1) + 0(-1) + 2(1) + 0(-1)]/4 &&= 0.5\\
\tilde{x}(3) &= [0(1) + 0(j) + 2(-1) + 0(-j)]/4 &&= -0.5
\end{aligned}$$

MATLAB verification:

```
Xtilde4 = [00,0,2,0]; N = length(Xtilde4);
[xtilde4] = real(idfs(Xtilde4,N))
xtilde4 =
    0.5000   -0.5000    0.5000   -0.5000
```

(e) Periodic DFS sequence: $\tilde{X}_5(k) = \{0, j, -2j, -j\}$, $N = 4$. Now,

$$\tilde{x}_5(n) = \frac{1}{N}\sum_{k=0}^{N-1} \tilde{X}_5(k)\, W_N^{-nk};\; N = 4;\; W_4^{-1} = e^{j2\pi/4} = j$$

Hence,

$$\begin{aligned}
\tilde{x}_5(0) &= [0(1)+j(1)-2j(1)-j(1)]/4 &&= -j0.5 \\
\tilde{x}_5(1) &= [0(1)+j(j)-2j(-1)-j(-j)]/4 &&= -0.5+j0.5 \\
\tilde{x}_5(2) &= [0(1)+j(-1)-2j(1)-2j(-1)]/4 &&= -j0.25 \\
\tilde{x}_5(3) &= [0(1)+j(-j)-2j(-1)-j(j)]/4 &&= 0.5+j0.5
\end{aligned}$$

MATLAB verification:

```
Xtilde5 = [0,j,-2*j,-j]; N5 = length(Xtilde5);
[xtilde5] = real(idfs(Xtilde5,N1))
xtilde1 =
         0 - 0.5000i  -0.5000 + 0.5000i   0.0000 - 0.2500i   0.5000 + 0.5000i
```

P5.3 Periodic $\tilde{x}_1(n)$ with fundamental period $N = 50$:

$$\tilde{x}_1(n) = \left\{ \begin{array}{cc} ne^{-0.3n}, & 0 \le n \le 25 \\ 0, & 26 \le n \le 49 \end{array} \right\}_{\text{PERIODIC}}$$

Periodic $\tilde{x}_2(n)$ with fundamental period $N = 100$:

$$\tilde{x}_2(n) = \left\{ \begin{array}{cc} ne^{-0.3n}, & 0 \le n \le 25 \\ 0, & 26 \le n \le 99 \end{array} \right\}_{\text{PERIODIC}}$$

(a) Computation of $\tilde{X}_1(k)$ by using MATLAB:

```
clear; close all;
% (a) DFS Xtilde1(k)
Hf_1 = figure('Units','normalized','position',[0.1,0.1,0.8,0.8],'color',[0,0,0]);
set(Hf_1,'NumberTitle','off','Name','P5.3a');
n1 = [0:49]; xtilde1 = [n1(1:26).*exp(-0.3*n1(1:26)),zeros(1,24)]; N1 = length(n1);
[Xtilde1] = dft(xtilde1,N1); k1 = n1;
mag_Xtilde1 = abs(Xtilde1); pha_Xtilde1 = angle(Xtilde1)*180/pi;
zei = find(mag_Xtilde1 < 1000*eps);
pha_Xtilde1(zei) = zeros(1,length(zei));
subplot(3,1,1); stem(n1,xtilde1); axis([-1,N1,min(xtilde1),max(xtilde1)]);
title('One period of the periodic sequence xtilde1(n)'); ylabel('xtilde1');
ntick = [n1(1):2:n1(N1),N1]';
set(gca,'XTickMode','manual','XTick',ntick,'FontSize',10)
subplot(3,1,2); stem(k1,mag_Xtilde1); axis([-1,N1,min(mag_Xtilde1),...
                                                    max(mag_Xtilde1)]);
title('Magnitude of Xtilde1(k)'); ylabel('|Xtilde1|')
ktick = [k1(1):2:k1(N1),N1]';
set(gca,'XTickMode','manual','XTick',ktick,'FontSize',10)
subplot(3,1,3); stem(k1,pha_Xtilde1); axis([-1,N1,-180,180]);
title('Phase of Xtilde1(k)'); xlabel('k'); ylabel('Angle in Deg')
ktick = [k1(1):2:k1(N1),N1]';
set(gca,'XTickMode','manual','XTick',ktick,'FontSize',10)
set(gca,'YTickMode','manual','YTick',[-180;-90;0;90;180])
```

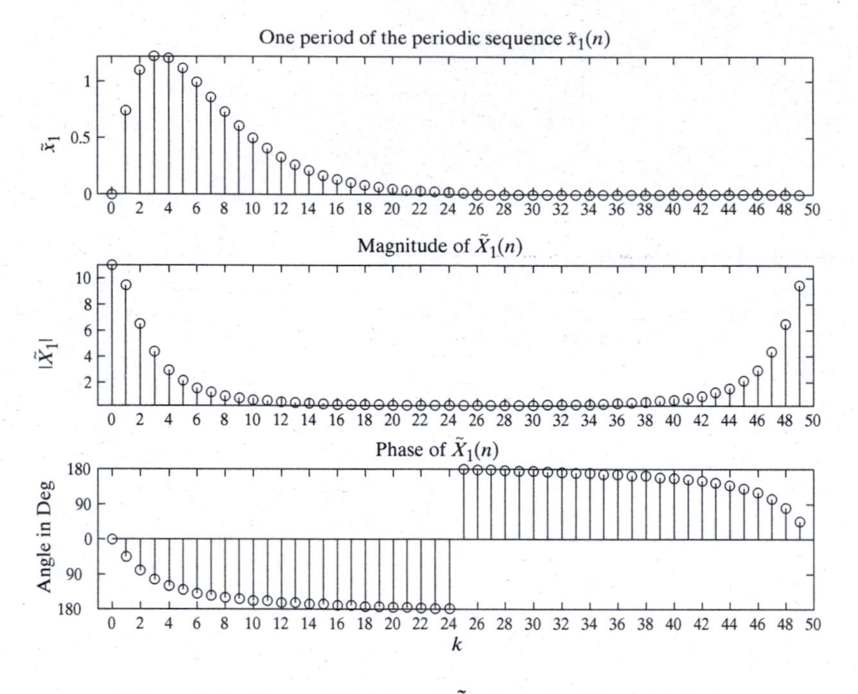

Figure D.1: Plots of $\tilde{x}_1(n)$ and $\tilde{X}_1(k)$ in Problem P5.3(a)

Plots of $\tilde{x}_1(n)$ and $\tilde{X}_1(k)$ are shown in Figure D.1.

(b) Computation of $\tilde{X}_2(k)$ by using MATLAB:

```
% (b) DFS Xtilde2(k)
Hf_2 = figure('Units','normalized','position',[0.1,0.1,0.8,0.8],'color',[0,0,0]);
set(Hf_2,'NumberTitle','off','Name','P5.3b');
n2 = [0:99]; xtilde2 = [xtilde1, zeros(1,50)]; N2 = length(n2);
[Xtilde2] = dft(xtilde2,N2); k2 = n2;
mag_Xtilde2 = abs(Xtilde2); pha_Xtilde2 = angle(Xtilde2)*180/pi;
zei = find(mag_Xtilde2 < 1000*eps);
pha_Xtilde2(zei) = zeros(1,length(zei));
subplot(3,1,1); stem(n2,xtilde2); axis([-1,N2,min(xtilde2),max(xtilde2)]);
title('One period of the periodic sequence xtilde2(n)'); ylabel('xtilde2');
ntick = [n2(1):5:n2(N2),N2]';
set(gca,'XTickMode','manual','XTick',ntick)
subplot(3,1,2); stem(k2,mag_Xtilde2); axis([-1,N2,min(mag_Xtilde2),...
                                           max(mag_Xtilde2)]);
title('Magnitude of Xtilde2(k)'); ylabel('|Xtilde2|')
ktick = [k2(1):5:k2(N2),N2]';
set(gca,'XTickMode','manual','XTick',ktick)
subplot(3,1,3); stem(k2,pha_Xtilde2); axis([-1,N2,-180,180]);
title('Phase of Xtilde2(k)'); xlabel('k'); ylabel('Degrees')
ktick = [k2(1):5:k2(N2),N2]';
set(gca,'XTickMode','manual','XTick',ktick)
set(gca,'YTickMode','manual','YTick',[-180;-90;0;90;180])
```

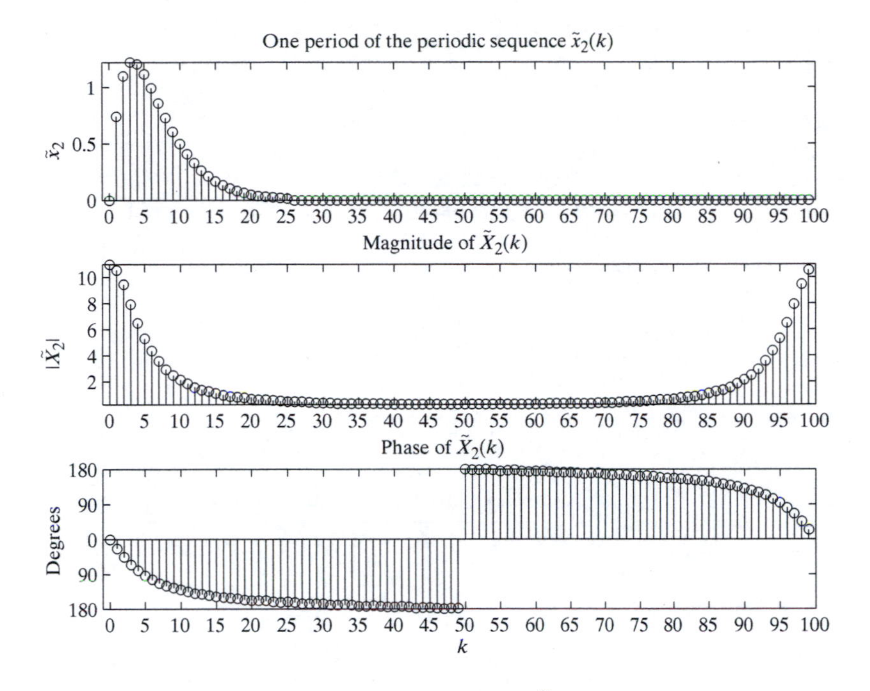

Figure D.2: Plots of magnitude and phase of $\tilde{X}_2(k)$ in Problem P5.3(b)

Plots of $\tilde{x}_2(n)$ and $\tilde{X}_2(k)$ are shown in Figure D.2.

(c) Changing the period from $N = 50$ to $N = 100$ resulted in a lower frequency-sampling interval (higher frequency resolution)ω_1 (i.e., in (0a)$\omega_1 = \pi/25$, but in (0b)$\omega_2 = \pi/50$). Hence, there are more terms in the DFS expansion of $\tilde{x}_2(n)$. The shape of the DTFT begins to fill in with $N = 100$.

P5.4 New periodic sequence $\tilde{x}_3(n)$ with period $N = 100$:

$$\tilde{x}_3(n) = \left[\tilde{x}_1(n), \tilde{x}_2(n)\right]_{\text{PERIODIC}}$$

(a) Computation of $\tilde{X}_3(k)$ by using MATLAB:

```
clear; close all;
% (a) DFS Xtilde3(k)
Hf_1 = figure('Units','normalized','position',[0.1,0.1,0.8,0.8],'color',[0,0,0]);
set(Hf_1,'NumberTitle','off','Name','P5.4a');
n1 = [0:49]; xtilde1 = [n1(1:26).*exp(-0.3*n1(1:26)),zeros(1,24)]; N1 = length(n1);
n3 = [0:99]; xtilde3 = [xtilde1, xtilde1]; N3 = length(n3);
[Xtilde3] = dft(xtilde3,N3); k3 = n3;
mag_Xtilde3 = abs(Xtilde3); pha_Xtilde3 = angle(Xtilde3)*180/pi;
zei = find(mag_Xtilde3 < 0.00001);
pha_Xtilde3(zei) = zeros(1,length(zei));
subplot(3,1,1); stem(n3,xtilde3); axis([-1,N3,min(xtilde3),max(xtilde3)]);
title('One period of the periodic sequence xtilde3(n)'); ylabel('xtilde3');
ntick = [n3(1):5:n3(N3),N3]';
```

```
set(gca,'XTickMode','manual','XTick',ntick)
subplot(3,1,2); stem(k3,mag_Xtilde3); axis([-1,N3,min(mag_Xtilde3),...
                                                  max(mag_Xtilde3)]);
title('Magnitude of Xtilde3(k)'); ylabel('|Xtilde3|')
ktick = [k3(1):5:k3(N3),N3]';
set(gca,'XTickMode','manual','XTick',ktick)
subplot(3,1,3); stem(k3,pha_Xtilde3); axis([-1,N3,-180,180]);
title('Phase of Xtilde3(k)')
xlabel('k'); ylabel('Degrees')
ktick = [k3(1):5:k3(N3),N3]';
set(gca,'XTickMode','manual','XTick',ktick)
set(gca,'YTickMode','manual','YTick',[-180;-90;0;90;180])
```

Plots of $\tilde{x}_3(n)$ and $\tilde{X}_3(k)$ are shown in Figure D.3.

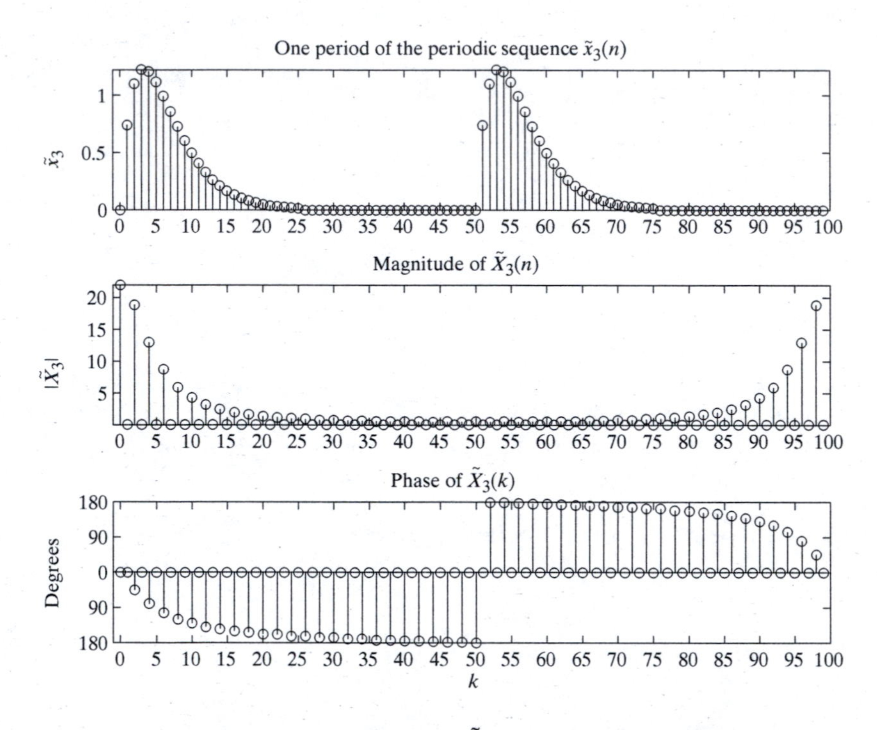

Figure D.3: Plots of $\tilde{x}_3(n)$ and $\tilde{X}_3(k)$ in Problem P5.4(a)

(b) Comparing the magnitude plot above with that of $\tilde{X}_1(k)$ in Problem P5.3 we observe that these plots are essentially similar. Plots of $\tilde{X}_3(k)$ have one zero between every sample of $\tilde{X}_1(k)$. (In general, for phase plots, we do get nonzero phase values when the magnitudes are zero. Clearly, these phase values have no meaning and should be ignored. This happens because of a particular algorithm used by MATLAB. We can avoid this problem by using the `find` function.) This makes sense, because sequences $\tilde{x}_1(n)$ and $\tilde{x}_3(n)$, when viewed over $-\infty < n < \infty$ interval, look exactly the same. The effect of periodicity doubling is in the doubling of magnitude of each sample.

(c) We can now generalize this argument. If

$$\tilde{x}_M(n) = \Big\{\underbrace{\tilde{x}_1(n), \tilde{x}_1(n), \ldots, \tilde{x}_1(n)}_{M \text{ times}}\Big\}_{\text{PERIODIC}}$$

then there will be $(M-1)$ zeros between samples of $\tilde{X}_M(k)$. The magnitudes of nonzero samples of $\tilde{X}_M(k)$ will be M times the magnitudes of the samples of $\tilde{X}_1(k)$:

$$\begin{aligned} \tilde{X}_M(Mk) &= M\tilde{X}_1(k) && , k = 0, 1, \ldots, N-1 \\ \tilde{X}_M(k) &= 0 && , k \neq 0, 1, \ldots, MN \end{aligned}$$

P5.5 $X(e^{j\omega})$ is a DTFT of the 10-point sequence $x(n) = \{2, 5, 3, -4, -2, 6, 0, -3, -3, 2\}$.

(a) $y_1(n)$ is a 3-point IDFS of three samples of $X(\omega)$ on the unit circle, so it can be obtained as a 3-point aliasing operation on $x(n)$. Thus,

$$y_1(n) = \{2 + (-4) + 0 + 2, 5 + (-2) + (-3), 3 + 6 + (-3)\} = \{0, 0, 6\}_{\text{periodic}}$$

MATLAB verification:

```
clear; close all;
x = [2,5,3,-4,-2,6,0,-3,-3,2]; n = 0:9;
% (a) y1(n) = 3-point IDFS{X(0),X(2*pi/3),X(4*pi/3)}
N = 3; k = 0:N-1; w = 2*pi*k/N;
[Y1] = dtft(x,n,w); y1 = real(idfs(Y1,N))
y1 =
    0.0000    0.0000    6.0000
```

(b) $y_2(n)$ is a 20-point IDFS of 20 samples of $X(\omega)$ on the unit circle, so, from the frequency-sampling theorem, there will not be any aliasing of $x(n)$ and , $y_2(n)$ will be a zero-padded version of $x(n)$. Thus,

$$y_2(n) = \{2, 5, 3, -4, -2, 6, 0, -3, -3, 2, 0, 0, 0, 0, 0, 0, 0, 0, 0, 0\}_{\text{periodic}}$$

MATLAB verification:

```
% (b) y1(n) = 20-point IDFS{X(0),X(2*pi/20),...,X(28*pi/20)}
N = 20; k = 0:N-1; w = 2*pi*k/N;
[Y2] = dtft(x,n,w); y2 = real(idfs(Y2,N))
y2 =
  Columns 1 through 7
    2.0000    5.0000    3.0000   -4.0000   -2.0000    6.0000    0.0000
  Columns 8 through 14
   -3.0000   -3.0000    2.0000    0.0000    0.0000    0.0000    0.0000
  Columns 15 through 20
    0.0000    0.0000    0.0000    0.0000    0.0000    0.0000
```

P5.6 A 12-point sequence $x(n) = \{1, 2, 3, 4, 5, 6, 6, 5, 4, 3, 2, 1\}$.

(a) DFT $X(k)$:

```
clear; close all;
%
xn = [1,2,3,4,5,6,6,5,4,3,2,1]; N = length(xn); % given signal x(n)
Xk = dft(xn,N); k = 0:N-1;                      % DFT of x(n)
mag_Xk = abs(Xk); pha_Xk = angle(Xk)*180/pi;    % Mag and Phase of X(k)
zei = find(mag_Xk < 0.00001);                   % Set phase values to
pha_Xk(zei) = zeros(1,length(zei));             %  zero when mag is zero
Hf_1 = figure('Units','normalized','position',[0.1,0.1,0.8,0.8],'color',[0,0,0]);
set(Hf_1,'NumberTitle','off','Name','HS020500');
subplot(2,1,1); stem(k,mag_Xk); axis([0,N,0,40])
set(gca,'XTickMode','manual','XTick',[0:1:N]);
set(gca,'YTickMode','manual','YTick',[0;20;40]);
xlabel('k'); ylabel('magnitude'); title('Magnitude plots of DFT and DTFT')
hold on
subplot(2,1,2); stem(k,pha_Xk); axis([0,N,-180,180])
set(gca,'XTickMode','manual','XTick',[0:1:N]);
set(gca,'YTickMode','manual','YTick',[-180;-90;0;90;180]);
xlabel('k'); ylabel('Degrees'); title('Phase plots of DFT and DTFT')
hold on
```

The stem plot of $X(k)$ is shown in Figure D.4.

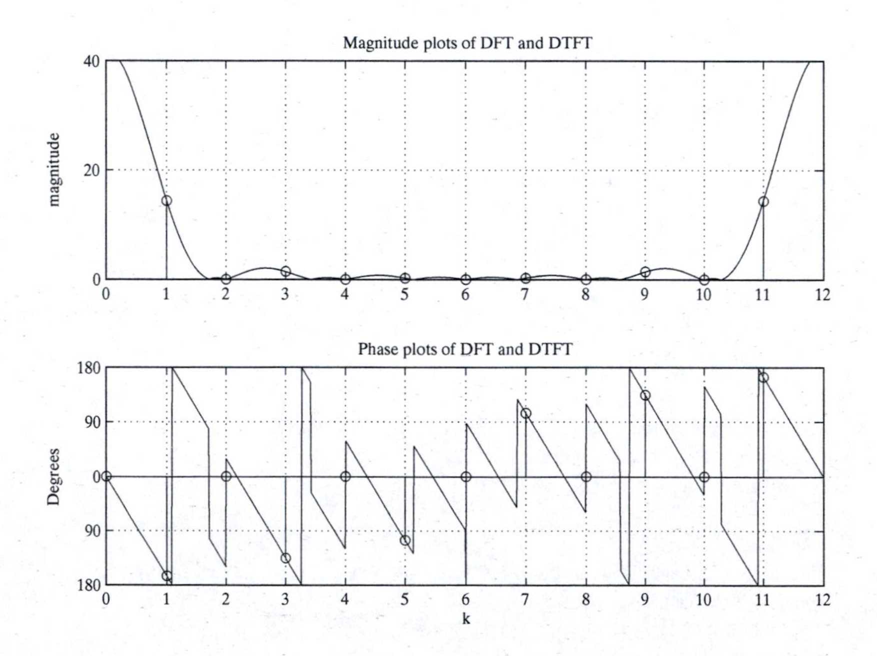

Figure D.4: Plots of DTFT and DFT of signal in Problem P5.6

(b) DTFT $X(\omega)$:

```
[X,w] = freqz(xn,1,1000,'whole');              % DTFT of xn
mag_X = abs(X); pha_X = angle(X)*180/pi;       % mag and phase of DTFT
Dw = (2*pi)/N;                                 % frequency resolution
subplot(2,1,1); plot(w/Dw,mag_X); grid
hold off
subplot(2,1,2); plot(w/Dw,pha_X); grid
hold off
```

The continuous plot of $X(\omega)$ is also shown in Figure D.4.

(c) Clearly, the DFT in part (0a) is the sampled version of $X(\omega)$.

(d) It is possible to reconstruct the DTFT from the DFT if the length of the DFT is greater than or equal to the length of the sequence $x(n)$. We can reconstruct the DTFT by using the complex interpolation formula

$$X(\omega) = \sum_{k=0}^{N-1} X(k)\phi\left(\omega - \frac{2\pi k}{N}\right), \quad \text{where} \quad \phi(\omega) = e^{-j\omega(N-1)/2}\frac{\sin(\omega N/2)}{N\sin(\omega/2)}$$

For $N = 12$, we have

$$X(\omega) = \sum_{k=0}^{11} X(k)e^{-j(5.5)\omega}\frac{\sin(6\omega)}{12\sin(\omega/2)}$$

P5.7 This problem is done via MATLAB.

(a) $x_1(n) = 2\cos(0.2\pi n)\,[u(n) - u(n-10)]$.

```
% (a) x1(n) = 2*cos(0.2*pi*n)*[u(n)-u(n-10)]
Hf_1 = figure('Units','normalized','position',[0.1,0.1,0.8,0.8],'color',[0,0,0]);
set(Hf_1,'NumberTitle','off','Name','P5.7a');
n1 = [0:9]; x1 = 2*cos(0.2*pi*n1); N1 = length(n1); N = 100;   % Length of DFT
[X1] = dft([x1, zeros(1,N-N1)],N);
mag_X1 = abs(X1(1:N/2+1)); w = (0:N/2)*2*pi/N;
subplot(2,1,1); stem(n1,x1); axis([-1,N1,min(x1),max(x1)]);
title('Sequence x1(n)'); ylabel('x1'); ntick = [n1(1):5:n1(N1),N1]';
set(gca,'XTickMode','manual','XTick',ntick); xlabel('n');
subplot(2,1,2); plot(w/pi,mag_X1); axis([0,1,0,max(mag_X1)]);
title('Magnitude of DTFT X1(omega)'); ylabel('|X1|');
xlabel('frequency in pi units');
```

The plots are shown in Figure D.5.

(b) $x_2(n) = \sin(0.45\pi n)\sin(0.55\pi n), \; 0 \le n \le 50$.

```
% (b) x2(n) = sin(0.45*pi*n)*sin(0.55*pi*n), 0 <= n <= 50
Hf_2 = figure('Units','normalized','position',[0.1,0.1,0.8,0.8],'color',[0,0,0]);
set(Hf_2,'NumberTitle','off','Name','P5.7b');
n2 = [0:50]; x2 = sin(0.45*pi*n2).*sin(0.55*pi*n2); N2 = length(n2);  N = 300;
[X2] = dft([x2, zeros(1,N-N2)],N);
mag_X2 = abs(X2(1:N/2+1)); w = (0:N/2)*2*pi/N;
```

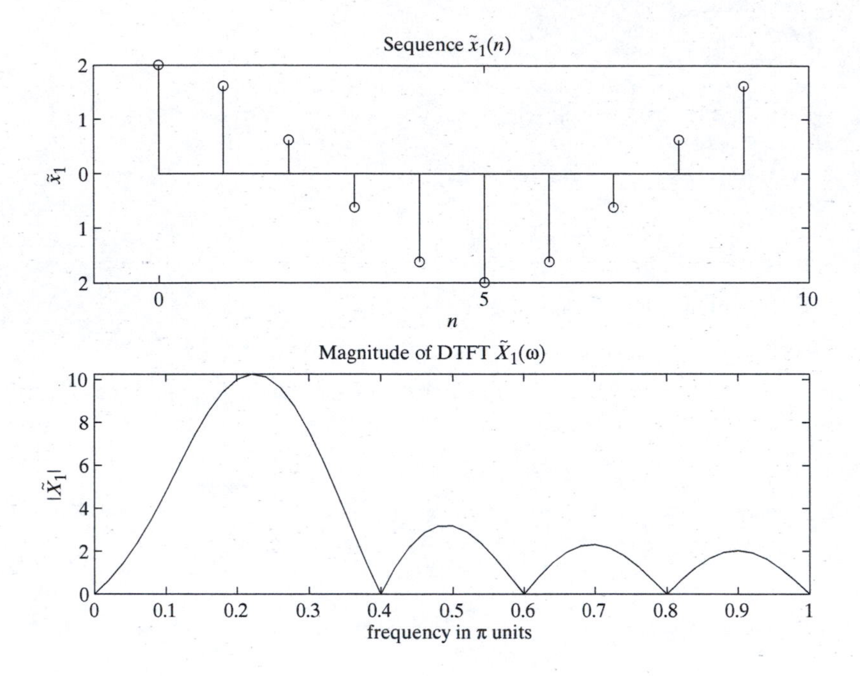

Figure D.5: Sequence plot and its DTFT magnitude plot in Problem P5.7(a)

```
subplot(2,1,1); stem(n2,x2); axis([-1,N2,min(x2),max(x2)]);
title('Sequence x2(n)'); ylabel('x2'); ntick = [n2(1):10:n2(N2)]';
set(gca,'XTickMode','manual','XTick',ntick); xlabel('n');
subplot(2,1,2); plot(w/pi,mag_X2); axis([0,1,0,max(mag_X2)]);
title('Magnitude of DTFT X2(omega)'); ylabel('|X2|');
xlabel('frequency in pi units');
```

The plots are shown in Figure D.6.

(c) $x_3(n) = 3\,(2)^n\,,\ -10 \leq n \leq 10$. This problem is done via MATLAB.

```
% (c) x3(n) = 3*(2)^n*[u(n+10)-u(n-11)]
Hf_1 = figure('Units','normalized','position',[0.1,0.1,0.8,0.8],'color',[0,0,0]);
set(Hf_1,'NumberTitle','off','Name','P5.7c');
n3 = [-10:10]; x3 = 3*(2).^n3; N3 = length(n3); N = 100;   % Length of DFT
[X3] = dft([x3, zeros(1,N-N3)],N);
mag_X3 = abs(X3(1:N/2+1)); w = (0:N/2)*2*pi/N;
subplot(2,1,1); stem(n3,x3); axis([-11,11,min(x3),max(x3)]);
title('Sequence x3(n)'); ylabel('x3'); ntick = [n3(1):5:n3(N3),N3]';
set(gca,'XTickMode','manual','XTick',ntick); xlabel('n');
subplot(2,1,2); plot(w/pi,mag_X3); axis([0,1,0,max(mag_X3)]);
title('Magnitude of DTFT X3(omega)'); ylabel('|X3|');
xlabel('frequency in pi units');
```

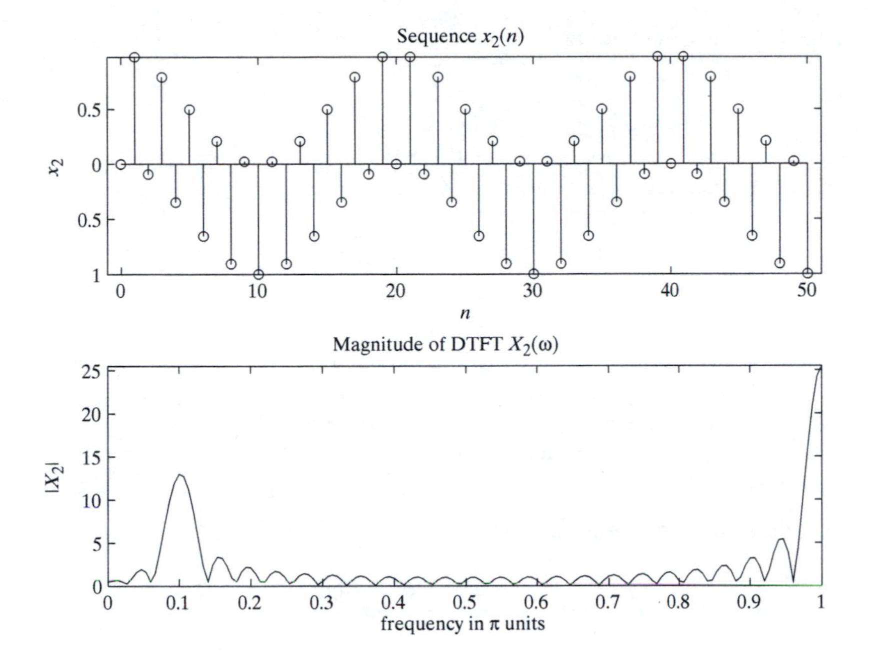

Figure D.6: Sequence plot and its DTFT magnitude plot in Problem P5.7(b)

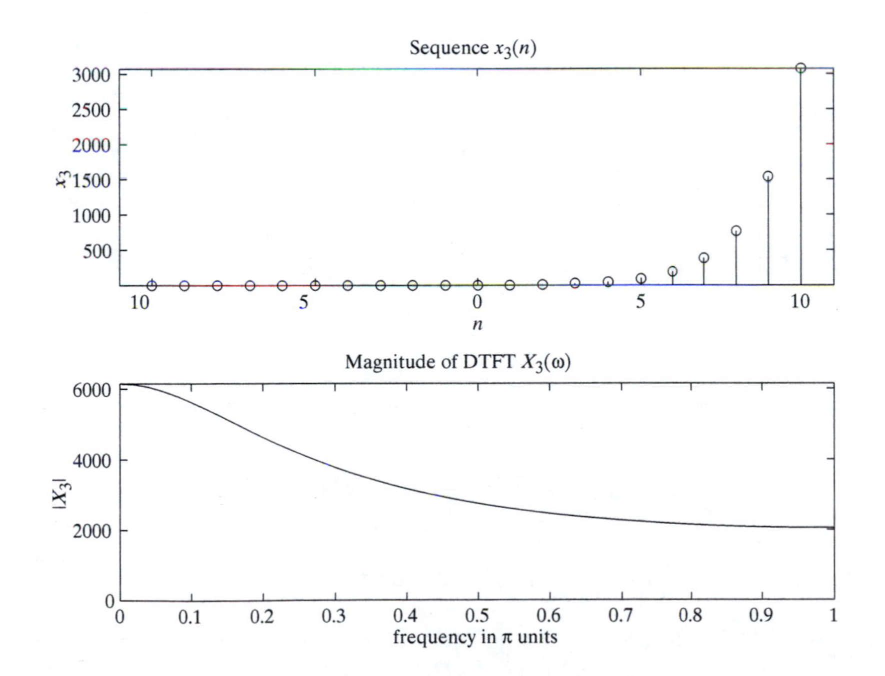

Figure D.7: Plots in Problem P5.7(c)

The plots are shown in Figure D.7.

(d) $x_5(n) = 5\left(0.9e^{j\pi/4}\right)^n u(n)$.

```
% (e) x5(n) = 5*(0.9*exp(j*pi/4)).^n*u(n)
Hf_1 = figure('Units','normalized','position',[.1,.1,.8,.8],'color',[0,0,0]);
set(Hf_1,'NumberTitle','off','Name','P5.7e');
n5 = [0:99]; x5 = 5*(0.9*exp(j*pi/4)).^n5;
N5 = length(n5);   % Length of DFT
[X5] = dft(x5,N5); x5 = real(x5);
mag_X5 = abs(X5(1:N5/2+1)); w = (0:N5/2)*2*pi/N5;
subplot(2,1,1); stem(n5,x5); %axis([-11,11,min(x5),max(x5)]);
title('Sequence real(x5(n))'); ylabel('real(x5)'); ntick = [n5(1):10:n5(N5),N5]';
set(gca,'XTickMode','manual','XTick',ntick); xlabel('n');
subplot(2,1,2); plot(w/pi,mag_X5); axis([0,1,0,max(mag_X5)]);
title('Magnitude of DTFT X5(omega)'); ylabel('|X5|');
xlabel('frequency in pi units');
```

The plots are shown in Figure D.8.

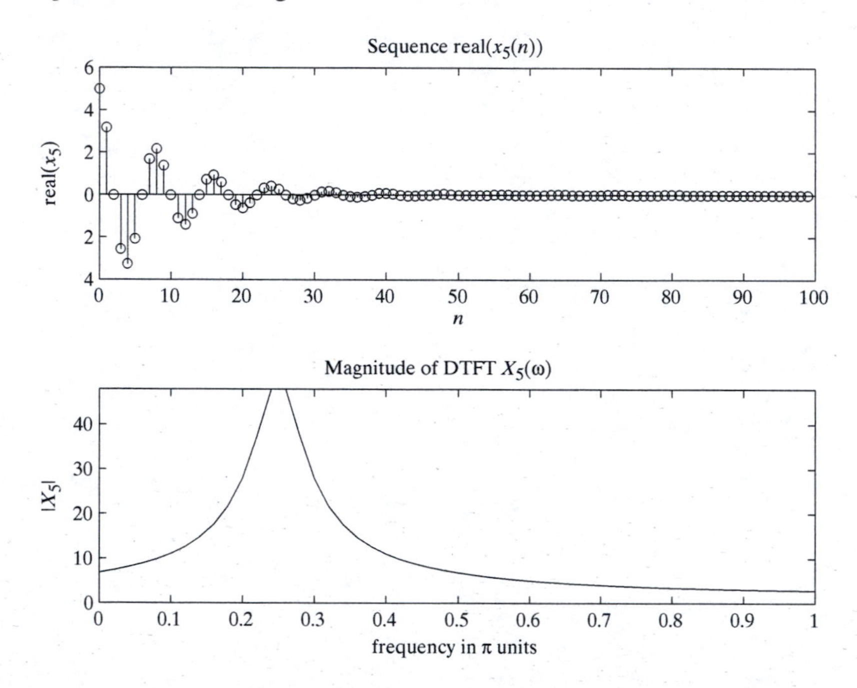

Figure D.8: DTFT plot in Problem P5.7(d)

P5.8 The impulse response $h(n)$ is causal and real-valued.

(a) It is known that $\text{Re}\{H(\omega)\} = \sum_{k=0}^{5} (0.5)^k \cos \omega k$. Consider

$$\text{Re}\{H(\omega)\} = \text{Re}\left\{\sum_{k=0}^{\infty} h(k)e^{-j\omega k}\right\} = \sum_{k=0}^{\infty} h(k)\,\text{Re}\left\{e^{-j\omega k}\right\} = \sum_{k=0}^{\infty} h(k)\cos \omega k$$

Comparing with the given expression, we obtain

$$h(n) = \begin{cases} (0.5)^k, & 0 \le k \le 5 \\ 0, & \text{else} \end{cases}$$

(b) It is known that $\text{Im}\{H(\omega)\} = \sum_{\ell=0}^{5} 2\ell \sin \omega\ell$ and $\int_{-\pi}^{\pi} H(\omega)d\omega = 0$. From the second condition,

$$\int_{-\pi}^{\pi} H(\omega)d\omega = h(0) = 0$$

Consider

$$\text{Im}\{H(\omega)\} = \text{Im}\left\{\sum_{\ell=0}^{\infty} h(\ell)\, e^{-j\omega\ell}\right\} = \sum_{\ell=0}^{\infty} h(\ell)\, \text{Im}\left\{e^{-j\omega\ell}\right\} = -\sum_{\ell=0}^{\infty} h(\ell) \sin \omega\ell$$

Comparing against the given expression, we obtain

$$h(n) = \begin{cases} -2\ell, & 0 \le \ell \le 5 \\ 0, & \text{else} \end{cases}$$

P5.9 An N-point sequence $x(n)$.

(a) The N-point DFT of $x(n)$: $X(k) = \sum_{m=0}^{N-1} x(m)\, W_N^{mk}$. The N-point DFT of $X(k)$: $y(n) = \sum_{k=0}^{N-1} X(k)\, W_N^{kn}$; hence,

$$\begin{aligned} y(n) &= \sum_{k=0}^{N-1} \left\{\sum_{m=0}^{N-1} x(m)\, W_N^{mk}\right\} W_N^{kn} = \sum_{m=0}^{N-1} x(m) \sum_{k=0}^{N-1} W_N^{mk} W_N^{kn},\ 0 \le n \le N-1 \\ &= \sum_{m=0}^{N-1} x(m) \sum_{k=0}^{N-1} W_N^{(m+n)k} = \sum_{m=0}^{N-1} x(m) \sum_{r=-\infty}^{\infty} N\delta(m+n-rN),\, 0 \le n \le N-1 \\ &= N \sum_{r=-\infty}^{\infty} x(-n+rN) = Nx((-n))_N,\, 0 \le n \le N-1 \end{aligned}$$

This means that $y(n)$ is a "circularly folded and amplified (by N)" version of $x(n)$. Continuing further, if we take two more DFTs of $x(n)$, then

$x(n) \longrightarrow$ [N-point DFT] $\longrightarrow$ [N-point DFT] $\longrightarrow$ [N-point DFT] $\longrightarrow$ [N-point DFT] $\longrightarrow N^2 x(n)$

Therefore, if a given DFT function is working correctly, then four successive applications of this function on any arbitrary signal will produce the same signal (multiplied by N^2). This approach can be used to verify a DFT function.

(b) MATLAB function for circular folding:

```
function x2 = circfold(x1,N)
% Circular folding with respect to N
% ----------------------------------
% function x2 = circfold(x1,N)
%         x2(n) = x1((-n) mod N)
%
x2 = real(dft(dft(x1,N),N))/N;
```

(c) MATLAB verification:

```
x = [1,2,3,4,5,6,6,5,4,3,2,1], N = length(x);
x =
     1     2     3     4     5     6     6     5     4     3     2     1
y = circfold(x,N)
y =
  Columns 1 through 7
    1.0000    1.0000    2.0000    3.0000    4.0000    5.0000    6.0000
  Columns 8 through 12
    6.0000    5.0000    4.0000    3.0000    2.0000
```

Clearly, the circular folding of the sequence is evident.

P5.10 Circular even–odd decomposition of a complex sequence.

$$x_{ec}(n) \triangleq \frac{1}{2}\left[x(n) + x^*\left((-n)\right)_N\right]$$
$$x_{oc}(n) \triangleq \frac{1}{2}\left[x(n) + x^*\left((-n)\right)_N\right]$$

(a) Using the DFT properties of conjugation and circular folding, we obtain

$$\text{DFT}\left[x_{ec}(n)\right] = \frac{1}{2}\left\{\text{DFT}\left[x(n)\right] + \text{DFT}\left[x^*\left((-n)\right)_N\right]\right\}$$
$$= \frac{1}{2}\left\{X(k) + \tilde{X}^*\left((-k)\right)_N\right\}, \text{ where } \tilde{X}(k) = \text{DFT}\left[x\left((-n)\right)_N\right]$$
$$= \frac{1}{2}\left\{X(k) + X^*(k)\right\} = \text{Re}\left[X(k)\right] = \text{Re}\left[X\left((-k)\right)_N\right]$$

Similarly, we can show that

$$\text{DFT}\left[x_{oc}(n)\right] = j\,\text{Im}\left[X(k)\right] = j\,\text{Im}\left[X\left((-k)\right)_N\right]$$

(b) The modified `circevod` function:

```
function [xec, xoc] = circevod(x)
% Complex-valued signal decomposition into circular-even and circular-odd parts
% -----------------------------------------------------------------------------
% [xec, xoc] = circecod(x)
%
N = length(x); n = 0:(N-1);
xec = 0.5*(x + conj(x(mod(-n,N)+1)));
xoc = 0.5*(x - conj(x(mod(-n,N)+1)));
```

(c) MATLAB verification:

```
% (c) Matlab Verification
n = 0:19; x = (0.9*exp(j*pi/3)).^n; N = length(x);
[xec, xoc] = circevod(x);
X = dft(x,N); Xec = dft(xec,N); Xoc = dft(xoc,N);
Hf_1 = figure('Units','normalized','position',[0.1,0.1,0.8,0.8],'color',[0,0,0]);
set(Hf_1,'NumberTitle','off','Name','P5.10');
subplot(2,2,1); stem(n,real(X)); axis([-0.5,20.5,-1,7]);
title('Real{DFT[x(n)]}'); ylabel('Re{X(k)}');
subplot(2,2,3); stem(n,real(Xec)); axis([-0.5,20.5,-1,7]);
title('DFT[xec(n)]'); ylabel('Xec(k)'); xlabel('k');
subplot(2,2,2); stem(n,imag(X)); axis([-0.5,20.5,-5,5]);
title('Imag{DFT[x(n)]}'); ylabel('Im{X(k)}');
subplot(2,2,4); stem(n,imag(Xoc)); axis([-0.5,20.5,-5,5]);
title('DFT[xoc(n)]'); ylabel('Xoc(k)'); xlabel('k');
```

The plots are shown in Figure D.9.

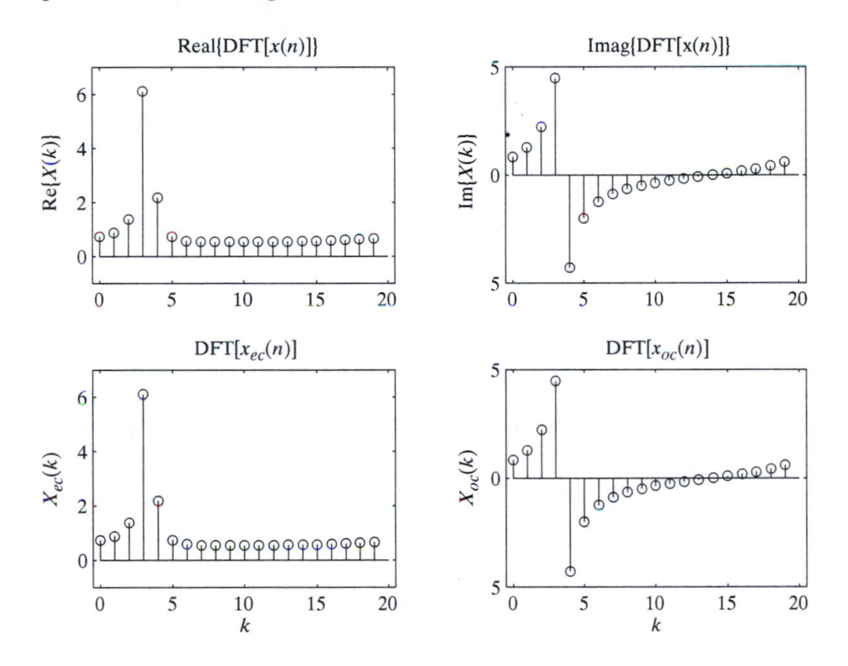

Figure D.9: Plots in Problem P5.10

P5.11 $x(n)$ is a real-valued sequence, soits DFT is conjugate symmetric. Thus,

$$X(k) = \{0.25, 0.125 - j0.3, 0, 0.125 - j0.06, 0.5, 0.125 + j0.06, 0, 0.125 + j0.3\}$$

From this, we can compute the 8-point sequence $x(n)$. MATLAB script:

```
clear, close all;
N = 8; X = [0.25,0.125-j*0.3,0,0.125-j*0.06,0.5];
```

```
X = [X,conj(X(4:-1:2))]              % Reconstruct the entire X(k)
X =
  Columns 1 through 4
   0.2500             0.1250 - 0.3000i        0              0.1250 - 0.0600i
  Columns 5 through 8
   0.5000             0.1250 + 0.0600i        0              0.1250 + 0.3000i
x = real(idft(X,N))
x =
  Columns 1 through 7
    0.1562    0.0324    0.1538    0.0324    0.0312   -0.0949    0.0337
  Column 8
   -0.0949
```

(a) Consider $x_1(n) = x\left((2-n)\right)_8 = x\left((-[n-2])\right)_8$. It is obtained by first, circular folding, and then, circular shifting by 2, of $x(n)$. Hence. using properties of the DFT,

$$X_1(k) = \text{DFT}\left[x\left((2-n)\right)_8\right] = W_8^{-2k} X_{⑧}\left((-k)\right)_8$$

MATLAB script:

```
% (a) x1(n) = x((2-n))_8; Circ folding followed by circ shifting by 2
N = 8; WN = exp(-j*2*pi/N); k = 0:N-1; m = 2;
X1 = circfold(X,N)
X1 =
  Columns 1 through 4
    0.2500 + 0.0000i   0.1250 + 0.3000i   0.0000 - 0.0000i   0.1250 + 0.0600i
  Columns 5 through 8
    0.5000 + 0.0000i   0.1250 - 0.0600i   0.0000 - 0.0000i   0.1250 - 0.3000i
X1 = (WN.^(m*k)).*X1
X1 =
  Columns 1 through 4
    0.2500 + 0.0000i   0.3000 - 0.1250i   0.0000 + 0.0000i  -0.0600 + 0.1250i
  Columns 5 through 8
    0.5000 + 0.0000i  -0.0600 - 0.1250i   0.0000 + 0.0000i   0.3000 + 0.1250i
% Matlab verification
x1 = circfold(x,N); x1 = cirshftt(x1,m,N);
X12 = dft(x1,N)
X12 =
  Columns 1 through 4
    0.2500 + 0.0000i   0.3000 - 0.1250i   0.0000 + 0.0000i  -0.0600 + 0.1250i
  Columns 5 through 8
    0.5000 + 0.0000i  -0.0600 - 0.1250i   0.0000 - 0.0000i   0.3000 + 0.1250i
difference = max(abs(X1-X12))
difference =
   4.7692e-016
```

(b) Let $X_{⑩}(k)$ be the 10-point DFT of $x(n)$. Then

$$\text{DFT}\left[x\left((n+5)\right)_{10}\right] = W_{10}^{5k} X_{⑩}(k)$$

Note that $x\left((n+5)\right)_{10}$ is a 10-point circularly shifted sequence of $x(n)$, shifted to the left by 5 samples. MATLAB script:

```
% (b) x2(n) = x((n+5))_{10}
N = 10; WN = exp(-j*2*pi/N); k = 0:N-1; m = -5;
X2 = (WN.^(m*k)).*dft([x,0,0],N)
X2 =
  Columns 1 through 4
   0.2500             -0.2916 + 0.2848i   0.0176 - 0.0487i  -0.0863 - 0.1368i
  Columns 5 through 8
   0.0108 + 0.0047i  -0.5000 + 0.0000i   0.0108 - 0.0047i  -0.0863 + 0.1368i
  Columns 9 through 10
   0.0176 + 0.0487i  -0.2916 - 0.2848i
% Matlab verification
x2 = cirshftt(x,m,N); X22 = dft(x2,N)
X22 =
  Columns 1 through 4
   0.2500             -0.2916 + 0.2848i   0.0176 - 0.0487i  -0.0863 - 0.1368i
  Columns 5 through 8
   0.0108 + 0.0047i  -0.5000 - 0.0000i   0.0108 - 0.0047i  -0.0863 + 0.1368i
  Columns 9 through 10
   0.0176 + 0.0487i  -0.2916 - 0.2848i
difference = max(abs(X2-X22))
difference =
  8.2523e-016
```

(c) Consider $x_3(n) = x^2(n) = x(n)\,x(n)$. Hence, the DFT of $x_3(n)$ is given by

$$X_3(k) = \frac{1}{N} X(k) \textcircled{N} X(k), \quad N = 8$$

MATLAB script:

```
% (c) x3(n) = x(n)*x(n);
N = 8;
X3 = circonvf(X,X,N)/N;
% Matlab verification
x3 = x.*x;
X32 = dft(x3,N);
difference = max(abs(X3-X32))
difference =
  4.5103e-017
```

(d) Let $X_{\textcircled{10}}(k)$ be the 10-point DFT of $x(n)$. Then

$$\text{DFT}\left[x(n) \textcircled{8} x\left((-n)\right)_8\right] = X(k) X\left((-k)\right)_8$$

Note that $x((-n))_8$ and $X((-k))_8$ are the 8-point circular foldings of their respective sequences. MATLAB script:

```
clear, close all;
N = 8; X = [0.25,0.125-j*0.3,0,0.125-j*0.06,0.5]; k = 0:N-1; n = k;
```

```
X = [X,conj(X(4:-1:2))]              % Reconstruct the entire X(k)
X =
  Columns 1 through 4
   0.2500           0.1250 - 0.3000i        0           0.1250 - 0.0600i
  Columns 5 through 8
   0.5000           0.1250 + 0.0600i        0           0.1250 + 0.3000i
x = real(idft(X,N))
x =
  Columns 1 through 7
    0.1562    0.0324    0.1538    0.0324    0.0312   -0.0949    0.0337
  Column 8
   -0.0949
%
% (d) x4(n) = x(n)(8)x((-n))_8
X0 = X(mod(-k,N)+1);                 % DFT of x((-n))_8
X4 = X .* X0                         % DFT of x(n) (8) x((-n))_8
X4 =
  Columns 1 through 7
    0.0625    0.1056         0    0.0192    0.2500    0.0192         0
  Column 8
    0.1056
% Verification
x4 = circonvt(x,x(mod(-n,N)+1),N); X44 = dft(x4,N);
difference = max(abs(X4-X44))
difference =
  2.1330e-016
```

P5.12 Two real DFTs using one complex DFT.

Suppose

$$x(n) = x_R(n) + jx_I(n);\ \ x_R(n) \text{ and } x_I(n) \text{ are real sequences;}$$

then

$$\begin{aligned} x_R(n) &= \frac{1}{2}\{x(n) + x^*(n)\} \\ jx_I(n) &= \frac{1}{2}\{x(n) - x^*(n)\} \end{aligned}$$

(a) Consider

$$\begin{aligned} X_R(k) &\stackrel{\text{def}}{=} \text{DFT}[x_R(n)] = \frac{1}{2}\left\{\text{DFT}[x(n)] + \text{DFT}\left[x^*(n)\right]\right\} \\ &= \frac{1}{2}\left\{X(k) + X^*((-k))_N\right\} \stackrel{\text{def}}{=} X_{ec} \end{aligned}$$

Similarly,

$$\left.\begin{aligned} jX_I(k) &\stackrel{\text{def}}{=} \text{DFT}[jx_I(n)] = \frac{1}{2}\left\{\text{DFT}[x(n)] - \text{DFT}\left[x^*(n)\right]\right\} \\ &= \frac{1}{2}\left\{X(k) - X^*((-k))_N\right\} \stackrel{\text{def}}{=} X_{oc} \end{aligned}\right\} \Rightarrow X_I = \frac{X_{oc}}{j} = -jX_{oc}$$

(b) MATLAB function `real2dft`:

```
function [X1,X2] = real2dft(x1,x2,N)
% DFTs of two real sequences
% [X1,X2] = real2dft(x1,x2,N)
%  X1 = N-point DFT of x1
%  X2 = N-point DFT of x2
%  x1 = real-valued sequence of length <= N
%  x2 = real-valued sequence of length <= N
%   N = length of DFT
%

% Check for length of x1 and x2
if length(x1) > N
        error('*** N must be >= the length of x1 ***')
end
if length(x2) > N
        error('*** N must be >= the length of x2 ***')
end
N1 = length(x1); x1 = [x1 zeros(1,N-N1)];
N2 = length(x2); x2 = [x2 zeros(1,N-N2)];
x = x1 + j*x2;
X = dft(x,N);
[X1, X2] = circevod(X); X2 = X2/j;
```

We will also need the `circevod` function for complex sequences (from Problem P5.10). This can be obtained from the one given in the text by two simple changes.

```
function [xec, xoc] = circevod(x)
% Complex signal decomposition into circular-even and circular-odd parts
% ----------------------------------------------------------------------
% [xec, xoc] = circecod(x)
%
N = length(x); n = 0:(N-1);
xec = 0.5*(x + conj(x(mod(-n,N)+1)));
xoc = 0.5*(x - conj(x(mod(-n,N)+1)));
```

(c) MATLAB verification:

```
clear; close all;
N = 64; n = 0:N-1; x1 = cos(0.25*pi*n); x2 = sin(0.75*pi*n);
[X1,X2] = real2dft(x1,x2,N);
X11 = dft(x1,N); X21 = dft(x2,N);
difference = max(abs(X1-X11))
difference =
  1.4918e-013
difference = max(abs(X2-X21))
difference =
  1.4914e-013
```

P5.13 Circular shifting.

(a) The MATLAB routine `cirshftf.m` to implement circular shift is written by using the frequency-domain property

$$y(n) \triangleq x\left((n-m)\right)_N = \text{IDFT}\left[X(k)W_N^{mk}\right]$$

This routine will be used in the next problem to generate a circulant matrix and has the following features: If m is a scalar, then $y(n)$ is circularly shifted sequence (or array). If m is a vector, then $y(n)$ is a matrix, each row of which is a circular shift in $x(n)$ corresponding to entries in the vector m.

```
function y = cirshftf(x,m,N)
%  Circular shift of m samples wrt size N in sequence x: (freq domain)
%  -------------------------------------------------------------------
% function y=cirshift(x,m,N)
%        y : output sequence containing the circular shift
%        x : input sequence of length <= N
%        m : sample shift
%        N : size of circular buffer
%
%  Method: y(n) = idft(dft(x(n))*WN^(mk))
%
%   If m is a scalar then y is a sequence (row vector)
%   If m is a vector then y is a matrix, each row is a circular shift
%       in x corresponding to entries in vector m
%   M and x should not be matrices
%
% Check whether m is scalar, vector, or matrix
[Rm,Cm] = size(m);
if Rm > Cm
    m = m'; % make sure that m is a row vector
end
[Rm,Cm] = size(m);
if Rm > 1
    error('*** m must be a vector ***') % stop if m is a matrix
end
% Check whether x is scalar, vector, or matrix
[Rx,Cx] = size(x);
if Rx > Cx
    x = x'; % make sure that x is a row vector
end
[Rx,Cx] = size(x);
if Rx > 1
    error('*** x must be a vector ***') % stop if x is a matrix
end
% Check for length of x
if length(x) > N
        error('N must be >= the length of x')
end
x=[x zeros(1,N-length(x))];
```

```
X=dft(x,N);
X=ones(Cm,1)*X;
WN=exp(-2*j*pi/N);
k=[0:1:N-1];
Y=(WN.^(m' * k)).*X;
y=real(conj(dfs(conj(Y),N)))/N;
```

(b) MATLAB verification:

```
n = [0:1:10]; x = 11*ones(1,length(n))-n;
y = cirshftf(x,10,15)
y =
  Columns 1 through 7
    6.0000    5.0000    4.0000    3.0000    2.0000    1.0000    0.0000
  Columns 8 through 14
    0.0000    0.0000    0.0000   11.0000   10.0000    9.0000    8.0000
  Column 15
    7.0000
```

P5.14 Parseval's relation for the DFT:

$$\begin{aligned}\sum_{n=0}^{N-1}|x(n)|^2 &= \sum_{n=0}^{N-1}x(n)x^*(n) = \sum_{n=0}^{N-1}\left\{\frac{1}{N}\sum_{k=0}^{N-1}X(k)\,W_N^{-nk}\right\}x^*(n)\\ &= \frac{1}{N}\sum_{k=0}^{N-1}X(k)\left\{\sum_{n=0}^{N-1}x^*(n)W_N^{-nk}\right\} = \frac{1}{N}\sum_{k=0}^{N-1}X(k)\left\{\sum_{n=0}^{N-1}x(n)W_N^{nk}\right\}^*\end{aligned}$$

Therefore,

$$\sum_{n=0}^{N-1}|x(n)|^2 = \frac{1}{N}\sum_{k=0}^{N-1}X(k)X^*(k) = \frac{1}{N}\sum_{k=0}^{N-1}|X(k)|^2$$

MATLAB verification:

```
x = [1,2,3,4,5,6,6,5,4,3,2,1]; N = length(x);
% power of x(n) in the time-domain
power_x = sum(x.*conj(x))
power_x =
   182
X = dft(x,N);
power_X = (1/N)*sum(X.*conj(X))
power_X =
  182.0000
```

P5.15 MATLAB function circonvf:

```
function y = circonvf(x1,x2,N)
%
%function y=circonvf(x1,x2,N)
%
```

```
% N-point circular convolution between x1 and x2: (freq domain)
% -------------------------------------------------------------
%       y  : output sequence containing the circular convolution
%       x1 : input sequence of length N1 <= N
%       x2 : input sequence of length N2 <= N
%       N  : size of circular buffer
%
% Method: y(n) = idft(dft(x1)*dft(x2))

% Check for length of x1
if length(x1) > N
        error('N must be >= the length of x1')
end

% Check for length of x2
if length(x2) > N
        error('N must be >= the length of x2')
end

x1=[x1 zeros(1,N-length(x1))];
x2=[x2 zeros(1,N-length(x2))];

X1=fft(x1); X2=fft(x2);
y=real(ifft(X1.*X2));
```

P5.16 Circular convolution by using circulant matrix operation.

$$x_1(n) = \{1, 2, 2\}, \; x_2(n) = \{1, 2, 3, 4\}, \; x_3(n) \triangleq x_1(n) \textcircled{4} x_2(n)$$

(a) Using the results from Example 5.13, we can express the above signals as

$$\mathbf{x}_1 = \begin{bmatrix} 1 \\ 2 \\ 2 \\ 0 \end{bmatrix}, \quad \mathbf{X}_2 = \begin{bmatrix} 1 & 4 & 3 & 2 \\ 2 & 1 & 4 & 3 \\ 3 & 2 & 1 & 4 \\ 4 & 3 & 2 & 1 \end{bmatrix}$$

The matrix $\mathbf{X}_2$ has the property that its every row (column) can be obtained from the previous row (column) by using circular shift. Such a matrix is called a *circulant* matrix. It is completely described by the first column or row. The following MATLAB function, `circulnt`, uses the `mod` function to generate a circulant matrix:

```
function C = circulnt(h,N)
% Circulant matrix generation using vector data values
% ----------------------------------------------------
% function C = circulnt(h,N)
%
%  C : Circulant matrix
%  h : input sequence of length <= N
%  N : size of the circular buffer
%  Method: C = h((n-m) mod N); n : col vec, m : row vec
```

```
  Mh = length(h); h = reshape(h,Mh,1);    % reshape h into column vector
   h = [h; zeros(N-Mh,1)];                % zero-pad h
   C = zeros(N,N);                        % establish size of C
   m = 0:N-1; n=m';                       % indices n and m
  nm = mod((n*ones(1,N)-ones(N,1)*m),N);  % (n-m) mod N in matrix form
C(:) = h(nm+1);                           % h((n-m) mod N)
```

(b) Circular convolution:

$$\mathbf{x}_3 = \mathbf{X}_2\mathbf{x}_1 = \begin{bmatrix} 1 & 4 & 3 & 2 \\ 2 & 1 & 4 & 3 \\ 3 & 2 & 1 & 4 \\ 4 & 3 & 2 & 1 \end{bmatrix} \begin{bmatrix} 1 \\ 2 \\ 2 \\ 0 \end{bmatrix} = \begin{bmatrix} 15 \\ 12 \\ 9 \\ 14 \end{bmatrix}$$

MATLAB script:

```
% Chapter 5: Problem~P5.16: Circular convolution using Circulant matrix
clear, close all;
N = 4; x1 = [1,2,2,0]; x2 = [1,2,3,4];

% (a) Circulant matrix
X2 = circulnt(x2,N)
X2 =
     1     4     3     2
     2     1     4     3
     3     2     1     4
     4     3     2     1

% (b) Circular Convolution
x3 = X2*x1'; x3 = x3'
x3 =
    15    12     9    14
```

P5.17 MATLAB function circulnt:

```
function C = circulnt(x,N)
%  Circulant matrix generation using vector data values
%  ----------------------------------------------------
%  function C = circulnt(h,N)
%
%   C : Circulant matrix
%   x : input sequence of length <= N
%   N : size of the circular buffer
%   Method: C = h((n-m) mod N);

  Mx = length(x);                         % length of x
   x = [x, zeros(N-Mx,1)];                % zero-pad x
   C = zeros(N,N);                        % establish size of C
   m = 0:N-1;                             % indices n and m
   x = circfold(x,N);                     % Circular folding
   C = cirshift(x,m,N);                   % Circular shifting
```

MATLAB verification on sequences in Problem P5.16:

```
clear, close all;
N = 4; x1 = [1,2,2,0]; x2 = [1,2,3,4];
% (a) Circulant matrix
X2 = circulnt(x2,N)
X2 =
    1.0000    4.0000    3.0000    2.0000
    2.0000    1.0000    4.0000    3.0000
    3.0000    2.0000    1.0000    4.0000
    4.0000    3.0000    2.0000    1.0000

% (b) Circular Convolution
x3 = X2*x1'; x3 = x3'
x3 =
   15.0000   12.0000    9.0000   14.0000
```

P5.18 (a) Circular convolution:

$$x_1(n) = \{1, 1, 1, 1\}, \; x_2(n) = \cos(\pi n/4)\,\mathcal{R}_N(n), \; N = 8; \quad x_3(n) = x_1(n) \textcircled{N} x_2(n)$$

Circular convolutions using MATLAB:

```
clear, close all;
%(a) x1(n) = [1,1,1,1]; x2(n) = cos(pi*n/4); N = 8;
N = 8; n = 0:N-1;
x1 = [1,1,1,1,0,0,0,0]; x2 = cos(pi*n/4);
x3 = circonvt(x1,x2,N)
x3 =
  Columns 1 through 7
    1.0000    2.4142    2.4142    1.0000   -1.0000   -2.4142   -2.4142
  Column 8
   -1.0000
```

(b) Circular convolution:

$$x_1(n) = n\mathcal{R}_N, \; x_2(n) = (N - n)\,\mathcal{R}_N(n), \; N = 10; \quad x_3(n) = x_1(n) \textcircled{N} x_2(n)$$

Circular convolutions using MATLAB:

```
N = 10; n = [0:N-1]; x1 = n; x2 = (N-n);
x3 = real(idft(dft(x1,N).*dft(x2,N),N))
x3 =
  Columns 1 through 7
  285.0000  250.0000  225.0000  210.0000  205.0000  210.0000  225.0000
  Columns 8 through 10
  250.0000  285.0000  330.0000
```

P5.19 (a) Error between linear and circular convolutions:

$$x_1(n) = \cos(2\pi n/N)\,\mathcal{R}_{16}(n), \; x_2(n) = \sin(2\pi n/N)\,\mathcal{R}_{16}(n); \quad N = 32$$

MATLAB script:

```
% (b) x1(n) = cos(2*pi*n/N)*R16, x2(n) = cos(2*pi*n/N)*R16; N = 32;
N = 32; n = 0:15;
x1 = cos(2*pi*n/N); x2 = sin(2*pi*n/N);
x3 = circonvf(x1,x2,N)
x3 =
  Columns 1 through 7
    0.0000    0.1951    0.5740    1.1111    1.7678    2.4944    3.2336
  Columns 8 through 14
    3.9231    4.5000    4.9039    5.0813    4.9888    4.5962    3.8890
  Columns 15 through 21
    2.8701    1.5607    0.0000   -1.3656   -2.4874   -3.3334   -3.8891
  Columns 22 through 28
   -4.1573   -4.1575   -3.9231   -3.5000   -2.9424   -2.3097   -1.6629
  Columns 29 through 32
   -1.0607   -0.5556   -0.1913    0.0000
x4 = conv(x1,x2), x4 = [x4, zeros(1,33)]; N4 = length(x4);
x4 =
  Columns 1 through 7
         0    0.1951    0.5740    1.1111    1.7678    2.4944    3.2336
  Columns 8 through 14
    3.9231    4.5000    4.9039    5.0813    4.9888    4.5962    3.8890
  Columns 15 through 21
    2.8701    1.5607    0.0000   -1.3656   -2.4874   -3.3334   -3.8891
  Columns 22 through 28
   -4.1573   -4.1575   -3.9231   -3.5000   -2.9424   -2.3097   -1.6629
  Columns 29 through 31
   -1.0607   -0.5556   -0.1913
 e = round(x3 - x4(1:N))
e =
  Columns 1 through 12
     0     0     0     0     0     0     0     0     0     0     0     0
  Columns 13 through 24
     0     0     0     0     0     0     0     0     0     0     0     0
  Columns 25 through 32
     0     0     0     0     0     0     0     0
x4(N+1:N4)
ans =
  Columns 1 through 12
     0     0     0     0     0     0     0     0     0     0     0     0
  Columns 13 through 24
     0     0     0     0     0     0     0     0     0     0     0     0
  Columns 25 through 32
     0     0     0     0     0     0     0     0
```

(b) Error between linear and circular convolutions:

$$x_1(n) = \{1, -1, 1, -1\}, \; x_2(n) = \{1, 0, -1, 0\}; \quad N = 5$$

MATLAB script:

```
% (e) x1(n) = {1,-1,1,-1}, x2(n) = {1,0,-1,0}; N = 5
x1 = [1,-1,1,-1]; x2 = [1,0,-1,0]; N = 5;
x3 = circonvf(x1,x2,N)
x3 =
    2.0000   -1.0000    0.0000    0.0000   -1.0000
x4 = conv(x1,x2), N4 = length(x4);
x4 =
     1    -1     0     0    -1     1     0
 e = x3 - x4(1:N)
e =
    1.0000    0.0000    0.0000    0.0000    0.0000
x4(N+1:N4)
ans =
     1     0
```

P5.20 MATLAB script:

```
nx = 0:10^6; m = 0:10; x = ((0.9).^m)*cos(0.1*pi*m'*n);
nh = 0:100; h = cos(0.5*pi*nh);
```

(a) MATLAB script:

```
t0 = clock; y1 = conv(x,h); t_conv = etime(clock,t0);
disp(sprintf('\n Computation time for convolution is %g',t_conv));
 Computation time for convolution is 5.05
```

(b) MATLAB function hsolpsav:

```
function [y] =  hsolpsav(x,h,N)
% High-speed Overlap-Save method of block convolutions using FFT
% --------------------------------------------------------------
% [y] = hsolpsav(x,h,N)
% y = output sequence
% x = input sequence
% h = impulse response
% N = block length (must be a power of two)
%
N = 2^(ceil(log10(N)/log10(2)));
Lenx = length(x); M = length(h);
M1 = M-1; L = N-M1;
h = fft(h,N);
%
x = [zeros(1,M1), x, zeros(1,N-1)];
K = floor((Lenx+M1-1)/(L)); % # of blocks
Y = zeros(K+1,N);
for k=0:K
    xk = fft(x(k*L+1:k*L+N));
    Y(k+1,:) = real(ifft(xk.*h));
```

```
end
Y = Y(:,M:N)'; y = (Y(:))';
```

MATLAB script:

```
% Highspeed convolution N = 1024
t0 = clock; y2 = hsolpsav(x,h,1024);  t_convhs = etime(clock,t0);
diff1 = max(abs(y1(1:10^6) - y2(1:10^6)));
disp(sprintf('\n Computation time for high-speed convolution (N=1024)...
                                                  is %g',t_convhs));
disp(sprintf(' Computation difference: %g',diff1));
 Computation time for high-speed convolution (N=1024) is 4.12
 Computation difference: 4.26326e-014

% Highspeed convolution N = 2048
t0 = clock; y2 = hsolpsav(x,h,2048);  t_convhs = etime(clock,t0);
diff1 = max(abs(y1(1:10^6) - y2(1:10^6)));
disp(sprintf('\n Computation time for high-speed convolution (N=2048)...
                                                  is %g',t_convhs));
disp(sprintf(' Computation difference: %g',diff1));
 Computation time for high-speed convolution (N=2048) convolution is 5.33
 Computation difference: 4.26326e-014
```

(c) From the above screen displays, the overlap-and-save method gives identical results with $N = 1024$ block size somewhat faster than the other two (note that these results depend on the cpu used).

Appendix E

Digital-Filter Structures

P6.1 A causal LTI system is described by

$$y(n) = \sum_{k=0}^{5} \left(\frac{1}{2}\right)^k x(n-k) + \sum_{\ell=1}^{5} \left(\frac{1}{3}\right)^k y(n-\ell)$$

and is driven by the input

$$x(n) = u(n), \; 0 \le n \le 100$$

A MATLAB script to produce various realizations is as follows:

```
clear; close all;
%% Direct Forms %%%%%%%%%%%%%%%%%%%%%%%%%%%%%%%%%%%%%%%%%%%%
disp('*** Direct Form Coefficients ***')
*** Direct Form Coefficients ***
k = 0:5; b = (0.5).^k
b =
    1.0000    0.5000    0.2500    0.1250    0.0625    0.0312
l = 1:5; a = [1, -(1/3).^l]
a =
    1.0000   -0.3333   -0.1111   -0.0370   -0.0123   -0.0041
% Output Response
[x,n] = stepseq(0,0,100);
y_DF = filter(b,a,x);

%% Cascade Form: %%%%%%%%%%%%%%%%%%%%%%%%%%%%%%%%%%%%%%%%%%%
disp('*** Cascade Form Coefficients ***')
*** Cascade Form Coefficients ***
[b0,B,A] = dir2cas(b,a)
```

```
b0 =
     1
B =
    1.0000    0.5000    0.2500
    1.0000   -0.5000    0.2500
    1.0000    0.5000         0
A =
    1.0000    0.4522    0.0745
    1.0000   -0.1303    0.0843
    1.0000   -0.6553         0
% Output Response
y_CF = casfiltr(b0,B,A,x);

%% Parallel Form: %%%%%%%%%%%%%%%%%%%%%%%%%%%%%%%%%%%%%%
disp('*** Parallel Form Coefficients ***')
*** Parallel Form Coefficients ***
[C,B,A] = dir2par(b,a)
C =
   -7.5938
B =
    3.4526    0.8052
    3.3193   -0.0367
    1.8219         0
A =
    1.0000    0.4522    0.0745
    1.0000   -0.1303    0.0843
    1.0000   -0.6553         0
% Output Response
y_PF = parfiltr(C,B,A,x);

%% Filter response plots
Hf_1 = figure('Units','normalized','position',[0.1,0.1,0.8,0.8],'color',[0,0,0]);
set(Hf_1,'NumberTitle','off','Name','P6.1');

subplot(3,1,1); stem(n,y_DF); axis([-1,101,0,5])
set(gca,'XTickMode','manual','XTick',[0:25:100],'fontsize',10);
set(gca,'YTickMode','manual','YTick',[0;4]);
ylabel('y_DF'); title('Direct Form Filter Response')

subplot(3,1,2); stem(n,y_CF); axis([-1,101,0,5])
set(gca,'XTickMode','manual','XTick',[0:25:100],'fontsize',10);
set(gca,'YTickMode','manual','YTick',[0;4]);
ylabel('y_DF'); title('Cascade Form Filter Response')

subplot(3,1,3); stem(n,y_PF); axis([-1,101,0,5])
set(gca,'XTickMode','manual','XTick',[0:25:100],'fontsize',10);
set(gca,'YTickMode','manual','YTick',[0;4]);
xlabel('n'); ylabel('y_DF'); title('Parallel Form Filter Response')
```

The block diagrams are shown in Figure E.1, and the output responses are shown in Figure E.2.

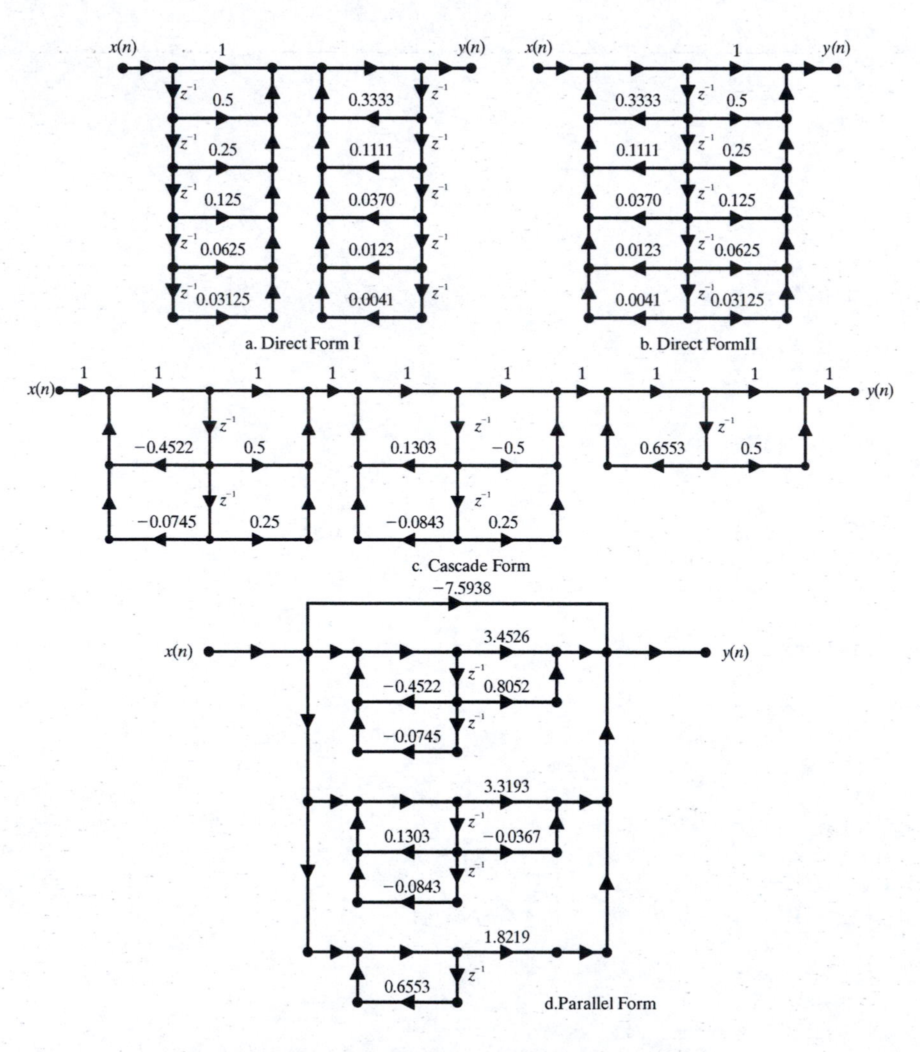

Figure E.1: Block diagrams in Problem P6.1

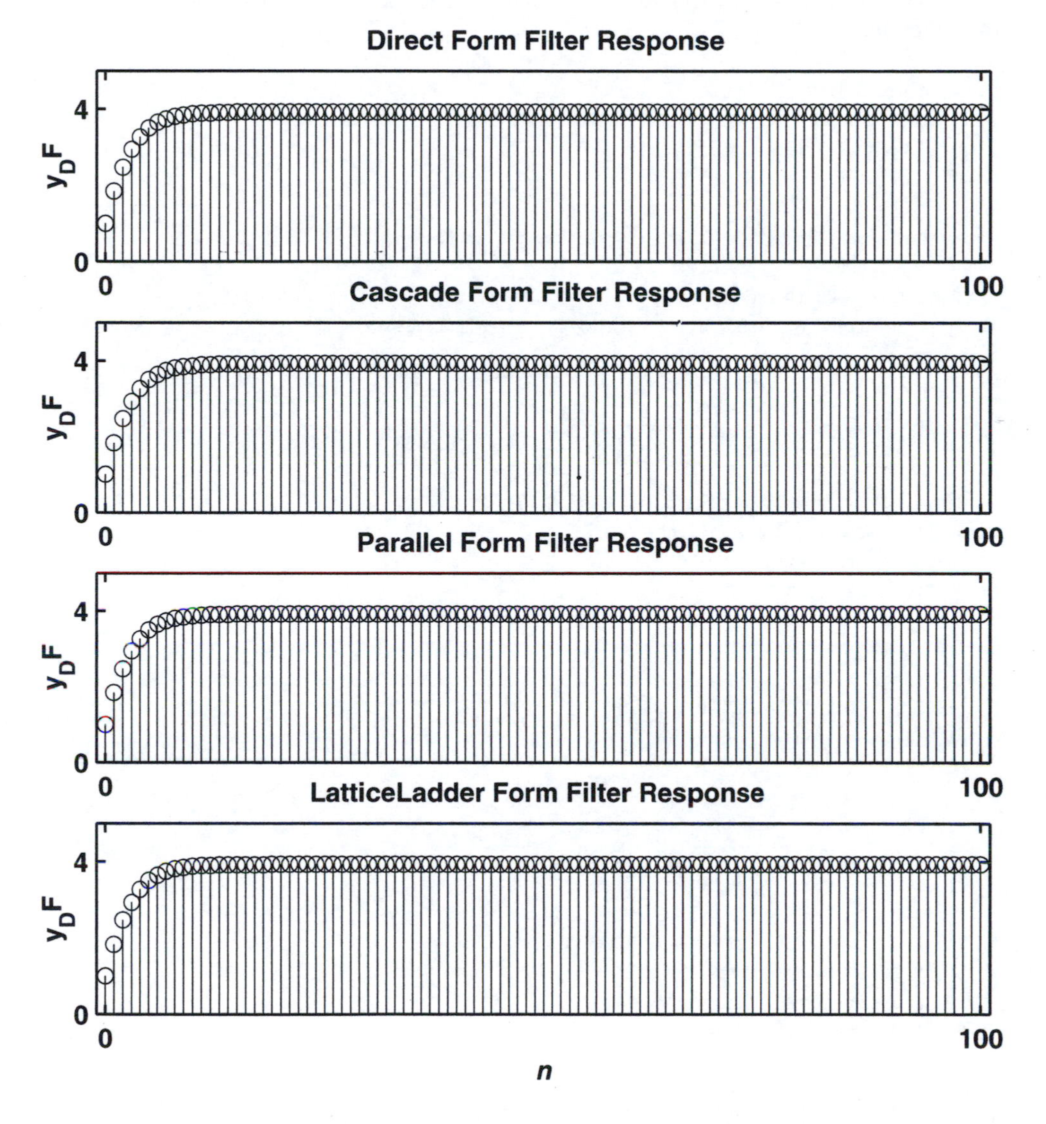

Figure E.2: Output responses in Problem P6.1

P6.2 The IIR filter is decribed by the system function

$$H(z) = 2\left(\frac{1+0z^{-1}+z^{-2}}{1-0.8z^{-1}+0.64z^{-2}}\right)\left(\frac{2-z^{-1}}{1-0.75z^{-1}}\right)\left(\frac{1+2z^{-1}+z^{-2}}{1+0.81z^{-2}}\right)$$

The given structure is a cascade form. A MATLAB script to produce various realizations is the following:

```
clear; close all;
%% Cascade Form Coefficients %%%%%%%%%%%%%%%%%%%%%%%%%%%
b0 = [2]; B = [1,0,1; 2,-1,0; 1,2,1];
A = [1,-0.8,0.64; 1,-0.75,0; 1,0,0.81];

%% Direct Forms %%%%%%%%%%%%%%%%%%%%%%%%%%%%%%%%%%%%%%%%%
disp('*** Direct Form Coefficients ***')
*** Direct Form Coefficients ***
[b,a] = cas2dir(b0,B,A); b = b(1:length(b)-1), a = a(1:length(a)-1)
b =
     4     6     4     4     0    -2
a =
    1.0000   -1.5500    2.0500   -1.7355    1.0044   -0.3888

%% Cascade Form: %%%%%%%%%%%%%%%%%%%%%%%%%%%%%%%%%%%%%%%%
disp('*** Cascade Form Coefficients ***')
*** Cascade Form Coefficients ***
[b0,B,A] = dir2cas(b,a)
b0 =
     4
B =
    1.0000    0.0000    1.0000
    1.0000    2.0000    1.0000
    1.0000   -0.5000         0
A =
    1.0000    0.0000    0.8100
    1.0000   -0.8000    0.6400
    1.0000   -0.7500         0

%% Parallel Form: %%%%%%%%%%%%%%%%%%%%%%%%%%%%%%%%%%%%%%%
disp('*** Parallel Form Coefficients ***')
*** Parallel Form Coefficients ***
[C,B,A] = dir2par(b,a)
C =
    5.1440
B =
    2.0137    0.1106
  -10.8732   15.0013
    7.7155         0
A =
    1.0000    0.0000    0.8100
    1.0000   -0.8000    0.6400
    1.0000   -0.7500         0
```

```
%% Lattice-Ladder Form: %%%%%%%%%%%%%%%%%%%%%%%%%%%%%%%%%%%%%%%
disp('*** Lattice-Ladder Form Coefficients ***')
*** Lattice-Ladder Form Coefficients ***
[K,C] = dir2ladr(b,a)
K =
   -0.3933    0.7034   -0.5916    0.4733   -0.3888
C =
    2.8172    6.8614    9.7493    3.8655   -3.1000   -2.0000
```

The block diagrams are shown in Figure E.3.

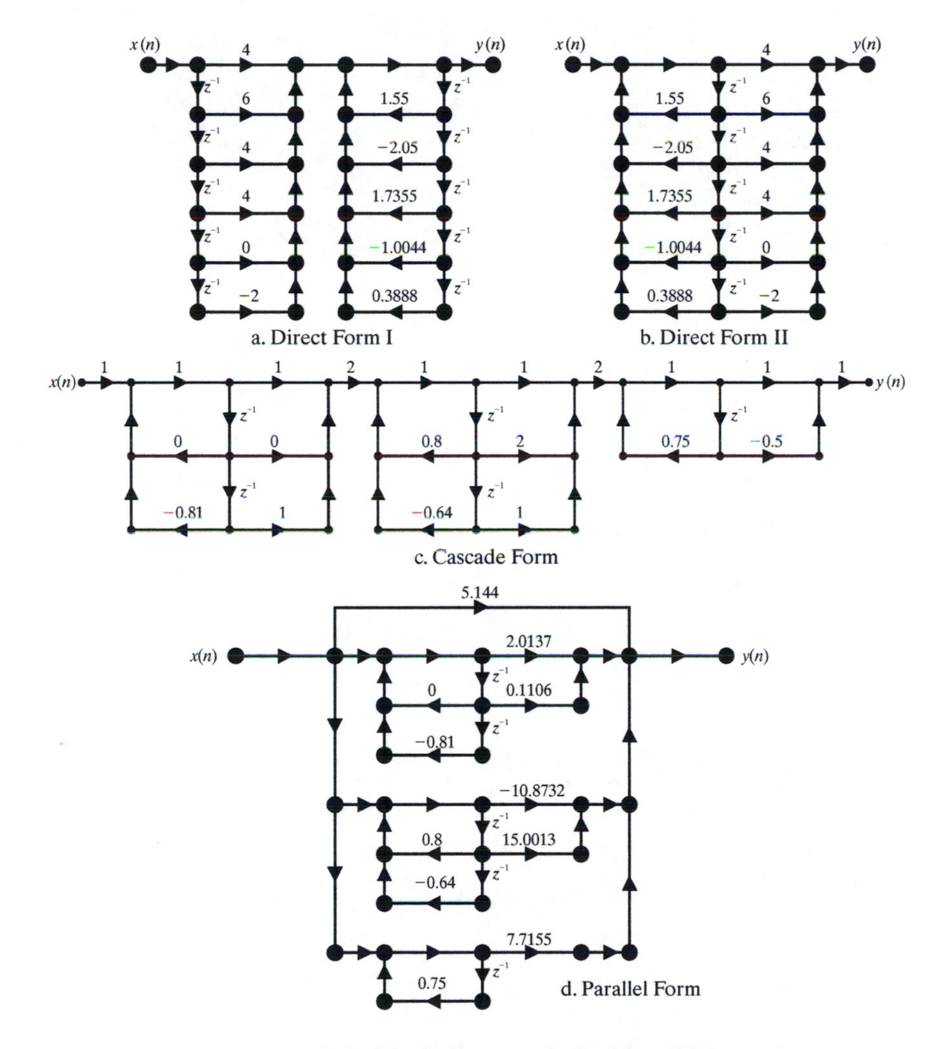

Figure E.3: Block diagrams in Problem P6.2

P6.3 The IIR filter is described by the system function

$$H(z) = \left(\frac{-14.75 - 12.9z^{-1}}{1 - \frac{7}{8}z^{-1} + \frac{3}{32}z^{-2}}\right) + \left(\frac{24.5 + 26.82z^{-1}}{1 - z^{-1} + 0.5z^{-2}}\right)$$

The given structure is a parallel form. A MATLAB script to produce various realizations is the following:

```
clear; close all;
%% Parallel Form Coefficients %%%%%%%%%%%%%%%%%%%%%%%%%%
C = []; B = [-14.75, -12.9; 24.5, 26.82];
A = [1, -7/8, 3/32; 1, -1, 1/2];

%% Direct Forms %%%%%%%%%%%%%%%%%%%%%%%%%%%%%%%%%%%%%%%%
disp('*** Direct Form Coefficients ***')
*** Direct Form Coefficients ***
[b,a] = par2dir(C,B,A)
b =
    9.7500    7.2325  -15.6456   -3.9356
a =
    1.0000   -1.8750    1.4688   -0.5312    0.0469

%% Cascade Form: %%%%%%%%%%%%%%%%%%%%%%%%%%%%%%%%%%%%%%%
disp('*** Cascade Form Coefficients ***')
*** Cascade Form Coefficients ***
[b0,B,A] = dir2cas(b,a)
b0 =
    9.7500
B =
    1.0000    1.8251    0.3726
    1.0000   -1.0834         0
A =
    1.0000   -1.0000    0.5000
    1.0000   -0.8750    0.0938

%% Parallel Form: %%%%%%%%%%%%%%%%%%%%%%%%%%%%%%%%%%%%%%
disp('*** Parallel Form Coefficients ***')
*** Parallel Form Coefficients ***
[C,B,A] = dir2par(b,a)
C =
     []
B =
   24.5000   26.8200
  -14.7500  -12.9000
A =
    1.0000   -1.0000    0.5000
    1.0000   -0.8750    0.0938
```

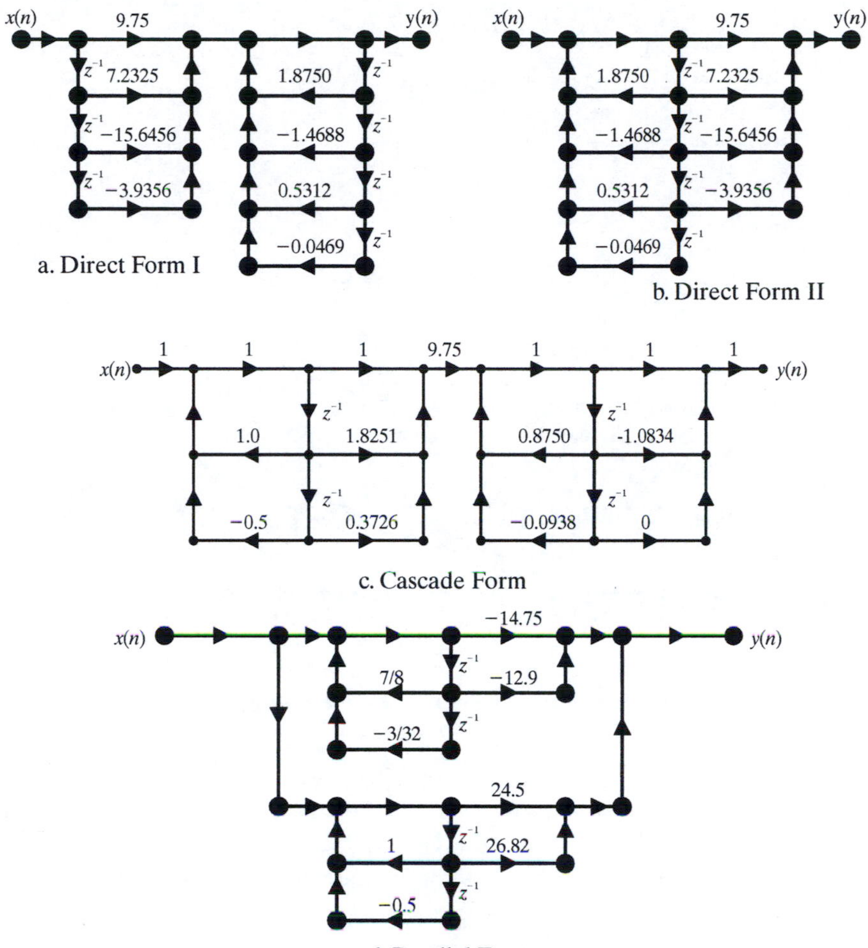

Figure E.4: Block diagrams in Problem P6.3

The block diagrams are shown in Figure E.4.

P6.4 A linear shift-invariant system has the system function

$$H(z) = \frac{0.5\left(1+z^{-1}\right)^6}{1-\frac{3}{2}z^{-1}+\frac{7}{8}z^{-2}-\frac{13}{16}z^{-3}-\frac{1}{8}z^{-4}-\frac{11}{32}z^{-5}+\frac{7}{16}z^{-6}}$$

The given block-diagram structure is of a cascade form.

(a) Conversion to the cascade form: The MATLAB script is as follows:

```
clear; close all;
%% Direct Form %%%%%%%%%%%%%%%%%%%%%%%%%%%%%%%%%%%%%%%%%%%
b0 = 0.5; broots = -ones(1,6); b = b0*poly(broots);
```

```
a = [1,-3/2,7/8,-13/16,-1/8,-11/32,7/16];

%% Cascade Form: %%%%%%%%%%%%%%%%%%%%%%%%%%%%%%%%%%%%%%%%
disp('*** Cascade Form Coefficients ***')
*** Cascade Form Coefficients ***
[b0,B,A] = dir2cas(b,a)
b0 =
    0.5000
B =
    1.0000    2.0067    1.0067
    1.0000    2.0000    1.0000
    1.0000    1.9933    0.9934
A =
    1.0000    1.0000    0.5000
    1.0000   -0.5000    1.0000
    1.0000   -2.0000    0.8750
```

The block-diagram structure is shown in Figure E.5.

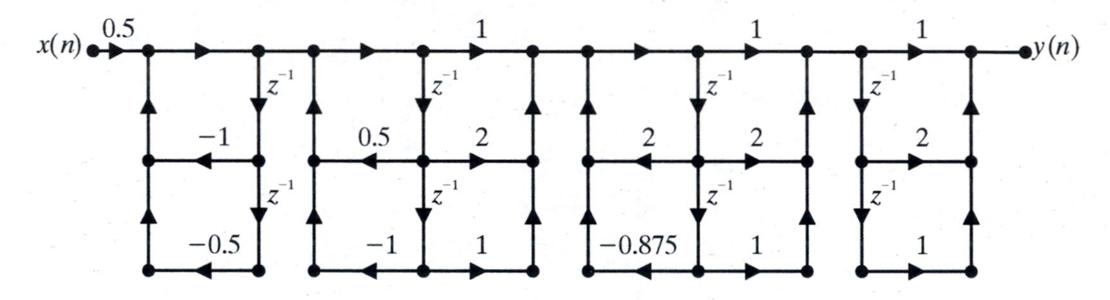

Figure E.5: Cascade-form structure in Problem P6.4

(b) The solution is not unique, because in the numerator and in the denominator the factors can be grouped differently.

P6.5 A linear shift-invariant system has the system function

$$H(z) = \frac{5 + 11.2z^{-1} + 5.44z^{-2} - 0.384z^{-3} - 2.3552z^{-4} - 1.2288z^{-5}}{1 + 0.8z^{-1} - 0.512z^{-3} - 0.4096z^{-4}}$$

The given structure is in a parallel form.

(a) Conversion to the parallel form: The MATLAB script is as follows:

```
clear; close all;
%% Direct Form %%%%%%%%%%%%%%%%%%%%%%%%%%%%%%%%%%%%%%%%%%%
b = [5, 11.2, 5.44, -0.384, -2.3552, -1.2288];
a = [1, 0.8, 0, -0.512, -0.4096];

%% Parallel Form: %%%%%%%%%%%%%%%%%%%%%%%%%%%%%%%%%%%%%%%%
disp('*** Parallel Form Coefficients ***')
*** Parallel Form Coefficients ***
```

```
[C,B,A] = dir2par(b,a)
C =
    2.0000    3.0000
B =
    1.0000    2.0000
    2.0000    3.0000
A =
    1.0000    0.8000    0.6400
    1.0000    0.0000   -0.6400
```

The block-diagram structure is shown in Figure E.6.

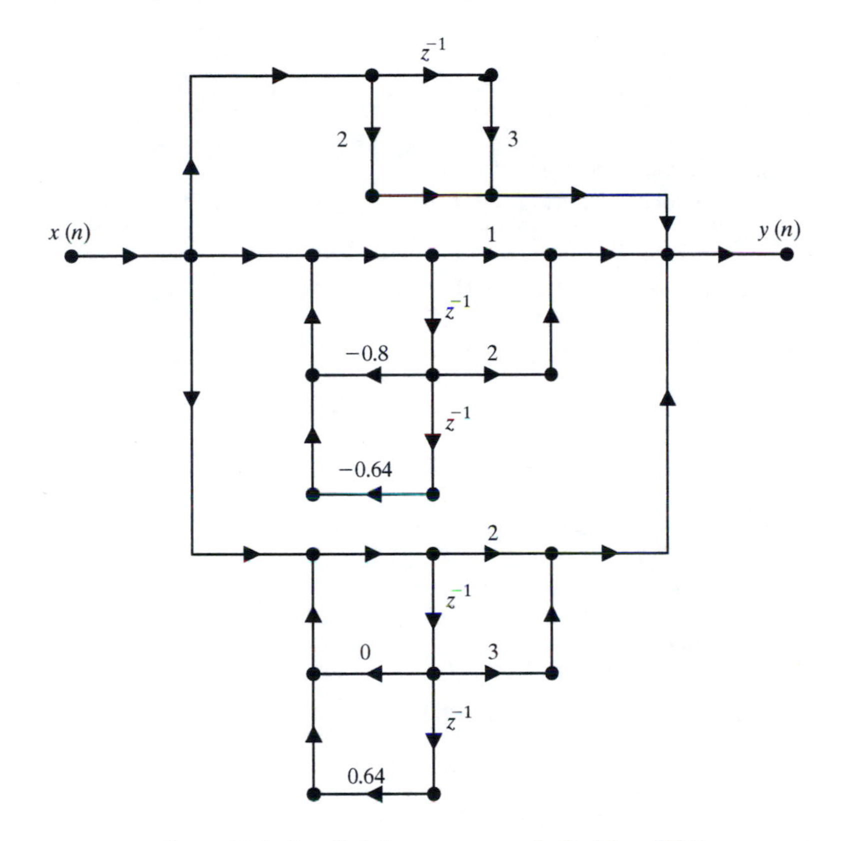

Figure E.6: Parallel-form structure in Problem P6.5

(b) The solution is unique because the numerator residues and the corresponding denominator poles must be grouped into a second-order section.

P6.6 A linear shift-invariant system has the system function

$$H(z) = \frac{0.5\left(1 + z^{-1}\right)^6}{1 - \frac{3}{2}z^{-1} + \frac{7}{8}z^{-2} - \frac{13}{16}z^{-3} - \frac{1}{8}z^{-4} - \frac{11}{32}z^{-5} + \frac{7}{16}z^{-6}}$$

The new structure has a parallel-of-cascades form.

(a) Conversion to the parallel-of-cascades form: A MATLAB script is the following:

```
clear; close all;
%% Direct Form %%%%%%%%%%%%%%%%%%%%%%%%%%%%%%%%%%%%%%%%%%%
b = 0.5*poly([-1,-1,-1,-1,-1,-1]);
a = [1, -3/2, 7/8, -13/16, -1/8, -11/32, 7/16];

%% Parallel Form: %%%%%%%%%%%%%%%%%%%%%%%%%%%%%%%%%%%%%%%%
disp('*** Parallel Form Coefficients ***')
*** Parallel Form Coefficients ***
[C,B,A] = dir2par(b,a)
C =
    1.1429
B =
   -0.0210   -0.0335
    0.9252   -3.6034
   -1.5470    9.9973
A =
    1.0000    1.0000    0.5000
    1.0000   -0.5000    1.0000
    1.0000   -2.0000    0.8750

%% Parallel of Cascade Forms: %%%%%%%%%%%%%%%%%%%%%%%%%%%
disp('*** Parallel of Cascade Forms Coefficients ***')
*** Parallel of Cascade Forms Coefficients ***
B = B(1:2,:), A = A(1:2,:)
B =
   -0.0210   -0.0335
    0.9252   -3.6034
A =
    1.0000    1.0000    0.5000
    1.0000   -0.5000    1.0000
[b1,a1] = par2dir(C,B,A)
b1 =
  Columns 1 through 4
   2.0470            -2.1298            -2.0023            -0.9780
  Column 5
   0.5714 + 0.0000i
a1 =
  Columns 1 through 4
   1.0000             0.5000             1.0000             0.7500
  Column 5
   0.5000 + 0.0000i
[b0,B1,A1] = dir2cas(b1,a1)
b0 =
    2.0470
B1 =
    1.0000    0.9985    0.5050
    1.0000   -2.0389    0.5527
A1 =
    1.0000    1.0000    0.5000
    1.0000   -0.5000    1.0000
```

The block-diagram structure is shown in Figure E.7.

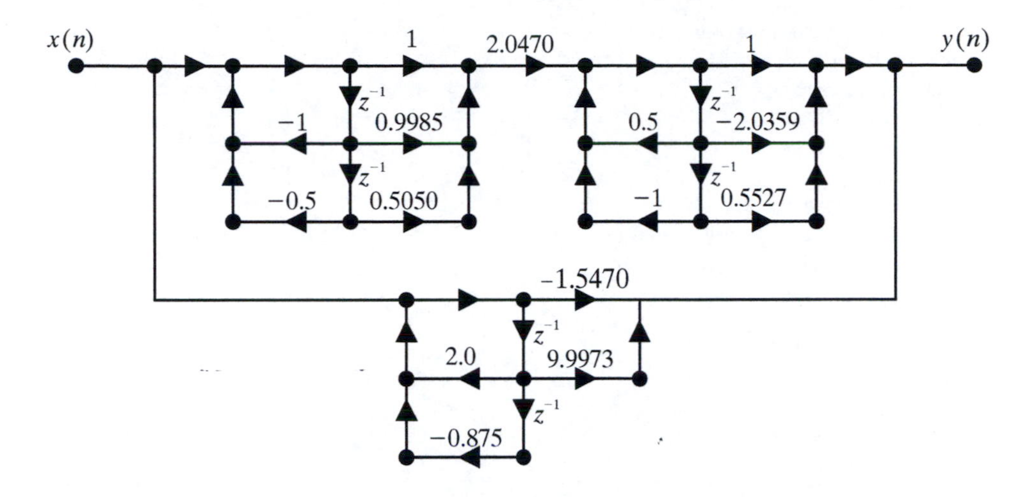

Figure E.7: Parallel-of-cascades structure in Problem P6.6

(b) The solution is unique because the numerator residues and the corresponding denominator poles must be grouped into a second-order section.

P6.7 A causal FIR filter is described by

$$y(n) = \sum_{k=0}^{10} \left(\frac{1}{2}\right)^{|5-k|} x(n-k)$$

A MATLAB script to produce various realizations is the following:

```
clear; close all;
%% Direct Form %%%%%%%%%%%%%%%%%%%%%%%%%%%%%%%%%%%%%%%%%%%%%%
disp('*** Direct Form Coefficients ***')
*** Direct Form Coefficients ***
k = 0:10; b = (0.5).^(abs(5-k)), a = 1;
b =

  Columns 1 through 7
    0.0312    0.0625    0.1250    0.2500    0.5000    1.0000    0.5000
  Columns 8 through 11
    0.2500    0.1250    0.0625    0.0312

%% Linear-Phase Form %%%%%%%%%%%%%%%%%%%%%%%%%%%%%%%%%%%%%%%%
disp('*** Linear-Phase Form Coefficients ***')
*** Linear-Phase Form Coefficients ***
M = length(b); L = (M-1)/2;
B = b(1:1:L+1)
B =
    0.0312    0.0625    0.1250    0.2500    0.5000    1.0000
```

```
%% Cascade Form: %%%%%%%%%%%%%%%%%%%%%%%%%%%%%%%%%%%%%%%%%%%%
disp('*** Cascade Form Coefficients ***')
*** Cascade Form Coefficients ***
[b0,B,A] = dir2cas(b,a)
b0 =
    0.0312
B =
    1.0000    1.7495    3.4671
    1.0000    0.5046    0.2884
    1.0000   -0.5506    0.2646
    1.0000   -2.0806    3.7789
    1.0000    2.3770    1.0000
A =
     1     0     0
     1     0     0
     1     0     0
     1     0     0
     1     0     0
```

The block diagrams are shown in Figure E.8.

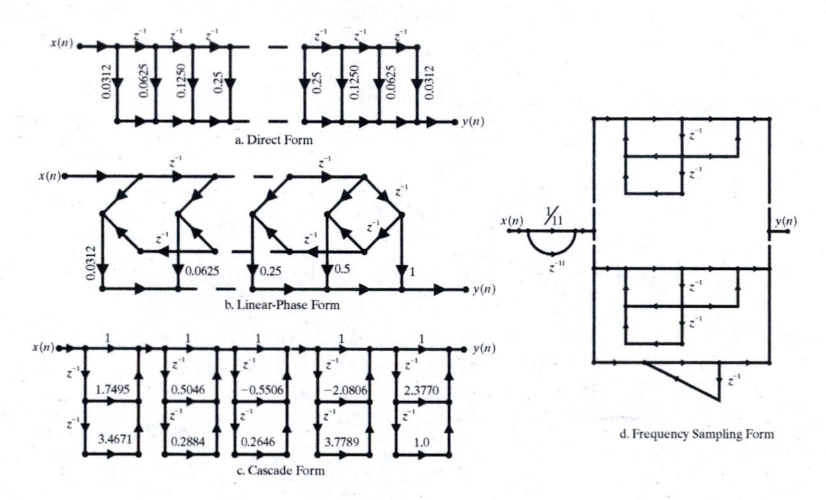

Figure E.8: Block diagrams in Problem P6.7

P6.8 The FIR filter is decribed by the system function

$$H(z) = \sum_{k=0}^{10} (2z)^{-k}$$

A MATLAB script to produce various realizations is the following:

```
clear; close all;
%% Direct Form %%%%%%%%%%%%%%%%%%%%%%%%%%%%%%%%%%%%%%%%%%%%%%
```

```
disp('*** Direct Form Coefficients ***')
*** Direct Form Coefficients ***
k = 0:10; b = (0.5).^k, a = 1;
b =
  Columns 1 through 7
    1.0000    0.5000    0.2500    0.1250    0.0625    0.0312    0.0156
  Columns 8 through 11
    0.0078    0.0039    0.0020    0.0010

%% Cascade Form: %%%%%%%%%%%%%%%%%%%%%%%%%%%%%%%%%%%%%%%%
disp('*** Cascade Form Coefficients ***')
*** Cascade Form Coefficients ***
[b0,B,A] = dir2cas(b,a)
b0 =
     1
B =
    1.0000    0.9595    0.2500
    1.0000    0.6549    0.2500
    1.0000    0.1423    0.2500
    1.0000   -0.4154    0.2500
    1.0000   -0.8413    0.2500
A =
     1     0     0
     1     0     0
     1     0     0
     1     0     0
     1     0     0

%% Lattice Form: %%%%%%%%%%%%%%%%%%%%%%%%%%%%%%%%%%%%%%%%
disp('*** Lattice-ladder Form Coefficients ***')
*** Lattice-ladder vForm Coefficients ***
[k] = dir2latc(b)
k =
  Columns 1 through 6
    1.0000    0.4000    0.1905    0.1941    0.0469    0.234
  Columns 7 through 11
    0.0117    0.0059    0.0029    0.0015    0.0010
```

The block diagrams are shown in Figure E.9.

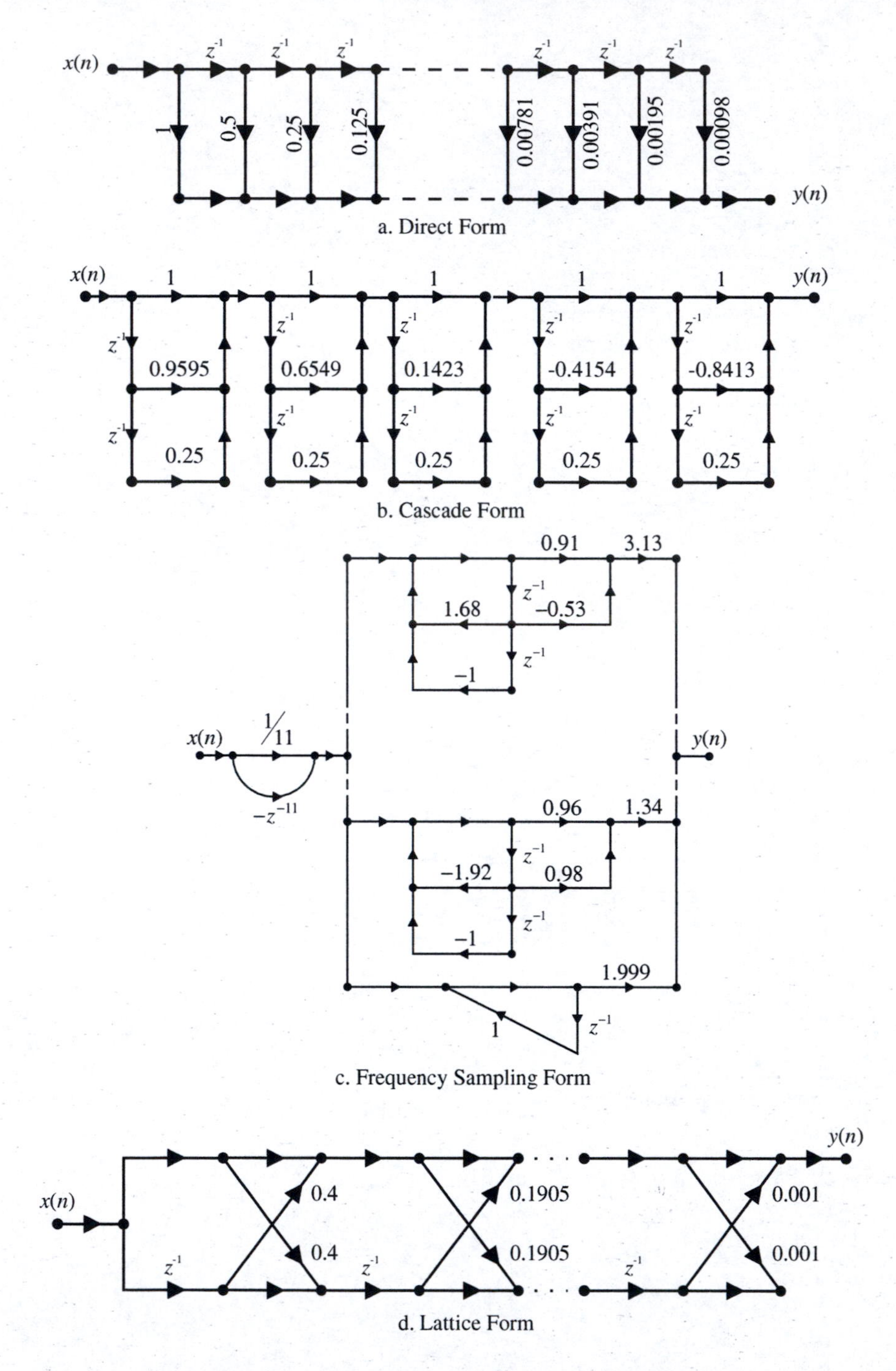

Figure E.9: Block diagrams in Problem P6.8

Appendix F

FIR Filter Design

P7.1 Amplitude-response function for **Type-2** LP FIR filter.

(a) Type-2 $\Rightarrow$ symmetric $h(n)$ and M – even:

$$h(n) = h(M-1-n),\ 0 \leq n \leq M-1; \quad \alpha = \frac{M-1}{2} \text{ is not an integer}$$

Consider

$$H(\omega) = \sum_{n=0}^{M-1} h(n)e^{-j\omega n} = \sum_{n=0}^{\frac{M}{2}-1} h(n)e^{-j\omega n} + \sum_{\frac{M}{2}}^{M-1} h(n)e^{-j\omega n}$$

By change of variable in the second sum,

$$n \to M-1-n \Rightarrow \frac{M}{2} \to \frac{M}{2} - 1,\ M-1 \to 0,\ \text{and}\ h(n) \to h\,(n)$$

Hence,

$$\begin{aligned}
H(\omega) &= \textstyle\sum_{n=0}^{\frac{M}{2}-1} h\,(n)\,e^{-j\omega n} + \sum_{n=0}^{\frac{M}{2}-1} h(n)e^{-j\omega(M-1-n)} \\
&= e^{-j\omega\left(\frac{M-1}{2}\right)} \textstyle\sum_{n=0}^{\frac{M}{2}-1} h(n) \left\{ e^{-j\omega n + j\omega\left(\frac{M-1}{2}\right)} + e^{-j\omega(M-1-n)+j\omega\left(\frac{M-1}{2}\right)} \right\} \\
&= e^{-j\omega\left(\frac{M-1}{2}\right)} \textstyle\sum_{n=0}^{\frac{M}{2}-1} h(n) \left\{ e^{+j\omega\left(\frac{M-1}{2}-n\right)} + e^{-j\omega\left(\frac{M-1}{2}-n\right)} \right\} \\
&= e^{-j\omega\left(\frac{M-1}{2}\right)} \textstyle\sum_{n=0}^{\frac{M}{2}-1} 2h(n) \cos\left[\left(\frac{M-1}{2} - n\right)\omega\right]
\end{aligned}$$

Change of variable yields

$$\frac{M}{2} - n \to n \Rightarrow n = 0 \to n = \frac{M}{2},\ n = \frac{M}{2} - 1 \to n = 1$$

and

$$\cos\left[\left(\frac{M-1}{2} - n\right)\omega\right] \to \cos\left[\omega\left(n - \frac{1}{2}\right)\right]$$

Hence,

$$H(\omega) = e^{-j\omega\left(\frac{M-1}{2}\right)} \sum_{n=1}^{\frac{M}{2}} 2h\left(\frac{M}{2} - n\right) \cos\left[\omega\left(n - \frac{1}{2}\right)\right]$$

Define $b(n) = 2h\left(\frac{M}{2} - n\right)$. Then,

$$H(\omega) = e^{-j\omega\left(\frac{M-1}{2}\right)} \sum_{n=1}^{\frac{M}{2}} b(n) \cos\left[\omega\left(n - \frac{1}{2}\right)\right] \Rightarrow H_r(\omega) = \sum_{n=1}^{\frac{M}{2}} b(n) \cos\left[\omega\left(n - \frac{1}{2}\right)\right]$$

(b) Now $\cos\left[\omega\left(n - \frac{1}{2}\right)\right]$ can be written as a linear combination of higher harmonics in $\cos\omega$ multiplied by $\cos\left(\frac{\omega}{2}\right)$:

$$\begin{aligned}
\cos\left(\frac{1\omega}{2}\right) &= \cos\frac{\omega}{2}\{\cos 0\omega\} \\
\cos\left(\frac{3\omega}{2}\right) &= \cos\frac{\omega}{2}\{2\cos\omega - 1\} \\
\cos\left(\frac{5\omega}{2}\right) &= \cos\frac{\omega}{2}\{\cos 0\omega - \cos\omega + \cos 2\omega\}
\end{aligned}$$

and so on. Note that the lowest harmonic frequency is zero and that the highest harmonic frequency is $(n-1)\,\omega$ in the $\cos\omega\left(n - \frac{1}{2}\right)$ expansion. Hence,

$$H_r(\omega) = \sum_{n=1}^{\frac{M}{2}} b(n) \cos\left[\omega\left(n - \frac{1}{2}\right)\right] = \cos\frac{\omega}{2} \sum_{0}^{\frac{M}{2}-1} \tilde{b}(n) \cos(\omega n)$$

where the $\tilde{b}(n)$ are related to the $b(n)$ through the trigonometric identities listed in (b).

P7.2 Type-3 $\Rightarrow$ antisymmetric $h(n)$ and M – odd:

$$h(n) = -h(M-1-n),\ 0 \le n \le M-1;\ h\left(\frac{M-1}{2}\right) = 0; \quad \alpha = \frac{M-1}{2} \text{ is an integer}$$

(a) Consider

$$H(\omega) = \sum_{n=0}^{M-1} h(n) e^{-j\omega n} = \sum_{n=0}^{\frac{M-3}{2}} h(n) e^{-j\omega n} + \sum_{\frac{M+1}{2}}^{M-1} h(n) e^{-j\omega n}$$

By change of variable in the second sum,

$$n \to M-1-n \Rightarrow \frac{M+1}{2} \to \frac{M-3}{2},\ M-1 \to 0, \text{ and } h(n) \to -h(n)$$

Hence,

$$
\begin{aligned}
H(\omega) &= \sum_{n=0}^{\frac{M-3}{2}} h(n)e^{-j\omega n} - \sum_{n=0}^{\frac{M-3}{2}} h(n)e^{-j\omega(M-1-n)} \\
&= e^{-j\left(\frac{M-1}{2}\right)} \sum_{n=0}^{\frac{M-3}{2}} h(n)\left\{e^{-j\omega n+j\omega\left(\frac{M-1}{2}\right)} - e^{-j\omega(M-1-n)+j\omega\left(\frac{M-1}{2}\right)}\right\} \\
&= e^{-j\left(\frac{M-1}{2}\right)} \sum_{n=0}^{\frac{M-3}{2}} h(n)\left\{e^{+j\omega\left(\frac{M-1}{2}-n\right)} - e^{-j\omega\left(\frac{M-1}{2}-n\right)}\right\} \\
&= je^{-j\left(\frac{M-1}{2}\right)} \sum_{n=0}^{\frac{M-3}{2}} 2h(n)\sin\left[\left(\frac{M-1}{2}-n\right)\omega\right]
\end{aligned}
$$

Change of variable:

$$
\frac{M-1}{2} - n \to n \Rightarrow n = 0 \to n = \frac{M-1}{2},\ n = \frac{M-3}{2} \to n = 1
$$

and

$$
\sin\left[\left(\frac{M-1}{2} - n\right)\omega\right] \to \sin(\omega n)
$$

Hence,

$$
H\left(e^{j\omega}\right) = je^{-j\omega\left(\frac{M-1}{2}\right)} \sum_{n=1}^{\frac{M-1}{2}} 2h\left(\frac{M-1}{2} - n\right)\sin(\omega n)
$$

Define $c(n) = 2h\left(\frac{M-1}{2} - n\right)$. Then,

$$
H(\omega) = je^{-j\omega\left(\frac{M-1}{2}\right)} \sum_{n=1}^{\frac{M-1}{2}} c(n)\sin(\omega n) \Rightarrow H_r(\omega) = \sum_{n=1}^{\frac{M}{2}} c(n)\sin(\omega n)
$$

(b) Now $\sin(\omega n)$ can be written as a linear combination of higher harmonics in $\cos\omega$ multiplied by $\sin\omega$:

$$
\begin{aligned}
\sin(\omega) &= \sin\omega\{\cos 0\omega\} \\
\sin(2\omega) &= \sin\omega\{2\cos\omega\} \\
\sin(3\omega) &= \sin\omega\{\cos 0\omega + 2\cos 2\omega\}
\end{aligned}
$$

and so on. Note that the lowest harmonic frequency is zero and that the highest harmonic frequency is $(n-1)\omega$ in the $\sin\omega n$ expansion. Hence,

$$
H_r(\omega) = \sum_{n=1}^{\frac{M-1}{2}} c(n)\sin(\omega n) = \sin\omega \sum_{0}^{\frac{M-3}{2}} \tilde{c}(n)\cos(\omega n)
$$

where the $\tilde{c}(n)$ are related to the $c(n)$ through the trigonometric identities shown in (b).

P7.3 MATLAB function for amplitude response:

```
function [Hr,w,P,L] = ampl_res(h);
%
% function [Hr,w,P,L] = ampl_res(h)
%
% Computes Amplitude response Hr(w) and its polynomial P of order L,
%   given a linear-phase FIR filter impulse response h.
% The type of filter is determined automatically by the subroutine.
%
% Hr = Amplitude Response
%  w = frequencies between [0 pi] over which Hr is computed
%  P = Polynomial coefficients
%  L = Order of P
%  h = Linear Phase filter impulse response
%

 M = length(h);
 L = floor(M/2);
if fix(abs(h(1:1:L))*10^10) ~= fix(abs(h(M:-1:M-L+1))*10^10)
    error('Not a linear-phase impulse response')
end

if 2*L ~= M
    if fix(h(1:1:L)*10^10) == fix(h(M:-1:M-L+1)*10^10)
        disp('*** Type-1 Linear-Phase Filter ***')
        [Hr,w,P,L] = hr_type1(h);
    elseif fix(h(1:1:L)*10^10) == -fix(h(M:-1:M-L+1)*10^10)
        disp('*** Type-3 Linear-Phase Filter ***')
        h(L+1) = 0;
        [Hr,w,P,L] = hr_type3(h);
    end
else
    if fix(h(1:1:L)*10^10) == fix(h(M:-1:M-L+1)*10^10)
        disp('*** Type-2 Linear-Phase Filter ***')
        [Hr,w,P,L] = hr_type2(h);
    elseif fix(h(1:1:L)*10^10) == -fix(h(M:-1:M-L+1)*10^10)
        disp('*** Type-4 Linear-Phase Filter ***')
        [Hr,w,P,L] = hr_type4(h);
    end
end
```

MATLAB verification:

```
clear; close all;
%% Matlab verification
h1 = [1 2 3 2 1]; [Hr1,w,P1,L1] = ampl_res(h1);
*** Type-1 Linear-Phase Filter ***
P1, L1,
P1 =
     3     4     2
```

```
L1 =
     2
%
h2 = [1 2 2 1]; [Hr2,w,P2,L2] = ampl_res(h2);
*** Type-2 Linear-Phase Filter ***
P2, L2,
P2 =
     4     2
L2 =
     2
%
h3 = [1 2 0 -2 -1]; [Hr3,w,P3,L3] = ampl_res(h3);
*** Type-3 Linear-Phase Filter ***
P3, L3,
P3 =
     0     4     2
L3 =
     2
%
h4 = [1 2 -2 -1]; [Hr4,w,P4,L4] = ampl_res(h4);
*** Type-4 Linear-Phase Filter ***
P4, L4,
P4 =
     4     2
L4 =
     2
%
%% Amplitude response plots
Hf_1 = figure('Units','normalized','position',[0.1,0.1,0.8,0.8],'color',[0,0,0]);
set(Hf_1,'NumberTitle','off','Name','P7.5');

subplot(2,2,1);plot(w/pi,Hr1);title('Type-1 FIR Filter');
ylabel('Hr(w)');
set(gca,'XTickMode','manual','XTick',[0:0.2:1],'fontsize',10);

subplot(2,2,2);plot(w/pi,Hr2);title('Type-2 FIR Filter');
ylabel('Hr(w)');
set(gca,'XTickMode','manual','XTick',[0:0.2:1],'fontsize',10);

subplot(2,2,3);plot(w/pi,Hr3);title('Type-3 FIR Filter');
ylabel('Hr(w)'); xlabel('frequency in pi units');
set(gca,'XTickMode','manual','XTick',[0:0.2:1],'fontsize',10);

subplot(2,2,4);plot(w/pi,Hr4);title('Type-4 FIR Filter');
ylabel('Hr(w)'); xlabel('frequency in pi units');
set(gca,'XTickMode','manual','XTick',[0:0.2:1],'fontsize',10);

% Super Title
suptitle('Problem~P7.3');
```

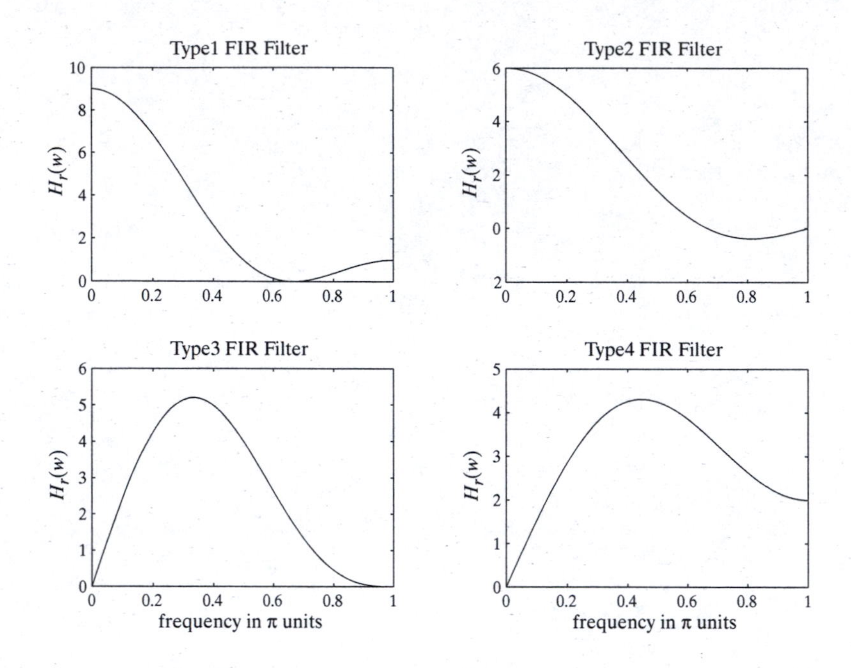

Figure F.1: Amplitude-response plots in Problem P7.3

The plots are shown in Figure F.1.

P7.4 The filter $H(z)$ has the following four zeros:

$$z_1 = re^{j\theta}, \quad z_2 = \frac{1}{r}e^{j\theta}, \quad z_3 = re^{-j\theta}, \quad z_4 = \frac{1}{r}e^{-j\theta}$$

The system function can be written as

$$\begin{aligned}
H(z) &= \left(1 - z_1 z^{-1}\right)\left(1 - z_2 z^{-1}\right)\left(1 - z_3 z^{-1}\right)\left(1 - z_4 z^{-1}\right) \\
&= \left(1 - re^{j\theta} z^{-1}\right)\left(1 - \tfrac{1}{r}e^{j\theta} z^{-1}\right)\left(1 - re^{-j\theta} z^{-1}\right)\left(1 - \tfrac{1}{r}e^{-j\theta} z^{-1}\right) \\
&= \left\{1 - (2r\cos\theta)\, z^{-1} + r^2 z^{-2}\right\}\left\{1 - \left(2r^{-1}\cos\theta\right) z^{-1} + r^{-2} z^{-2}\right\} \\
&= 1 - 2\cos\theta\left(r + r^{-1}\right) z^{-1} + \left(r^2 + r^{-2} + 4\cos^2\theta\right) z^{-2} - 2\cos\theta\left(r + r^{-1}\right) z^{-3} + z^{-4}
\end{aligned}$$

Hence, the impulse response of the filter is

$$h(n) = \left\{\underset{\uparrow}{1}, -2\cos\theta\left(r + r^{-1}\right), \left(r^2 + r^{-2} + 4\cos^2\theta\right), -2\cos\theta\left(r + r^{-1}\right), 1\right\}$$

which is a finite-duration symmetric-impulse response. This implies that the filter is a linear-phase FIR filter.

P7.5 The bandstop-filter specifications are

lower passband edge:	0.3π
lower stopband edge:	0.4π
upper stopband edge:	0.6π
upper passband edge:	0.7π
passband ripple:	0.5 dB
stopband attenuation:	40 dB

Hanning window design via MATLAB:

```
clear; close all;
%% Specifications:
wp1 = 0.3*pi; % lower passband edge
ws1 = 0.4*pi; % lower stopband edge
ws2 = 0.6*pi; % upper stopband edge
wp2 = 0.7*pi; % upper passband edge
Rp = 0.5;     % passband ripple
As = 40;      % stopband attenuation
%
% Select the min(delta1,delta2) since delta1=delta2 in window design
[delta1,delta2] = db2delta(Rp,As);
if (delta1 < delta2)
   delta2 = delta1; disp('Delta1 is smaller than delta2')
   [Rp,As] = delta2db(delta1,delta2)
end
%
tr_width = min((ws1-wp1),(wp2-ws2));
M = ceil(6.2*pi/tr_width); M = 2*floor(M/2)+1, % choose odd M
M =
    63
n = 0:M-1;
w_han = (hanning(M))';
wc1 = (ws1+wp1)/2; wc2 = (ws2+wp2)/2;
hd = ideal_lp(pi,M)+ideal_lp(wc1,M)-ideal_lp(wc2,M);
h = hd .* w_han;
[db,mag,pha,grd,w] = freqz_m(h,1);
delta_w = pi/500;
Asd = floor(-max(db((ws1/delta_w)+1:(ws2/delta_w)+1))),        % Actual Attn
Asd =
    43
Rpd = -min(db(1:(wp1/delta_w)+1)),                % Actual passband ripple
Rpd =
    0.0884
%
%% Filter Response Plots
Hf_1 = figure('Units','normalized','position',[0.1,0.1,0.8,0.8],'color',[0,0,0]);
set(Hf_1,'NumberTitle','off','Name','P7.7');

subplot(2,2,1); stem(n,hd); title('Ideal Impulse Response');
axis([-1,M,min(hd)-0.1,max(hd)+0.1]); xlabel('n'); ylabel('hd(n)')
```

```
set(gca,'XTickMode','manual','XTick',[0;M-1],'fontsize',10)

subplot(2,2,2); stem(n,w_han); title('Hanning Window');
axis([-1,M,-0.1,1.1]); xlabel('n'); ylabel('w_ham(n)')
set(gca,'XTickMode','manual','XTick',[0;M-1],'fontsize',10)
set(gca,'YTickMode','manual','YTick',[0;1],'fontsize',10)

subplot(2,2,3); stem(n,h); title('Actual Impulse Response');
axis([-1,M,min(hd)-0.1,max(hd)+0.1]); xlabel('n'); ylabel('h(n)')
set(gca,'XTickMode','manual','XTick',[0;M-1],'fontsize',10)

subplot(2,2,4); plot(w/pi,db); title('Magnitude Response in dB');
axis([0,1,-As-30,5]); xlabel('frequency in pi units'); ylabel('Decibels')
set(gca,'XTickMode','manual','XTick',[0;0.3;0.4;0.6;0.7;1])
set(gca,'XTickLabelMode','manual','XTickLabels',[' 0 ';'0.3';'0.4';'0.6';...
                                                   '0.7';' 1 '],'fontsize',10)
set(gca,'YTickMode','manual','YTick',[-40;0])
set(gca,'YTickLabelMode','manual','YTickLabels',[' 40';' 0 ']);grid

% Super Title
suptitle('Problem~P7.5: Bandstop Filter');
```

The filter-design plots are shown in Figure F.2.

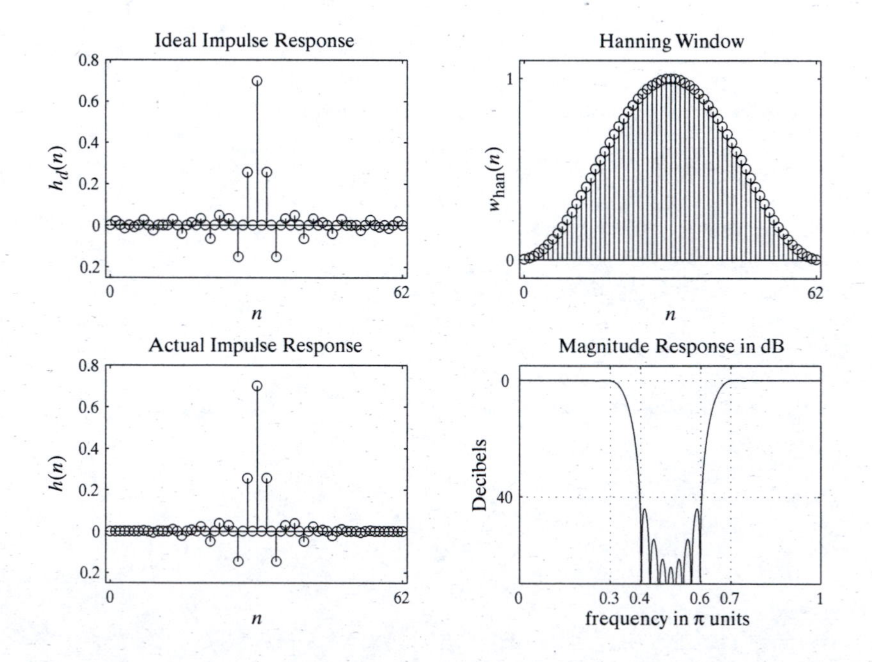

Figure F.2: Bandstop-filter design using Hanning window in Problem P7.5

P7.6 The bandpass-filter specifications are

lower stopband edge:	0.3π
lower passband edge:	0.4π
upper passband edge:	0.5π
upper stopband edge:	0.6π
passband ripple:	0.5 dB
stopband attenuation:	50 dB

Hamming window design via MATLAB:

```
clear; close all;
%% Specifications:
ws1 = 0.3*pi; % lower stopband edge
wp1 = 0.4*pi; % lower passband edge
wp2 = 0.5*pi; % upper passband edge
ws2 = 0.6*pi; % upper stopband edge
Rp = 0.5;     % passband ripple
As = 50;      % stopband attenuation
%
% Select the min(delta1,delta2) since delta1=delta2 in window design
[delta1,delta2] = db2delta(Rp,As);
if (delta1 < delta2)
   delta2 = delta1; disp('Delta1 is smaller than delta2')
   [Rp,As] = delta2db(delta1,delta2)
end
%
tr_width = min((wp1-ws1),(ws2-wp2));
M = ceil(6.6*pi/tr_width); M = 2*floor(M/2)+1, % choose odd M
M =
    67
n = 0:M-1;
w_ham = (hamming(M))';
wc1 = (ws1+wp1)/2; wc2 = (ws2+wp2)/2;
hd = ideal_lp(wc2,M)-ideal_lp(wc1,M);
h = hd .* w_ham;
[db,mag,pha,grd,w] = freqz_m(h,1);
delta_w = pi/500;
Asd = floor(-max(db([1:floor(ws1/delta_w)+1]))),          % Actual Attn
Asd =
    51
Rpd = -min(db(ceil(wp1/delta_w)+1:floor(wp2/delta_w)+1)), % Actual passband ripple
Rpd =
    0.0488
%
%% Filter Response Plots
Hf_1 = figure('Units','normalized','position',[0.1,0.1,0.8,0.8],'color',[0,0,0]);
set(Hf_1,'NumberTitle','off','Name','P7.8');

subplot(2,2,1); stem(n,hd); title('Ideal Impulse Response: Bandpass');
```

```
axis([-1,M,min(hd)-0.1,max(hd)+0.1]); xlabel('n'); ylabel('hd(n)')
set(gca,'XTickMode','manual','XTick',[0;M-1],'fontsize',10)

subplot(2,2,2); stem(n,w_ham); title('Hamming Window');
axis([-1,M,-0.1,1.1]); xlabel('n'); ylabel('w_ham(n)')
set(gca,'XTickMode','manual','XTick',[0;M-1],'fontsize',10)
set(gca,'YTickMode','manual','YTick',[0;1],'fontsize',10)

subplot(2,2,3); stem(n,h); title('Actual Impulse Response: Bandpass');
axis([-1,M,min(hd)-0.1,max(hd)+0.1]); xlabel('n'); ylabel('h(n)')
set(gca,'XTickMode','manual','XTick',[0;M-1],'fontsize',10)

subplot(2,2,4); plot(w/pi,db); title('Magnitude Response in dB');
axis([0,1,-As-30,5]); xlabel('frequency in pi units'); ylabel('Decibels')
set(gca,'XTickMode','manual','XTick',[0;0.3;0.4;0.5;0.6;1])
set(gca,'XTickLabelMode','manual','XTickLabels',[' 0 ';'0.3';'0.4';'0.5';...
                                                 '0.6';' 1 '],'fontsize',10)
set(gca,'YTickMode','manual','YTick',[-50;0])
set(gca,'YTickLabelMode','manual','YTickLabels',['-50';' 0 ']);grid
```

The filter-design plots are shown in Figure F.3.

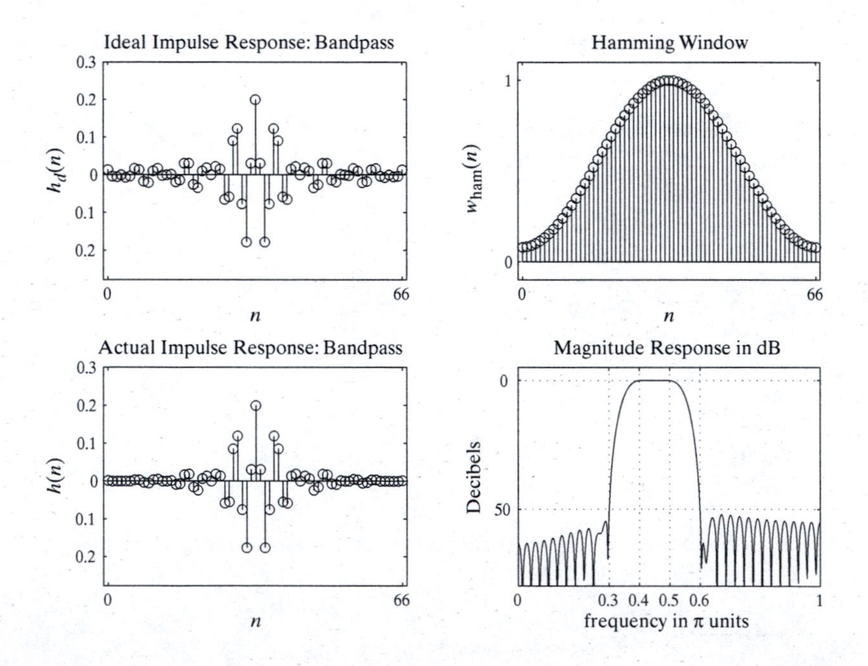

Figure F.3: Bandpass-filter design using Hamming window in Problem P7.6

P7.7 The highpass-filter specifications are

stopband edge:	0.4π
passband edge:	0.6π
passband ripple:	0.5 dB
stopband attenuation:	60 dB

Kaiser window design via MATLAB:

```
clear; close all;
%% Specifications:
ws = 0.4*pi;  % stopband edge
wp = 0.6*pi;  % passband edge
Rp = 0.5;     % passband ripple
As = 60;      % stopband attenuation
%
% Select the min(delta1,delta2) since delta1=delta2 in window design
[delta1,delta2] = db2delta(Rp,As);
if (delta1 < delta2)
   delta2 = delta1; disp('Delta1 is smaller than delta2')
   [Rp,As] = delta2db(delta1,delta2)
end
%
tr_width = wp-ws; M = ceil((As-7.95)/(14.36*tr_width/(2*pi))+1)+1;
M = 2*floor(M/2)+3, % choose odd M, Increased order by 2 to get Asd>60
M =
    41
n = [0:1:M-1]; beta = 0.1102*(As-8.7);
w_kai = (kaiser(M,beta))';                         % Kaiser Window
wc = (ws+wp)/2; hd = ideal_lp(pi,M)-ideal_lp(wc,M); % Ideal HP Filter
h = hd .* w_kai;                                   % Window design
[db,mag,pha,grd,w] = freqz_m(h,1);
delta_w = pi/500;
Asd = -floor(max(db(1:1:(ws/delta_w)+1))),         % Actual Attn
Asd =
    61
Rpd = -min(db((wp/delta_w)+1:1:501)),              % Actual passband ripple
Rpd =
    0.0148
%
%% Filter Response Plots
Hf_1 = figure('Units','normalized','position',[0.1,0.1,0.8,0.8],'color',[0,0,0]);
set(Hf_1,'NumberTitle','off','Name','P7.9');

subplot(2,2,1); stem(n,hd); title('Ideal Impulse Response');
axis([-1,M,min(hd)-0.1,max(hd)+0.1]); xlabel('n'); ylabel('hd(n)')
set(gca,'XTickMode','manual','XTick',[0;M-1],'fontsize',10)

subplot(2,2,2); stem(n,w_kai); title('Kaiser Window');
axis([-1,M,-0.1,1.1]); xlabel('n'); ylabel('w_kai(n)')
```

```
set(gca,'XTickMode','manual','XTick',[0;M-1],'fontsize',10)
set(gca,'YTickMode','manual','YTick',[0;1],'fontsize',10)

subplot(2,2,3); stem(n,h); title('Actual Impulse Response');
axis([-1,M,min(hd)-0.1,max(hd)+0.1]); xlabel('n'); ylabel('h(n)')
set(gca,'XTickMode','manual','XTick',[0;M-1],'fontsize',10)

subplot(2,2,4); plot(w/pi,db); title('Magnitude Response in dB');
axis([0,1,-As-30,5]); xlabel('frequency in pi units'); ylabel('Decibels')
set(gca,'XTickMode','manual','XTick',[0;0.4;0.6;1])
set(gca,'XTickLabelMode','manual','XTickLabels',[' 0 ';'0.4';'0.6';' 1 '],...
'fontsize',10)
set(gca,'YTickMode','manual','YTick',[-60;0])
set(gca,'YTickLabelMode','manual','YTickLabels',[' 60';' 0 ']);grid

% Super Title
suptitle('Problem~P7.7: Highpass Filter Design');
```

The filter-design plots are shown in Figure F.4.

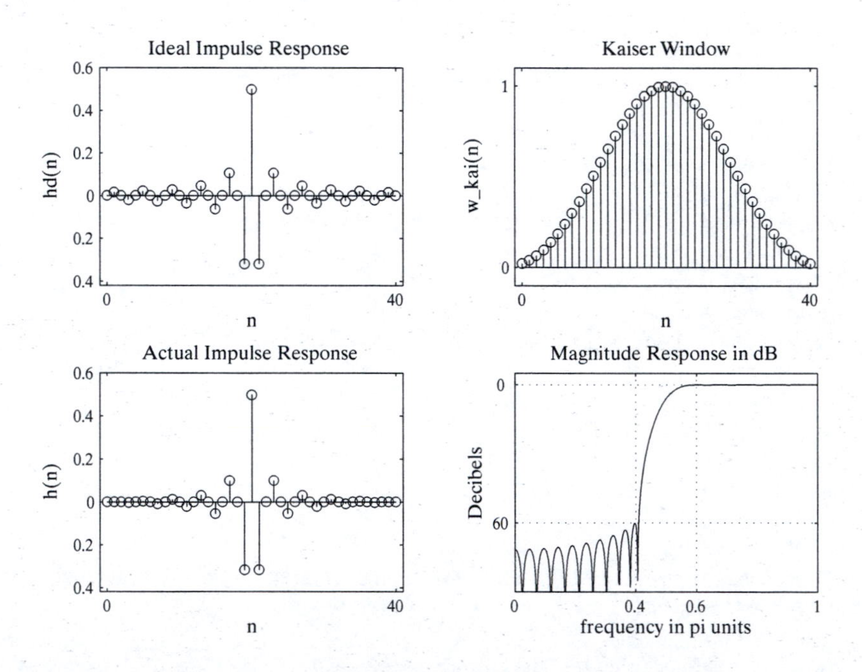

Figure F.4: Highpass-filter design using Kaiser window in Problem P7.7

P7.8 The bandpass-filter specifications are

lower stopband edge:	0.25π
lower passband edge:	0.35π
upper passband edge:	0.65π
upper stopband edge:	0.75π
passband tolerance:	0.05
stopband tolerance:	0.01

Kaiser window design via MATLAB:

```
clear; close all;
%% Specifications:
ws1 = 0.25*pi; % lower stopband edge
wp1 = 0.35*pi; % lower passband edge
wp2 = 0.65*pi; % upper passband edge
ws2 = 0.75*pi; % upper stopband edge
delta1 = 0.05; % passband ripple magnitude
delta2 = 0.01; % stopband attenuation magnitude
%

%% Determination of Rp and As in dB
[Rp,As] = delta2db(delta1,delta2)
Rp =
    0.8693
As =
   40.4238
%
%% Determination of Window Parameters
tr_width = min((wp1-ws1),(ws2-wp2));
M = ceil((As-7.95)/(14.36*tr_width/(2*pi))+1)+1;
M = 2*floor(M/2)+1, % Odd filter length
M =
    49
n=[0:1:M-1];
if As >= 50
        beta = 0.1102*(As-8.7);
elseif (As < 50) & (As > 21)
        beta = 0.5842*(As-21)^(0.4) + 0.07886*(As-21);
else
        error('As must be greater than 21')
end
w_kai = (kaiser(M,beta))';
wc1 = (ws1+wp1)/2; wc2 = (ws2+wp2)/2;
hd = ideal_lp(wc2,M)-ideal_lp(wc1,M);
h = hd .* w_kai;
[db,mag,pha,grd,w] = freqz_m(h,1);
delta_w = pi/500;
Asd = floor(-max(db([1:floor(ws1/delta_w)+1]))),          % Actual Attn
Asd =
```

```
    42
Rpd = -min(db(ceil(wp1/delta_w)+1:floor(wp2/delta_w)+1)), % Actual passband ripple
Rpd =
    0.1097
%
%% Filter Response Plots
Hf_1 = figure('Units','normalized','position',[0.1,0.1,0.8,0.8],'color',[0,0,0]);
set(Hf_1,'NumberTitle','off','Name','P7.10');

subplot(2,2,1); stem(n,hd); title('Ideal Impulse Response: Bandpass');
axis([-1,M,min(hd)-0.1,max(hd)+0.1]); xlabel('n'); ylabel('hd(n)')
set(gca,'XTickMode','manual','XTick',[0;M-1],'fontsize',10)

subplot(2,2,2); stem(n,w_kai); title('Kaiser Window');
axis([-1,M,-0.1,1.1]); xlabel('n'); ylabel('w_ham(n)')
set(gca,'XTickMode','manual','XTick',[0;M-1],'fontsize',10)
set(gca,'YTickMode','manual','YTick',[0;1],'fontsize',10)

subplot(2,2,3); stem(n,h); title('Actual Impulse Response: Bandpass');
axis([-1,M,min(hd)-0.1,max(hd)+0.1]); xlabel('n'); ylabel('h(n)')
set(gca,'XTickMode','manual','XTick',[0;M-1],'fontsize',10)

subplot(2,2,4); plot(w/pi,db); title('Magnitude Response in dB');
axis([0,1,-As-30,5]); xlabel('frequency in pi units'); ylabel('Decibels')
set(gca,'XTickMode','manual','XTick',[0;0.25;0.35;0.65;0.75;1])
set(gca,'XTickLabelMode','manual','XTickLabels',...
['  0 ';'0.25';'0.35';'0.65';'0.75';'  1 '], 'fontsize',10)
set(gca,'YTickMode','manual','YTick',[-40;0])
set(gca,'YTickLabelMode','manual','YTickLabels',['-40';' 0 ']);grid
```

The filter-design plots are shown in Figure F.5.

P7.9 The kai_hpf function:

```
function [h,M] = kai_hpf(ws,wp,As);
% [h,M] = kai_hpf(ws,wp,As);
% HighPass FIR filter design using Kaiser window
%
%  h = Impulse response of the designed filter
%  M = Length of h which is an odd number
% ws = Stopband edge in radians (0 < wp < ws < pi)
% wp = Passband edge in radians (0 < wp < ws < pi)
% As = Stopband attenuation in dB (As > 0)

if ws <= 0
    error('Stopband edge must be larger than 0')
end
if wp >= pi
    error('Passband edge must be smaller than pi')
end
if wp <= ws
```

```
    error('Passband edge must be larger than Stopband edge')
end

% Select the min(delta1,delta2) since delta1=delta2 in window design
tr_width = wp-ws; M = ceil((As-7.95)/(14.36*tr_width/(2*pi))+1)+1;
M = 2*floor(M/2)+1;                                  % choose odd M
if M > 255
    error('M is larger than 255')
end
n = [0:1:M-1]; beta = 0.1102*(As-8.7);
w_kai = (kaiser(M,beta))';                           % Kaiser Window
wc = (ws+wp)/2; hd = ideal_lp(pi,M)-ideal_lp(wc,M); % Ideal HP Filter
h = hd .* w_kai;                                     % Window design
```

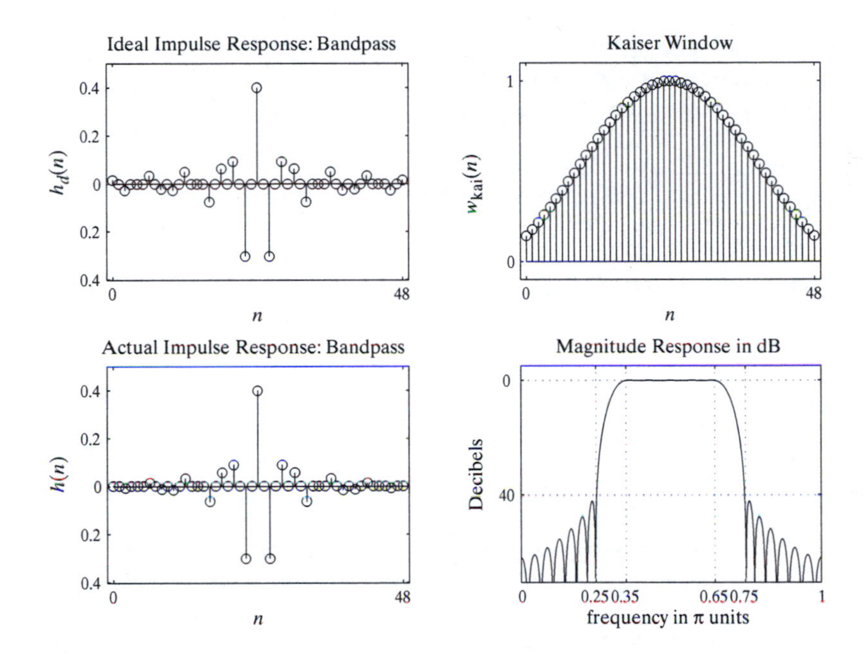

Figure F.5: Bandpass-filter design using Kaiser window in Problem P7.8

MATLAB verification:

```
clear; close all;
%% Specifications:
ws = 0.4*pi;  % stopband edge
wp = 0.6*pi;  % passband edge
As = 60;      % stopband attenuation
%
[h,M] = kai_hpf(ws,wp,As); n = 0:M-1;
[db,mag,pha,grd,w] = freqz_m(h,1);
delta_w = pi/500;
Asd = -floor(max(db(1:1:(ws/delta_w)+1))),          % Actual Attn
```

```
Asd =
    60
Rpd = -min(db((wp/delta_w)+1:1:501)),              % Actual passband ripple
Rpd =
    0.0147
%
%% Filter Response Plots
Hf_1 = figure('Units','normalized','position',[0.1,0.1,0.8,0.8],'color',[0,0,0]);
set(Hf_1,'NumberTitle','off','Name','P7.11b');

subplot(2,1,1); stem(n,h); title('Actual Impulse Response');
axis([-1,M,min(h)-0.1,max(h)+0.1]); xlabel('n'); ylabel('h(n)')
set(gca,'XTickMode','manual','XTick',[0;M-1],'fontsize',10)

subplot(2,1,2); plot(w/pi,db); title('Magnitude Response in dB');
axis([0,1,-As-30,5]); xlabel('frequency in pi units'); ylabel('Decibels')
set(gca,'XTickMode','manual','XTick',[0;0.4;0.6;1])
set(gca,'XTickLabelMode','manual','XTickLabels',[' 0 ';'0.4';'0.6';' 1 '],...
'fontsize',10)
set(gca,'YTickMode','manual','YTick',[-60;0])
set(gca,'YTickLabelMode','manual','YTickLabels',[' 60';' 0 ']);grid

% Super Title
suptitle('Problem~P7.9: Highpass Filter Design');
```

The filter-design plots are shown in Figure F.6.

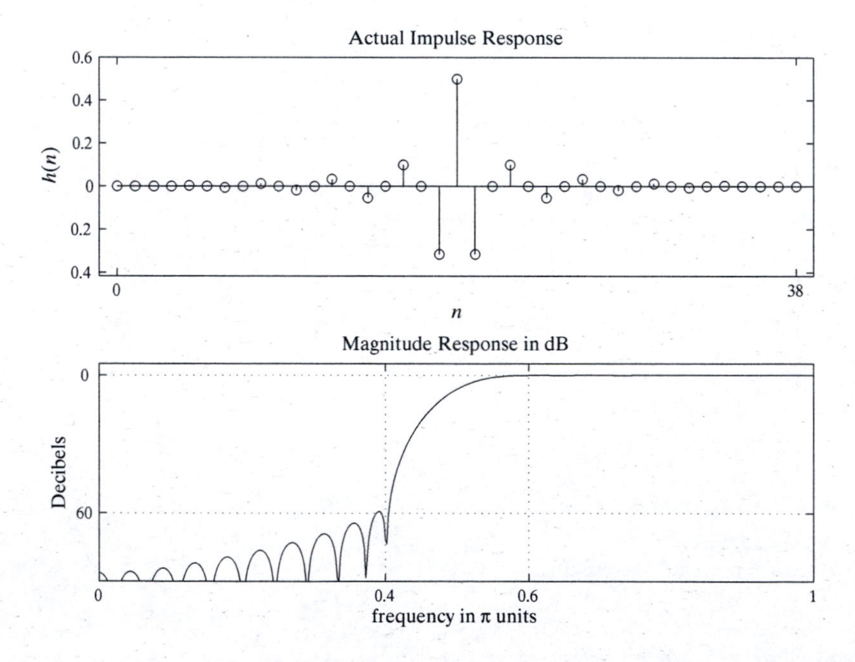

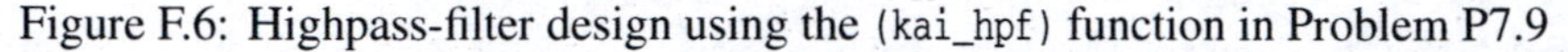
Figure F.6: Highpass-filter design using the (kai_hpf) function in Problem P7.9

P7.10 The staircase-filter specifications are

Band-1:	0	$\le$	ω	$\le$	0.3π	Ideal Gain $= 1$	δ_1	$=$	0.010
Band-2:	0.4π	$\le$	ω	$\le$	0.7π	Ideal Gain $= 0.5$	δ_2	$=$	0.005
Band-3:	0.8π	$\le$	ω	$\le$	π	Ideal Gain $= 0$	δ_3	$=$	0.001

Blackman window design via MATLAB:

```
clear; close all;
%% Specifications:
w1L = 0.0*pi; w1U = 0.3*pi; delta1 = 0.010; % Band-1 Specs
w2L = 0.4*pi; w2U = 0.7*pi; delta2 = 0.005; % Band-2 Specs
w3L = 0.8*pi; w3U = 1.0*pi; delta3 = 0.001; % Band-3 Specs
%% Determination of Rp and As in dB
As = -20*log10(delta3)
As =
   60.0000
%
%% Determination of Window Parameters
tr_width = min((w2L-w1U),(w3L-w2U));
M = ceil(11*pi/tr_width); M = 2*floor(M/2)+1, % choose odd M
M =
   111
n=[0:1:M-1];
w_bla = (blackman(M))';
wc1 = (w1U+w2L)/2; wc2 = (w2U+w3L)/2;
hd = 0.5*ideal_lp(wc1,M) + 0.5*ideal_lp(wc2,M);
h = hd .* w_bla;
[db,mag,pha,grd,w] = freqz_m(h,1);
delta_w = pi/500;
Asd = floor(-max(db([floor(w3L/delta_w)+1:501]))),          % Actual Attn
Asd =
    79
%
%% Filter Response Plots
Hf_1 = figure('Units','normalized','position',[0.1,0.1,0.8,0.8],'color',[0,0,0]);
set(Hf_1,'NumberTitle','off','Name','P7.12');

subplot(2,2,1); stem(n,w_bla); title('Blackman Window');
axis([-1,M,-0.1,1.1]); xlabel('n'); ylabel('w_ham(n)')
set(gca,'XTickMode','manual','XTick',[0;M-1],'fontsize',10)
set(gca,'YTickMode','manual','YTick',[0;1],'fontsize',10)

subplot(2,2,2); stem(n,h); title('Actual Impulse Response');
axis([-1,M,min(hd)-0.1,max(hd)+0.1]); xlabel('n'); ylabel('h(n)')
set(gca,'XTickMode','manual','XTick',[0;M-1],'fontsize',10)

subplot(2,2,3); plot(w/pi,mag); title('Magnitude Response');
axis([0,1,0,1]); xlabel('frequency in pi units'); ylabel('|H|')
set(gca,'XTickMode','manual','XTick',[0;0.3;0.4;0.7;0.8;1])
set(gca,'XTickLabelMode','manual','XTickLabels',['  0';'0.3';'0.4';'0.7';'0.8';'  1'],...
```

```
'fontsize',10)
set(gca,'YTickMode','manual','YTick',[0;0.5;1])
set(gca,'YTickLabelMode','manual','YTickLabels',[' 0 ';'0.5';' 1 ']);grid

subplot(2,2,4); plot(w/pi,db); title('Magnitude Response in dB');
axis([0,1,-As-30,5]); xlabel('frequency in pi units'); ylabel('Decibels')
set(gca,'XTickMode','manual','XTick',[0;0.3;0.4;0.7;0.8;1])
set(gca,'XTickLabelMode','manual','XTickLabels',['  0';'0.3';'0.4';'0.7';'0.8';'  1'],...
'fontsize',10)
set(gca,'YTickMode','manual','YTick',[-60;0])
set(gca,'YTickLabelMode','manual','YTickLabels',[' 60';' 0 ']);grid

% Super Title
suptitle('Problem~P7.10: Staircase Filter');
```

The filter-design plots are shown in Figure F.7.

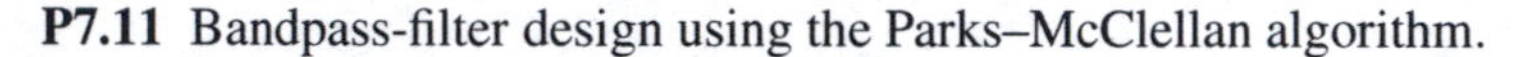

P7.11 Bandpass-filter design using the Parks–McClellan algorithm.

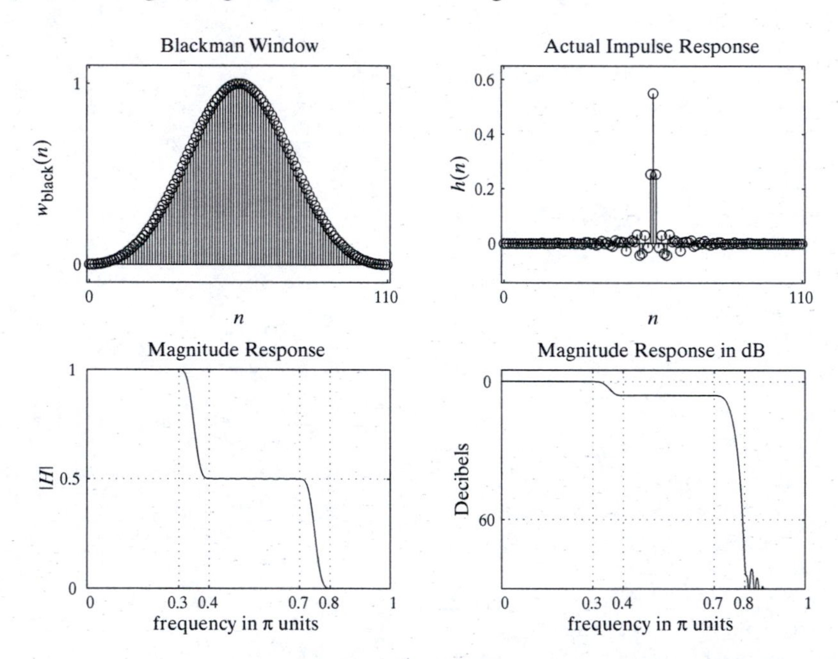

Figure F.7: Staircase-filter design using Blackman window in Problem P7.10

(a) We are given $M = 25$, $A_s = 50$ dB, and the ideal cutoff frequencies, $\omega_{c_1} = \pi/3$ and $\omega_{c_2} = 2\pi/3$, so we will have to compute the band-edge frequencies, ω_{s_1}, ω_{p_1}, ω_{p_2}, and ω_{s_2}, to implement the Parks–McClellan algorithm. Using the formula due to Kaiser, we can calculate the transition width as follows:

$$M - 1 \simeq \frac{A_s - 13}{14.36\,(\Delta\omega/2\pi)} \Rightarrow \Delta\omega \simeq 2\pi\,\frac{A_s - 13}{14.36\,(M-1)}$$

No additional information is given, so we will assume that the transition bandwidths are equal and that the tolerance values are also equal in each band:

$$\omega_{s_1} = \omega_{c_1} - \frac{\Delta\omega}{2},\ \omega_{p_1} = \omega_{c_1} + \frac{\Delta\omega}{2},\ \omega_{p_2} = \omega_{c_2} - \frac{\Delta\omega}{2},\ \omega_{s_2} = \omega_{c_2} + \frac{\Delta\omega}{2}$$

and

$$\delta_1 = \delta_2 = \delta_3$$

Using these values, we will run the remez algorithm and check for a stopband attenuation of 50 dB. If the actual attenuation is less than (more than)50 then we will increase (decrease)$\Delta\omega$ until the attenuation condition is met. In the following MATLAB script, the conditions were met at the computed value.

```
clear; close all;
%% Specifications
wc1 = pi/3;   % lower cutoff frequency
wc2 = 2*pi/3; % upper cutoff frequency
As = 50;      % stopband attenuation
M = 25;       % filter length
%
% (a) Design
tr_width = 2*pi*(As-13)/(14.36*(M-1)), % transition width in radians
tr_width =
    0.6746
ws1 = wc1-tr_width/2; wp1 = wc1+tr_width/2;
wp2 = wc2-tr_width/2; ws2 = wc2+tr_width/2;
f = [0,ws1/pi,wp1/pi,wp2/pi,ws2/pi,1];
m = [0,0,1,1,0,0];
n = 0:M-1;
h = remez(M-1,f,m);
[db,mag,pha,grd,w] = freqz_m(h,1);
delta_w = pi/500;
Asd = floor(-max(db([1:floor(ws1/delta_w)+1]))), % Actual Attn
Asd =
    50
Rpd = -min(db(ceil(wp1/delta_w)+1:floor(wp2/delta_w)+1)), % Actual ripple
Rpd =
    0.0518
%
%% Filter Response Plots
Hf_1 = figure('Units','normalized','position',[0.1,0.1,0.8,0.8],'color',[0,0,0]);
set(Hf_1,'NumberTitle','off','Name','P7.11a');
subplot(2,1,1); stem(n,h); title('Impulse Response: Bandpass');
axis([-1,M,min(h)-0.1,max(h)+0.1]); xlabel('n'); ylabel('h(n)')
set(gca,'XTickMode','manual','XTick',[0;12;24],'fontsize',10)

subplot(2,1,2); plot(w/pi,db); title('Magnitude Response in dB');
axis([0,1,-80,5]); xlabel('frequency in pi units'); ylabel('Decibels')
set(gca,'XTickMode','manual','XTick',f,'fontsize',10)
set(gca,'YTickMode','manual','YTick',[-50;0])
set(gca,'YTickLabelMode','manual','YTickLabels',['-50';' 0 '],'fontsize',10);grid
```

The impulse-response plot is shown in Figure F.8.

(b) Amplitude response:

```
% (b) Amplitude Response Plot
Hf_2 = figure('Units','normalized','position',[0.1,0.1,0.8,0.8],'color',[0,0,0]);
set(Hf_2,'NumberTitle','off','Name','P7.11b');

[Hr,w,a,L] = Hr_type1(h);
plot(w/pi,Hr); title('Amplitude Response in Problem~11(b)')
xlabel('frequency in pi units'); ylabel('Hr(w)')
axis([0,1,-0.1,1.1])
set(gca,'XTickMode','manual','XTick',f)
```

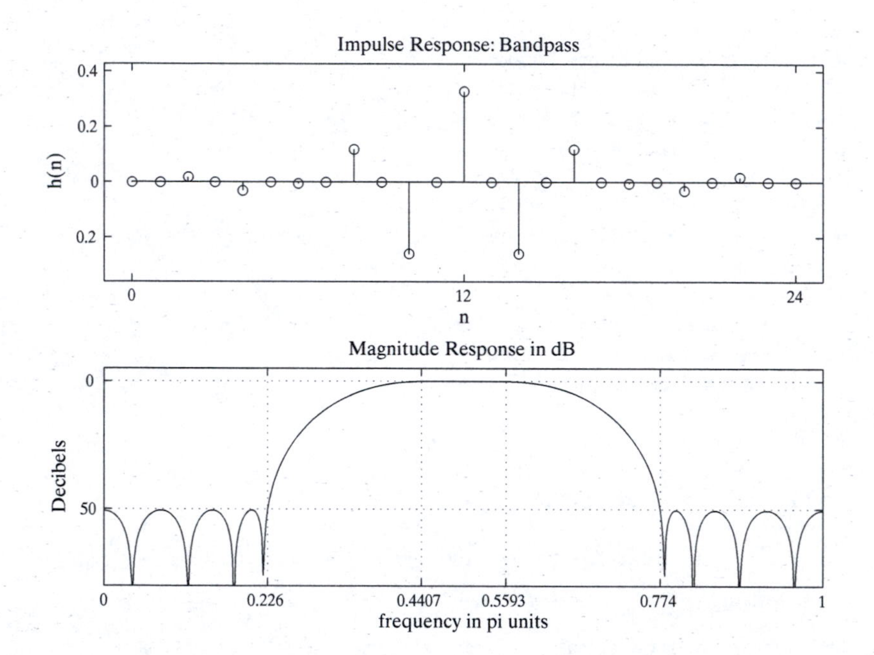

Figure F.8: Impulse-response plot in Problem P7.11(a)

The amplitude-response plot is shown in Figure F.10.

P7.12 Bandstop-filter design of Problem P7.5 by using the Parks–McClellan algorithm.

(a) MATLAB script:

```
clear; close all;
%% Specifications
wp1 = 0.3*pi; % lower passband edge
ws1 = 0.4*pi; % lower stopband edge
```

```
ws2 = 0.6*pi; % upper stopband edge
wp2 = 0.7*pi; % upper passband edge
Rp = 0.5;     % passband ripple
As = 40;      % stopband attenuation
%
% (a) Design
delta1 = (10^(Rp/20)-1)/(10^(Rp/20)+1);
delta2 = (1+delta1)*(10^(-As/20));
weights = [delta2/delta1, 1, delta2/delta1];
delta_f =min((wp2-ws2)/(2*pi), (ws1-wp1)/(2*pi));
M = ceil((-20*log10(sqrt(delta1*delta2))-13)/(14.6*delta_f)+1);
M = 2*floor(M/2)+1
M =
    33
f = [0, wp1/pi, ws1/pi, ws2/pi, wp2/pi, 1];
m = [1 1 0 0 1 1];
h = remez(M-1,f,m,weights);
[db,mag,pha,grd,w] = freqz_m(h,[1]);
delta_w = pi/500;
Asd = floor(-max(db([floor(ws1/delta_w)+1:floor(ws2/delta_w)]))), % Actual Attn
Asd =
    40
M = M+2
M =
    35
h = remez(M-1,f,m,weights);
[db,mag,pha,grd,w] = freqz_m(h,[1]);
delta_w = pi/500;
Asd = floor(-max(db([floor(ws1/delta_w)+1:floor(ws2/delta_w)]))), % Actual Attn
Asd =
    40
n = 0:M-1;
%
%%Filter Response Plots
Hf_1 = figure('Units','normalized','position',[0.1,0.1,0.8,0.8],'color',[0,0,0]);
set(Hf_1,'NumberTitle','off','Name','P7.12a');

subplot(2,1,1); stem(n,h); title('Impulse Response: Bandpass');
axis([-1,M,min(h)-0.1,max(h)+0.1]); xlabel('n'); ylabel('h(n)')
set(gca,'XTickMode','manual','XTick',[0;17;34],'fontsize',10)

subplot(2,1,2); plot(w/pi,db); title('Magnitude Response in dB');
axis([0,1,-60,5]); xlabel('frequency in pi units'); ylabel('Decibels')
set(gca,'XTickMode','manual','XTick',f,'fontsize',10)
set(gca,'YTickMode','manual','YTick',[-40;0])
set(gca,'YTickLabelMode','manual','YTickLabels',[' 40';' 0 '],'fontsize',10);grid

% Super Title
suptitle('Problem~P7.12(a): Bandstop Filter');
```

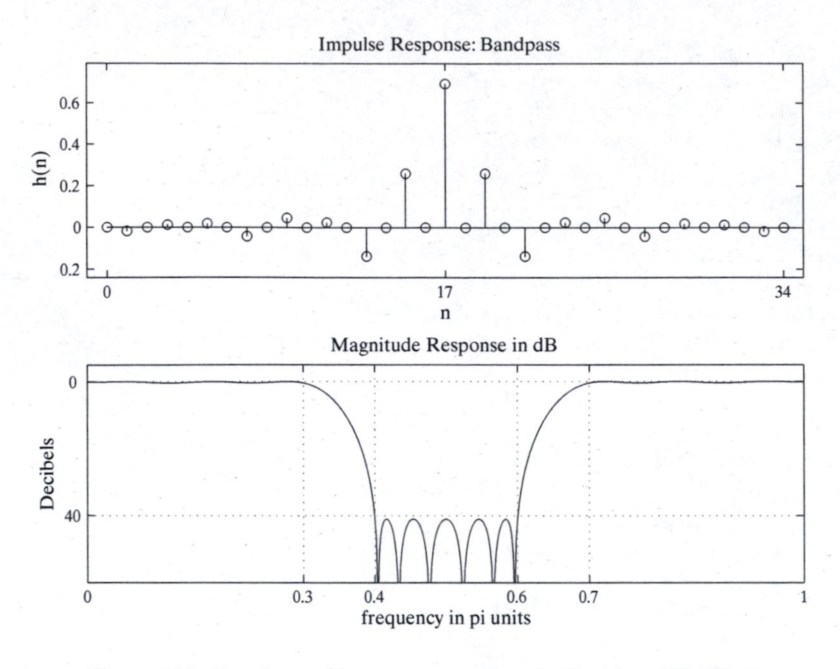

Figure F.9: Bandstop-filter response plots in Problem P7.12(a)

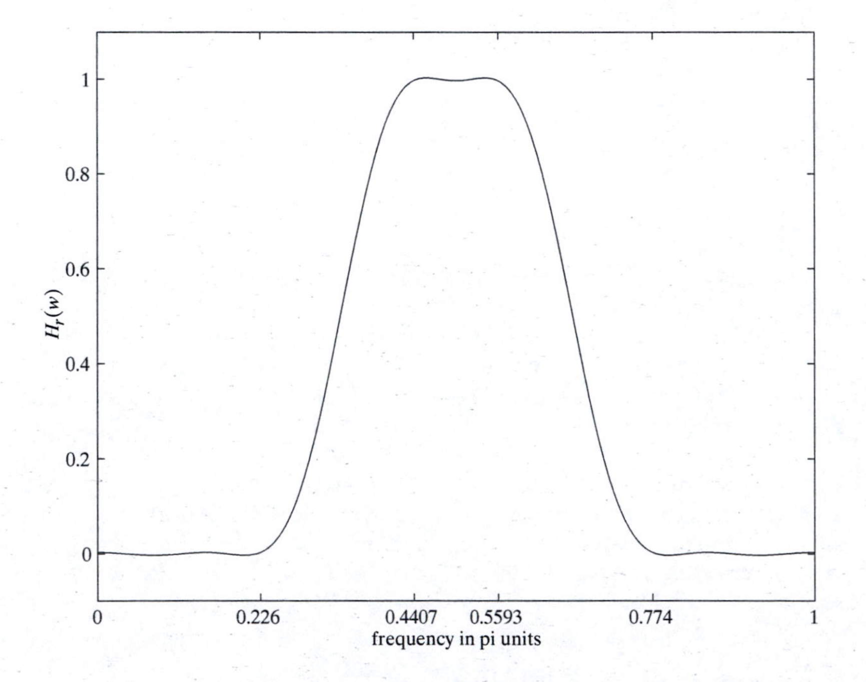

Figure F.10: Amplitude-response plot in Problem P7.11(b)

The filter-response plots are shown in Figure F.9.

(b) Amplitude response:

```
% (b) Amplitude Response Plot
Hf_2 = figure('Units','normalized','position',[0.1,0.1,0.8,0.8],'color',[0,0,0]);
set(Hf_2,'NumberTitle','off','Name','P7.12b');

[Hr,w,a,L] = Hr_type1(h);
plot(w/pi,Hr); title('Amplitude Response in Problem~P7.12(b)')
xlabel('frequency in pi units'); ylabel('Hr(w)')
axis([0,1,-0.1,1.1])
set(gca,'XTickMode','manual','XTick',f)
```

The amplitude-response plot is shown in Figure F.11.

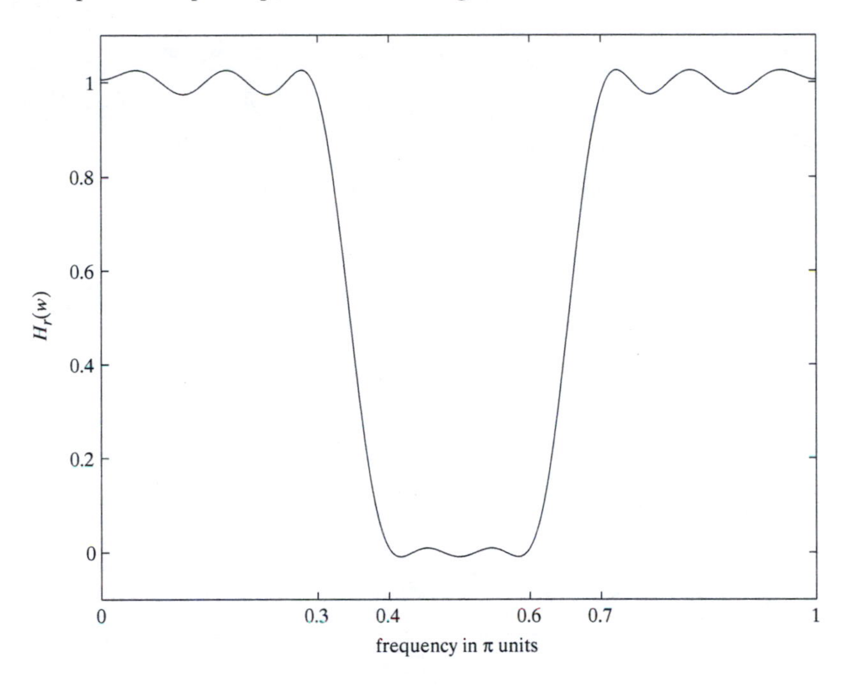

Figure F.11: Amplitude-response plot in Problem P7.12(b)

P7.13 Design of a 25-tap differentiator with unit slope by using the Parks–McClellan algorithm.

(a) MATLAB script:

```
clear; close all;
% Specifications
M = 25; w1 = 0.1*pi; w2 = 0.9*pi; % slope = 1 sam/cycle
%
% (a) Design
f = [w1/pi w2/pi]; m = [w1/(2*pi) w2/(2*pi)];
h = remez(M-1,f,m,'differentiator');
```

```
[db,mag,pha,grd,w] = freqz_m(h,1);
Hf_1 = figure('Units','normalized','position',[0.1,0.1,0.8,0.8],'color',[0,0,0]);
set(Hf_1,'NumberTitle','off','Name','P7.13a');
subplot(2,1,1);stem([0:1:M-1],h);title('Impulse Response'); axis([-1,25,-0.2,.2]);
xlabel('n','fontsize',10); ylabel('h(n)','fontsize',10);
set(gca,'XTickMode','manual','XTick',[0;12;24],'fontsize',10);
subplot(2,1,2);plot(w/(2*pi),mag);title('Magnitude Response');grid;
axis([0,0.5,0,0.5]);
xlabel('Normalized frequency in cycles/sam','fontsize',10)
set(gca,'XTickMode','manual','XTick',[0;w1/(2*pi);w2/(2*pi);0.5],'fontsize',10);
set(gca,'YTickMode','manual','YTick',[0;0.05;0.45;0.5],'fontsize',10);
```

The differentiator design plots are shown in Figure F.12, from which we observe that this filter provides a slope equal to 1 sam/cycle or $\pi/2$ sam/rad.

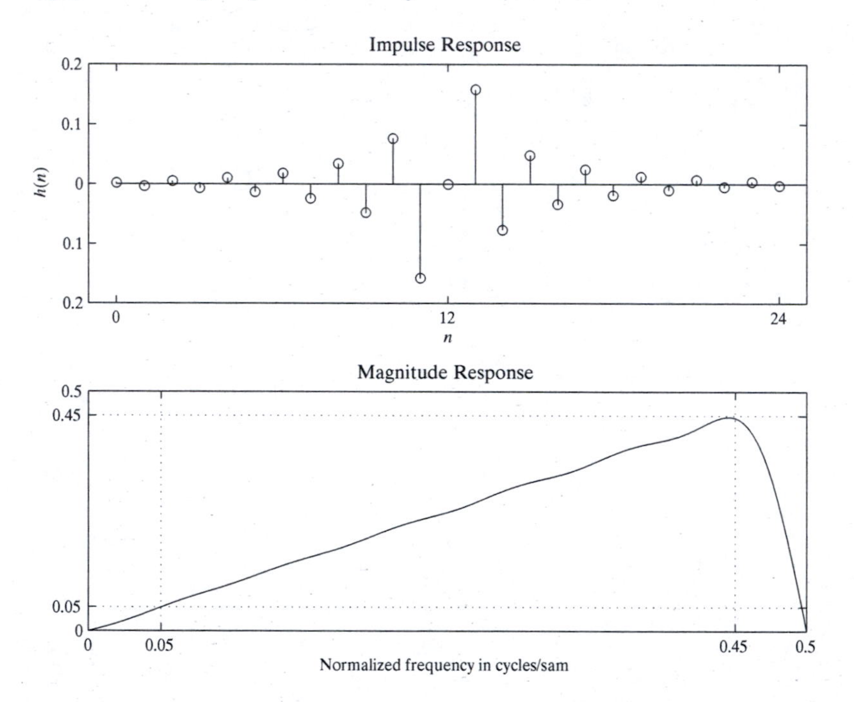

Figure F.12: Digital-differentiator design using P–M algorithm in Problem P7.13(a).

(b) Output when $x(n) = 3\sin(0.25\pi n)$, $0 \le n \le 100$: First find the sign of $h(n)$ and then appropriately convolve with $x(n)$.

```
% (b) Differentiator verification
Hf_2 = figure('Units','normalized','position',[0.1,0.1,0.8,0.8],'color',[0,0,0]);
set(Hf_2,'NumberTitle','off','Name','P7.13b');
[Hr,w,P,L] = ampl_res(h); subplot; plot(w/(2*pi), Hr);
*** Type-3 Linear-Phase Filter ***
title('Amplitude Response'); grid; axis([0,0.5,-0.5,0]);
set(gca,'XTickMode','manual','XTick',[0;w1/(2*pi);w2/(2*pi);0.5],'fontsize',10);
set(gca,'YTickMode','manual','YTick',[-0.5;-0.45;-0.05;0],'fontsize',10);
```

The amplitude-response plot is shown in Figure F.13. The sign of $h(n)$ from Figure F.13 is negative; hence, we will convolve $x(n)$ with $-h(n)$, then compare the input and output in the steady state (i.e., when $n > 25$) with output shifted by 12 to the left to account for the phase delay of the differentiator.

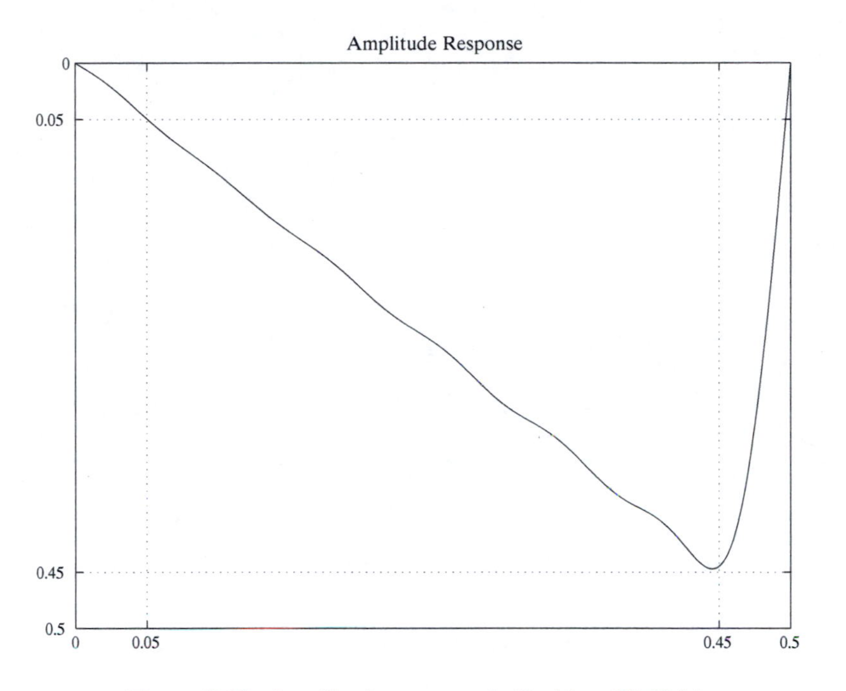

Figure F.13: Amplitude response in Problem P7.13(b).

```
Hf_3 = figure('Units','normalized','position',[0.1,0.1,0.8,0.8],'color',[0,0,0]);
set(Hf_3,'NumberTitle','off','Name','P7.13c');
n=[0:1:100]; x = 3*sin(0.25*pi*n); y = conv(x,-h);
m = [41:1:81];
plot(m,x(41:1:81),m,y(41+12:1:81+12));grid % add 12 sample delay to y
xlabel('n'); title('Input-Output Sequences'); axis([40,82,-4,4]);
set(gca,'XTickMode','manual','XTick',[41;81],'fontsize',10);
set(gca,'YTickMode','manual','YTick',[-3;0;3],'fontsize',10);
```

The input–output plots are shown in Figure F.14. Because the slope is $\pi/2$ sam/rad, the gain at $\omega = 0.25\pi$ is equal to 0.125. Therefore, the output (when properly shifted) should be

$$y(n) = 3\,(0.125)\cos\left(0.25\pi n\right) = 0.375\cos\left(0.25\pi n\right)$$

From Figure F.14 we can verify that $y(n)$(the lower curve) is indeed a cosine waveform with amplitude ≈ 0.4.

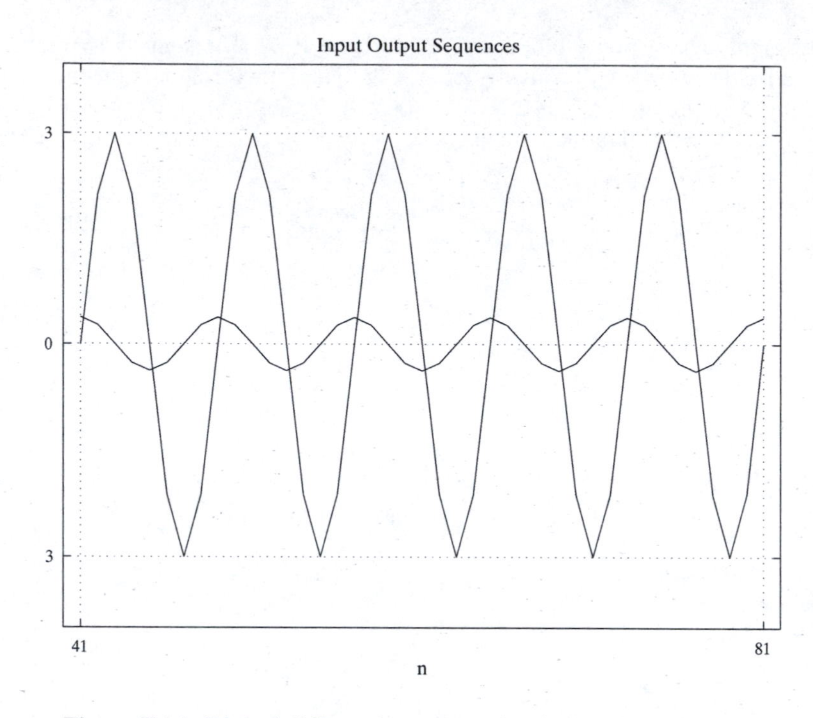

Figure F.14: Digital-differentiator operation in Problem P7.13(b).

P7.14 Multiband-filter design by using Parks–McClellan algorithm.

The filter specifications are

	Band-1	**Band-2**	**Band-3**
Lower band-edge	0	0.5	0.8
Upper band-edge	0.4	0.7	1
Ideal value	0.4	0.95	0.025
Tolerance	0.05	0.05	0.025

MATLAB Script:

```
clear; close all;
%% Specifications
f = [0,0.4,0.5,0.7,0.8,1];
m = [0.4,0.4,0.95,0.95,0.025,0.025];
delta1 = 0.05; delta2 = 0.05; delta3 = 0.025;
weights = [delta3/delta2, delta3/delta2, delta3/delta3];
As = -20*log10(0.05)
As =
   26.0206
%
%% Design
delta_f = 0.05; % Transition width in cycles per sample
M = ceil((-20*log10(sqrt(delta2*delta3))-13)/(14.6*delta_f)+1)
```

```
M =
    23
h = remez(M-1,f,m,weights);
[db,mag,pha,grd,w] = freqz_m(h,1);
delta_w = pi/500;
Asd = floor(-max(db([(0.8*pi/delta_w)+1:501]))), % Actual Attn
Asd =
    24
M = M+1
M =
    24
h = remez(M-1,f,m,weights);
[db,mag,pha,grd,w] = freqz_m(h,1);
Asd = floor(-max(db([(0.8*pi/delta_w)+1:501]))), % Actual Attn
Asd =
    25
M = M+1
M =
    25
h = remez(M-1,f,m,weights);
[db,mag,pha,grd,w] = freqz_m(h,1);
Asd = floor(-max(db([(0.8*pi/delta_w)+1:501]))), % Actual Attn
Asd =
    25
M = M+1
M =
    26
h = remez(M-1,f,m,weights);
[db,mag,pha,grd,w] = freqz_m(h,1);
Asd = floor(-max(db([(0.8*pi/delta_w)+1:501]))), % Actual Attn
Asd =
    25
M = M+1
M =
    27
h = remez(M-1,f,m,weights);
[db,mag,pha,grd,w] = freqz_m(h,1);
Asd = floor(-max(db([(0.8*pi/delta_w)+1:501]))), % Actual Attn
Asd =
    26
n = 0:M-1;
%
%% Impulse Response Plot
Hf_1 = figure('Units','normalized','position',[0.1,0.1,0.8,0.8],'color',[0,0,0]);
set(Hf_1,'NumberTitle','off','Name','P7.14a');

stem(n,h); title('Impulse Response Plot in Problem~P14');
axis([-1,M,min(h)-0.1,max(h)+0.1]); xlabel('n'); ylabel('h(n)')
set(gca,'XTickMode','manual','XTick',[0;13;26],'fontsize',12)

%
```

```
%% Amplitude Response Plot
Hf_2 = figure('Units','normalized','position',[0.1,0.1,0.8,0.8],'color',[0,0,0]);
set(Hf_2,'NumberTitle','off','Name','P7.14b');

[Hr,w,a,L] = Hr_type1(h);
plot(w/pi,Hr); title('Amplitude Response in Problem~P14')
xlabel('frequency in pi units'); ylabel('Hr(w)')
axis([0,1,0,1])
set(gca,'XTickMode','manual','XTick',f)
set(gca,'YTickMode','manual','YTick',[0;0.05;0.35;0.45;0.9;1]); grid
```

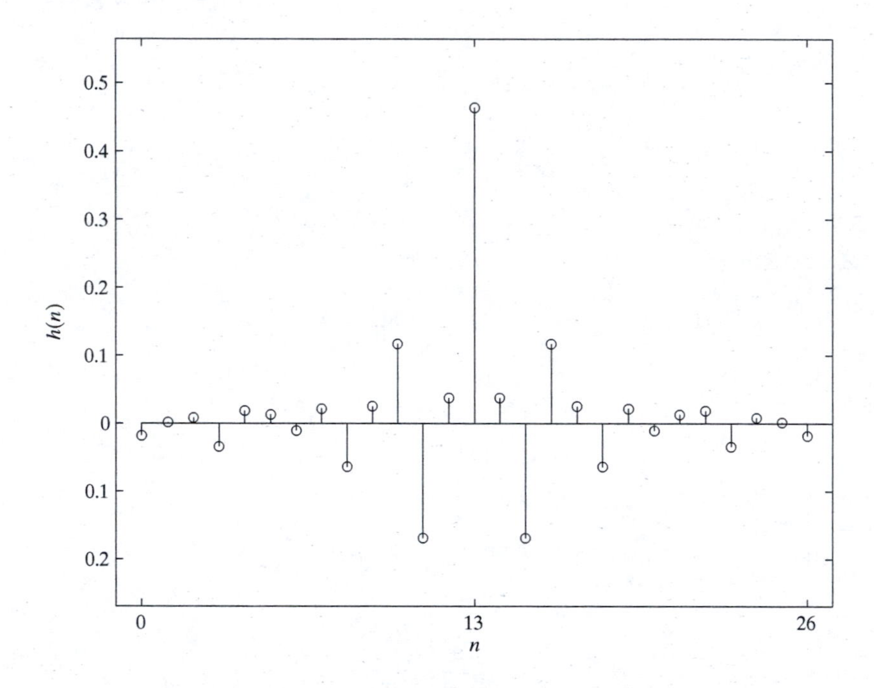

Figure F.15: Impulse-response plot in Problem P7.14

The impulse response is shown in Figure F.15, the amplitude-response plot in Figure F.16.

P7.15 Narrow-bandpass filter design by using Parks–McClellan algorithm.

We want to design a 50th-order narrowband bandpass filter to filter out a noise component having its center frequency at $\omega_c = \pi/2$, bandwidth of 0.02π, and stopband attenuation of 30 dB.

(a) In this design, we already know the order of the filter. The only parameters that we don't know are the stopband cutoff frequencies ω_{s_1} and ω_{s_2}. Let the transition bandwidth be $\Delta\omega$, and let the passband be symmetrical with respect to the center frequency ω_c. Then

$$\omega_{p_1} = \omega_c - 0.01\pi, \ \omega_{p_2} = \omega_c + 0.01\pi, \ \omega_{s_1} = \omega_{p_1} - \Delta\omega, \text{ and } \omega_{s_2} = \omega_{p_2} + \Delta\omega$$

We will also assume that the tolerance values in each band are equal. Now we will begin with an initial value for $\Delta\omega = 0.2\pi$ and run the `remez` algorithm to obtain the actual stopband

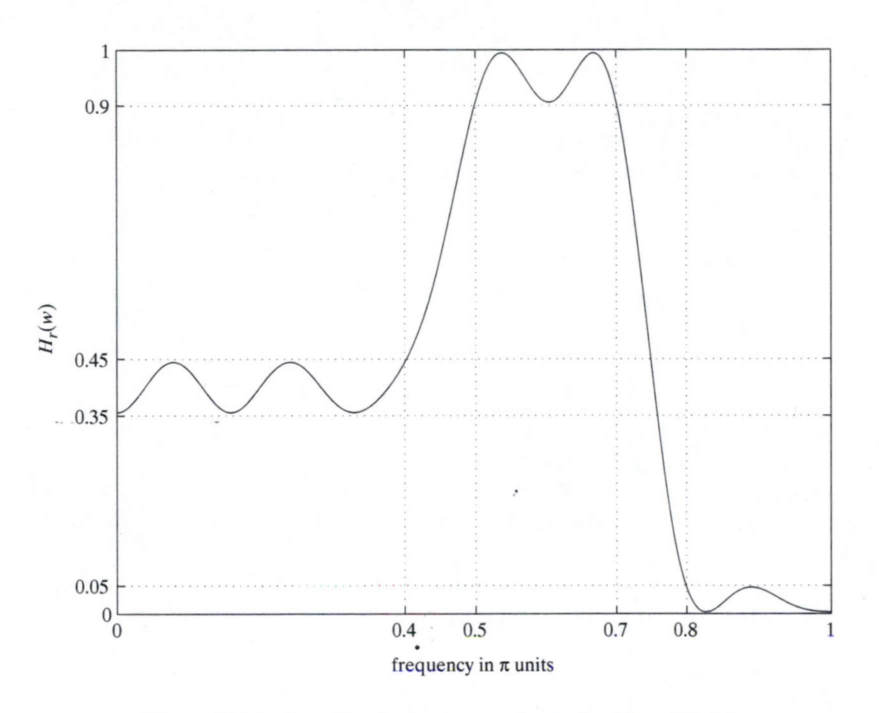

Figure F.16: Amplitude-response plot in Problem P7.14

attenuation. If it is smaller (larger) than the given 30 dB, then we will increase (decrease) $\Delta\omega$ and iterate to obtain the desired solution. The desired solution was found at $\Delta\omega = 0.5\pi$. MATLAB script:

```
clear; close all;
%% Specifications
        N = 50;                                   % Order of the filter
       w0 = 0.5*pi;                               % Center frequency
 Bandwidth = 0.02*pi;                             % Bandwidth
%
%   Deltaw = Transition bandwidth (iteration variable)
%
wp1 = w0-Bandwidth/2; wp2 = w0+Bandwidth/2;

% (a) Design
Deltaw = 0.02*pi;                                  % Initial guess
ws1=wp1-Deltaw; ws2=wp2+Deltaw;
F=[0, ws1, wp1, wp2, ws2, pi]/pi;
m=[0,0,1,1,0,0];
h=remez(50,F,m);
[db,mag,pha,grd,w]=freqz_m(h,1);
delta_w = pi/500;
Asd = floor(-max(db([1:floor(ws1/delta_w)]))), % Actual Attn
```

```
Asd =
    13
% Next iteration
Deltaw = Deltaw+0.01*pi;
ws1=wp1-Deltaw; ws2=wp2+Deltaw;
F=[0, ws1, wp1, wp2, ws2, pi]/pi;
h=remez(50,F,m);
[db,mag,pha,grd,w]=freqz_m(h,1);
delta_w = pi/500;
Asd = floor(-max(db([1:floor(ws1/delta_w)]))), % Actual Attn
Asd =
    20
% Next iteration
Deltaw = Deltaw+0.01*pi;
ws1=wp1-Deltaw; ws2=wp2+Deltaw;
F=[0, ws1, wp1, wp2, ws2, pi]/pi;
h=remez(50,F,m);
[db,mag,pha,grd,w]=freqz_m(h,1);
delta_w = pi/500;
Asd = floor(-max(db([1:floor(ws1/delta_w)]))), % Actual Attn
Asd =
    26
% Next iteration
Deltaw = Deltaw+0.01*pi;
ws1=wp1-Deltaw; ws2=wp2+Deltaw;
F=[0, ws1, wp1, wp2, ws2, pi]/pi;
h=remez(50,F,m);
[db,mag,pha,grd,w]=freqz_m(h,1);
delta_w = pi/500;
Asd = floor(-max(db([1:floor(ws1/delta_w)]))), % Actual Attn
Asd =
    30
Hf_1 = figure('Units','normalized','position',[0.1,0.1,0.8,0.8],'color',[0,0,0]);
set(Hf_1,'NumberTitle','off','Name','P7.15a');
plot(w/pi,db); axis([0,1,-50,0]); title('Log-Magnitude Response in P7.15a');
xlabel('frequency in pi units'); ylabel('DECIBELS')
set(gca,'XTickMode','manual','XTick',[0;ws1/pi;ws2/pi;1],'fontsize',10)
set(gca,'YTickMode','manual','YTick',[-30;0])
set(gca,'YTickLabelMode','manual','YTickLabels',[' 30';' 0 '],'fontsize',10);grid
```

The log-magnitude response is shown in Figure F.17.

(b) The time-domain response of the filter. MATLAB script:

```
% (b) Time-domain Response
n = [0:1:200]; x = 2*cos(pi*n/2)+randn(1,201); y = filter(h,1,x);
Hf_2 = figure('Units','normalized','position',[0.1,0.1,0.8,0.8],'color',[0,0,0]);
set(Hf_2,'NumberTitle','off','Name','P7.15b');
subplot(211);stem(n(101:201),x(101:201));title('Input sequence x(n)')
subplot(212);stem(n(101:201),y(101:201));title('Output sequence y(n)')
```

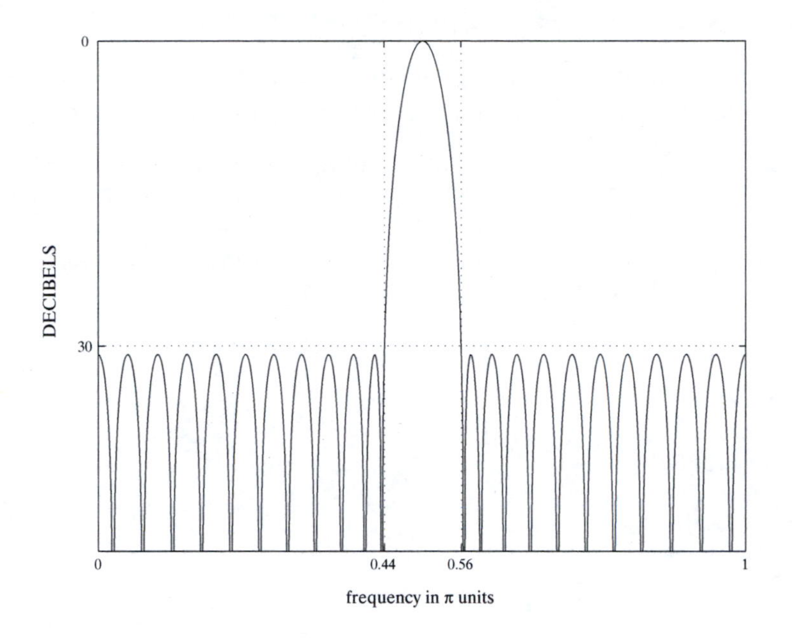

Figure F.17: Log-magnitude response plot in Problem P7.15(a)

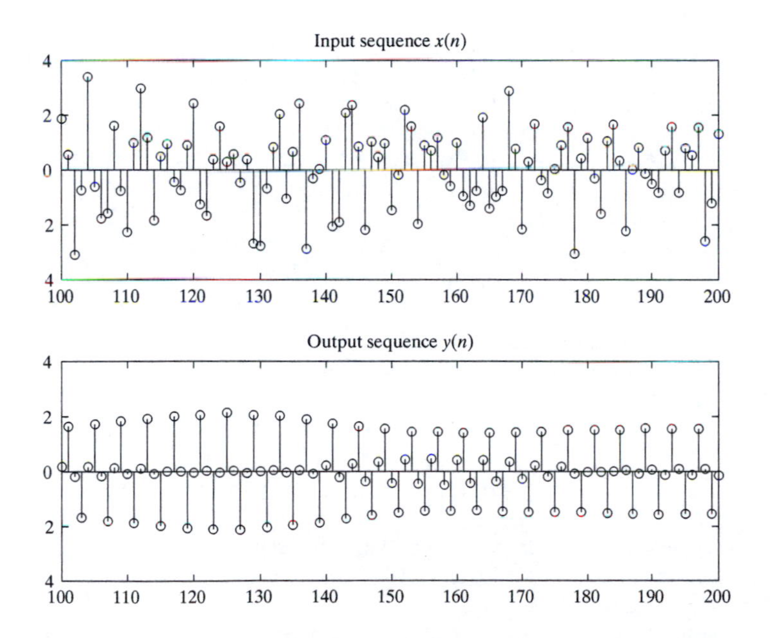

Figure F.18: The time-domain response in Problem P7.15(b)

The time-domain response is shown in Figure F.18.

P7.16 If we choose $M = 60$ then we have two samples in the transition band. Let the frequency samples in the lower transition band be T_1 and T_2. Then the samples of the amplitude response are

$$H_r(\omega) = [\underbrace{0, ..., 0}_{10}, T_1, T_2, \underbrace{1, ..., 1}_{4}, T_2, T_1, \underbrace{0, ..., 0}_{25}, T_1, T_2, \underbrace{1, ..., 1}_{4}, T_2, T_1, \underbrace{0, ..., 0}_{9}]$$

The optimum values of T_1 and T_2 for $M = 60$ and four samples in the passband are (from Table C.2 from P&M book with M even, $M = 64$, BW = 4)

$$T_1 = 0.11038818, \;\; T_2 = 0.59730067$$

The MATLAB script is:

```
clear; close all; clc;

%% Specifications:
ws1 = 0.3*pi;  % lower stopband edge
ws2 = 0.6*pi;  % upper stopband edge
wp1 = 0.4*pi;  % lower passband edge
wp2 = 0.5*pi;  % upper passband edge
 Rp = 0.5;     % passband ripple
 As = 50;      % stopband attenuation

% Design
M = 60; alpha = (M-1)/2; l = 0:M-1; wl = (2*pi/M)*l;
T1 = 0.11038818; T2 = 0.59730067 % From Table C.2 M even, M= 64, BW = 4
Hrs = [zeros(1,10),T1,T2,ones(1,4),T2,T1,zeros(1,25),T1,T2,ones(1,4),T2,T1,zeros(1,9)];
Hdr = [0,0,1,1,0,0]; wdl = [0,ws1,wp1,wp2,ws2,pi]/pi;
k1 = 0:floor((M-1)/2); k2 = floor((M-1)/2)+1:M-1;
angH = [-alpha*(2*pi)/M*k1, alpha*(2*pi)/M*(M-k2)];
H = Hrs.*exp(j*angH); h = real(ifft(H,M));
[db,mag,pha,grd,w] = freqz_m(h,1);
[Hr,ww,a,L] = Hr_Type2(h);

% Plots
Hf_P0715 = figure('Units','normalized','position',[0.05,0.05,0.9,0.85],...
   'color',[0,0,0], 'paperunits','inches','paperposition',[0,0,6,6]);
set(Hf_P0715,'NumberTitle','off','Name','P7.15');

subplot(1,1,1)
subplot(2,2,1);plot(wl(1:31)/pi,Hrs(1:31),'ro',wdl,Hdr,'g','linewidth',1.5);
axis([0,1,-0.1,1.1]);
title('Frequency Samples','fontsize',12,'fontweight','bold')
xlabel('frequency in \pi units','fontsize',12); ylabel('H_r(k)','fontsize',12)
set(gca,'XTickMode','manual','XTick',wdl)
set(gca,'YTickMode','manual','YTick',[0,0.109,0.59,1]); grid

subplot(2,2,2); plot([-1,M],[0,0],'w'); hold on;
stem(l,h,'r','filled'); axis([-1,M,-0.2,0.2])
```

```
title('Impulse Response','fontsize',12,'fontweight','bold');
xlabel('n','fontsize',12); ylabel('h(n)','fontsize',12); hold off;

subplot(2,2,3); plot(ww/pi,Hr,'r',wl(1:31)/pi,Hrs(1:31),'ro','linewidth',1.5);
axis([0,1,-0.1,1.1]);
title('Amplitude Response','fontsize',12,'fontweight','bold')
xlabel('frequency in \pi units','fontsize',12);
ylabel('H_r(e^{j\omega})','fontsize',12)
set(gca,'XTickMode','manual','XTick',wdl)
set(gca,'YTickMode','manual','YTick',[0,0.110,0.597,1]); grid

subplot(2,2,4);plot(w/pi,db,'r','linewidth',1.5); axis([0,1,-100,10]); grid
title('Magnitude Response','fontsize',12,'fontweight','bold');
xlabel('frequency in \pi units','fontsize',12);
ylabel('Decibels','fontsize',12);
set(gca,'XTickMode','manual','XTick',wdl)
set(gca,'YTickMode','Manual','YTick',[-50;0]);
set(gca,'YTickLabelMode','manual','YTickLabels',['50';' 0'])

suptitle('Bandpass Filter Design: M=60,T1=0.110, T2=0.597')
```

The plots in Figure F.19 show an acceptable bandpass filter design.

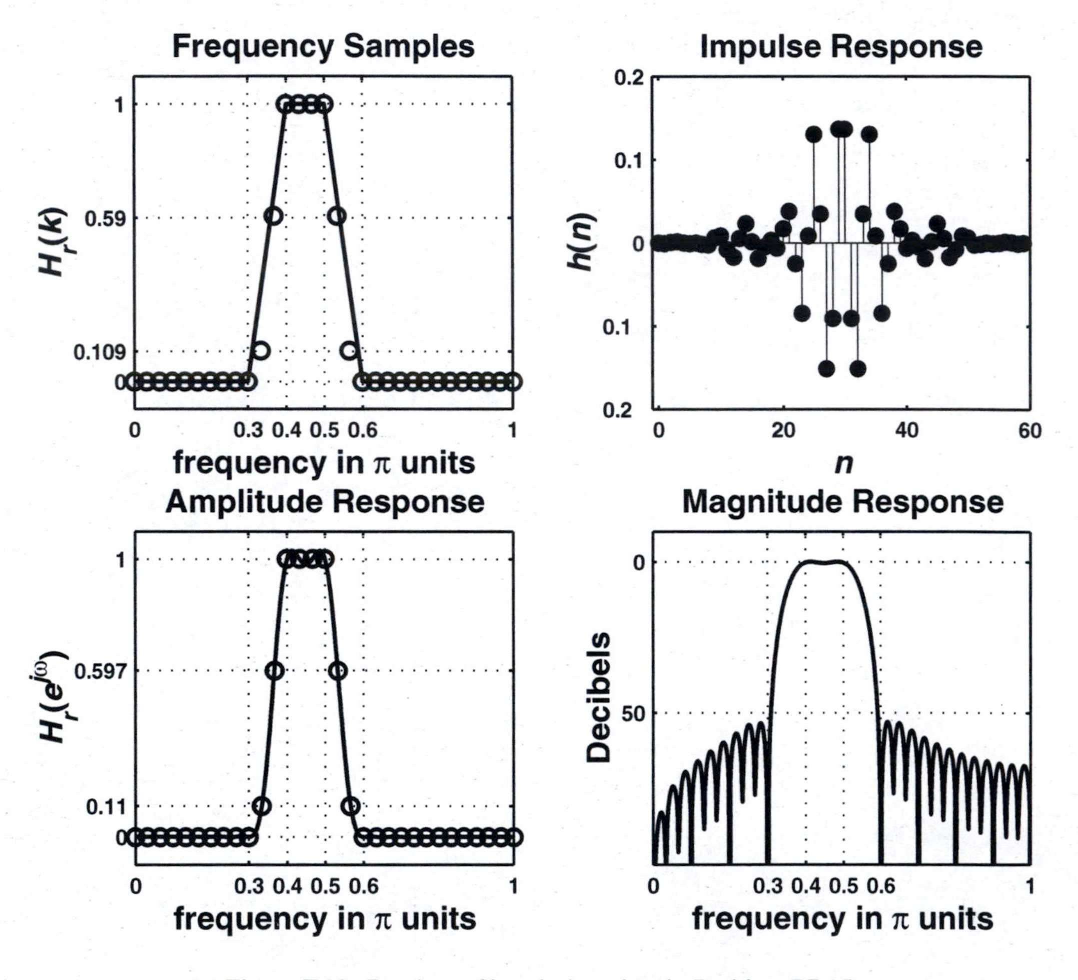

Figure F.19: Bandpass filter design plots in Problem P7.15

Appendix G

IIR Filter Design

P8.1 Analog Butterworth lowpass-filter design: $\Omega_p = 30$ rad/s, $R_p = 1$ dB, $\Omega_s = 40$ rad/s, $A_s = 30$ dB.
MATLAB script:

```
clear, close all;
% Filter Specifications
Wp = 30; Ws = 40; Rp = 1; As = 30;
% Filter Design
[b,a] = afd_butt(Wp,Ws,Rp,As); format short e

*** Butterworth Filter Order = 15
% Cascade Structure
[C,B,A] = sdir2cas(b,a)
C =
  2.8199e+022
B =
     0     0     1
A =
  1.0000e+000  6.1393e+001  9.8484e+002
  1.0000e+000  5.7338e+001  9.8484e+002
  1.0000e+000  5.0777e+001  9.8484e+002
  1.0000e+000  4.1997e+001  9.8484e+002
  1.0000e+000  3.1382e+001  9.8484e+002
  1.0000e+000  1.9395e+001  9.8484e+002
  1.0000e+000  6.5606e+000  9.8484e+002
            0  1.0000e+000  3.1382e+001
% Frequency Response
Wmax = 100; [db,mag,pha,w] = freqs_m(b,a,Wmax); pha = unwrap(pha);
% Impulse Response
%   The impulse response of the designed filter when computed by Matlab is numerically
%   unstable due to large coefficient values.  Hence we compute the impulse response
%   of the filter with Wp/10 and Ws/10 band edges to keep coefficient values small.
%   The actual impulse response is time-scaled and amplitude scaled version of the
%   computed impulse response.
[b,a] = afd_butt(Wp/10,Ws/10,Rp,As); [ha,x,t] = impulse(b,a);
```

```
*** Butterworth Filter Order = 15
t = t/10; ha = ha/10;
%
% Plots
Hf_1 = figure('Units','normalized','position',[0.1,0.1,0.8,0.8],'color',[0,0,0]);
set(Hf_1,'NumberTitle','off','Name','P8.1');
%
subplot(2,2,1);plot(w,mag); axis([0,Wmax,0,1.1]);
xlabel('Analog frequency in rad/sec','fontsize',10);
ylabel('Magnitude','fontsize',10); title ('Magnitude Response','fontsize',10);
set(gca,'XTickMode','manual','Xtick',[0;Wp;Ws;Wmax],'fontsize',10);
magRp = round(10^(-Rp/20)*100)/100;
set(gca,'YTickMode','manual','Ytick',[0;magRp;1],'fontsize',10);grid
%
subplot(2,2,2);plot(w,db); axis([0,Wmax,-100,0]);
xlabel('Analog frequency in rad/sec','fontsize',10);
ylabel('Decibels','fontsize',10); title ('Log-Magnitude Response','fontsize',10);
set(gca,'XTickMode','manual','Xtick',[0;Wp;Ws;Wmax],'fontsize',10);
set(gca,'YTickMode','manual','Ytick',[-100;-As;0],'fontsize',10);grid
AS = [' ',num2str(As)];
set(gca,'YTickLabelMode','manual','YTickLabels',['100';AS;'  0']);
%
minpha = floor(min(pha/pi)); maxpha = ceil(max(pha/pi));
subplot(2,2,3);plot(w,pha/pi); axis([0,Wmax,minpha,maxpha]);
xlabel('Analog frequency in rad/sec','fontsize',10);
ylabel('Phase in pi units','fontsize',10); title ('Phase Response','fontsize',10);
set(gca,'XTickMode','manual','Xtick',[0;Wp;Ws;Wmax],'fontsize',10);
phaWp = (round(pha(Wp/Wmax*500+1)/pi*100))/100;
phaWs = (round(pha(Ws/Wmax*500+1)/pi*100))/100;
set(gca,'YTickMode','manual','Ytick',[phaWs;phaWp;0],'fontsize',10); grid
%
subplot(2,2,4); plot(t,ha);  title ('Impulse Response','fontsize',10);
xlabel('t (sec)', 'fontsize',10); ylabel('ha(t)','fontsize',10);
```

The system function is given by

$$\begin{aligned} H_a(s) = {} & 2.8199 \times 10^{22} \left(\frac{1}{s^2 + 61.393s + 984.84}\right)\left(\frac{1}{s + 31.382}\right) \\ & \left(\frac{1}{s^2 + 57.338s + 984.84}\right)\left(\frac{1}{s^2 + 50.777s + 984.84}\right) \times \\ & \left(\frac{1}{s^2 + 41.997s + 984.84}\right)\left(\frac{1}{s^2 + 31.382s + 984.84}\right) \times \\ & \left(\frac{1}{s^2 + 19.395s + 984.84}\right)\left(\frac{1}{s^2 + 6.5606s + 984.84}\right) \end{aligned}$$

The filter-design plots are given in Figure G.1.

P8.2 Analog elliptic lowpass-filter design: $\Omega_p = 10$ rad/s, $R_p = 1$ dB, $\Omega_s = 15$ rad/s, $A_s = 40$ dB. MATLAB script:

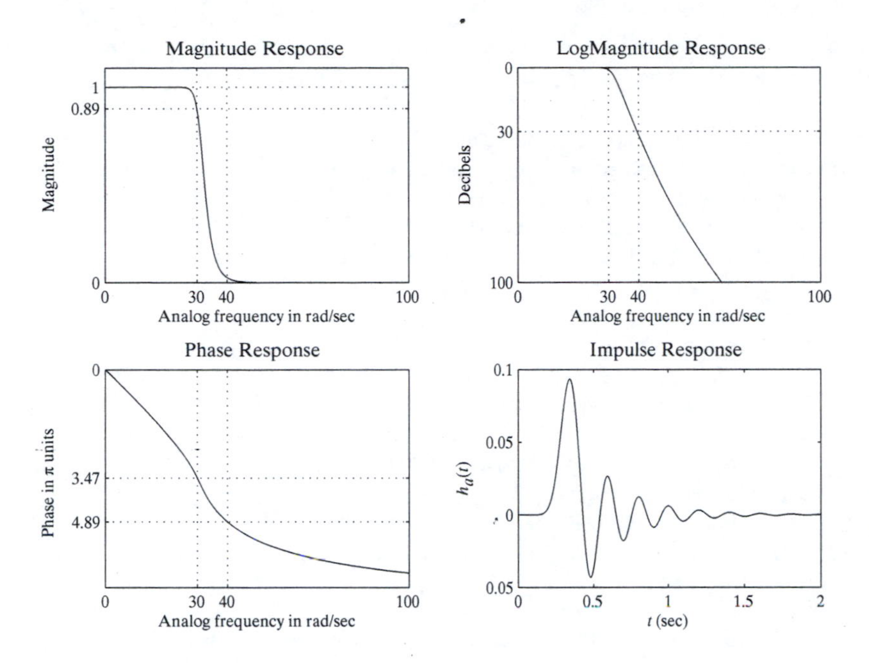

Figure G.1: Analog Butterworth lowpass-filter plots in Problem P8.1

```
clear; close all;
% Filter Specifications
Wp = 10; Ws = 15; Rp = 1; As = 40;
% Filter Design
[b,a] = afd_elip(Wp,Ws,Rp,As); format short e;

*** Elliptic Filter Order =  5
% Rational Function Form
a0 = a(1); b = b/a0, a = a/a0
b =
  4.6978e-001            0  2.2007e+002            0  2.2985e+004
a =
  1.0000e+000  9.2339e+000  1.8471e+002  1.1292e+003  7.8813e+003  2.2985e+004
% Frequency Response
Wmax = 30; [db,mag,pha,w] = freqs_m(b,a,Wmax); pha = unwrap(pha);
% Impulse Response
[ha,x,t] = impulse(b,a);
%
% Plots
Hf_1 = figure('Units','normalized','position',[0.1,0.1,0.8,0.8],'color',[0,0,0]);
set(Hf_1,'NumberTitle','off','Name','P8.2');
%
subplot(2,2,1);plot(w,mag); axis([0,Wmax,0,1]);
xlabel('Analog frequency in rad/sec','fontsize',10);
ylabel('Magnitude','fontsize',10); title ('Magnitude Response','fontsize',10);
set(gca,'XTickMode','manual','Xtick',[0;Wp;Ws;Wmax],'fontsize',10);
```

```
magRp = round(10^(-Rp/20)*100)/100;
set(gca,'YTickMode','manual','Ytick',[0;magRp;1],'fontsize',10);grid
%
subplot(2,2,2);plot(w,db); axis([0,Wmax,-100,0]);
xlabel('Analog frequency in rad/sec','fontsize',10);
ylabel('log-Magnitude in dB','fontsize',10);
title ('Log-Magnitude Response','fontsize',10);
set(gca,'XTickMode','manual','Xtick',[0;Wp;Ws;Wmax],'fontsize',10);
set(gca,'YTickMode','manual','Ytick',[-100;-As;0],'fontsize',10);grid
AS = [' ',num2str(As)];
set(gca,'YTickLabelMode','manual','YTickLabels',['100';AS;'  0'],'fontsize',10);
%
minpha = floor(min(pha/pi)); maxpha = ceil(max(pha/pi));
subplot(2,2,3);plot(w,pha/pi); axis([0,Wmax,minpha,maxpha]);
xlabel('Analog frequency in rad/sec','fontsize',10);
ylabel('Phase in pi units','fontsize',10);
title ('Phase Response','fontsize',10);
set(gca,'XTickMode','manual','Xtick',[0;Wp;Ws;Wmax],'fontsize',10);
phaWp = (round(pha(Wp/Wmax*500+1)/pi*100))/100;
phaWs = (round(pha(Ws/Wmax*500+1)/pi*100))/100;
set(gca,'YTickMode','manual','Ytick',[phaWs;phaWp;0],'fontsize',10); grid
%
subplot(2,2,4); plot(t,ha);  title ('Impulse Response','fontsize',10);
xlabel('t (sec)', 'fontsize',10); ylabel('ha(t)','fontsize',10);
```

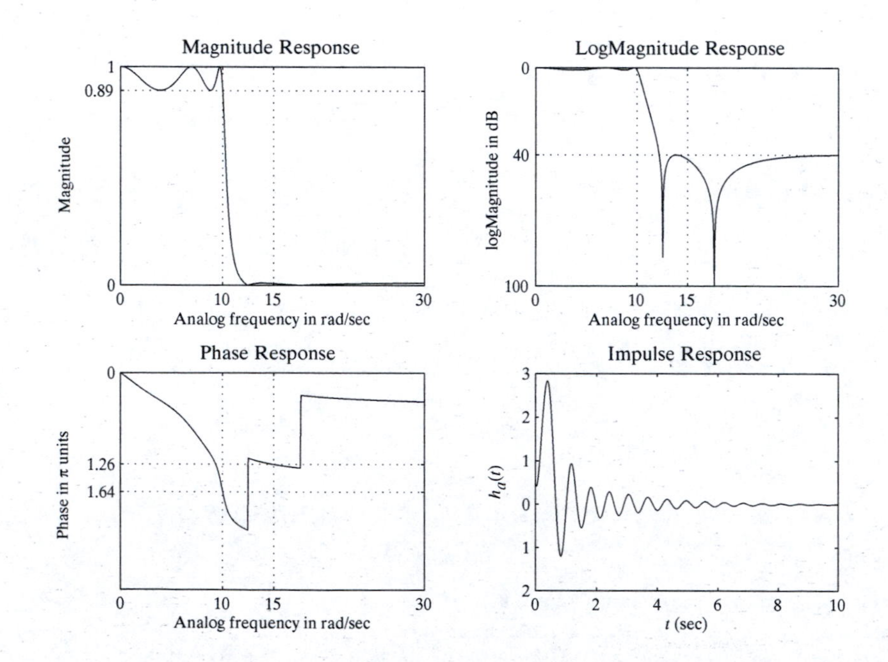

Figure G.2: Analog elliptic lowpass-filter design plots in P8.2

The system function is given by

$$H_a(s) = \frac{.46978s^4 + 220.07s^2 + 2298.5}{s^5 + 9.23s^4 + 184.71s^3 + 1129.2s^2 + 7881.3s + 22985}$$

The filter-design plots are given in Figure G.2.

P8.3 The filter passband must include the 100-Hz component; the stopband must include the 130-Hz component. To obtain a minimum-order filter, the transition band must be as large as possible. This means that the passband cutoff must be at 100 Hz and that the stopband cutoff must be at 130 Hz. Hence, the analog Chebyshev-I lowpass-filter specifications are $\Omega_p = 2\pi\ (100)$ rad/s, $R_p = 2$ dB, $\Omega_s = 2\pi\ (130)$ rad/s, $A_s = 50$ dB.
MATLAB script:

```
clear, close all;
% Filter Specifications
Fp = 100; Fs = 130; Rp = 2; As = 50;
Wp = 2*pi*Fp; Ws = 2*pi*Fs;
% Filter Design
[b,a] = afd_chb1(Wp,Ws,Rp,As); format short e

*** Chebyshev-1 Filter Order =  9
% Cascade Structure
[C,B,A] = sdir2cas(b,a)
C =
  7.7954e+022
B =
     0     0     1
A =
  1.0000e+000  1.4245e+002  5.1926e+004
  1.0000e+000  1.1612e+002  1.6886e+005
  1.0000e+000  7.5794e+001  3.0183e+005
  1.0000e+000  2.6323e+001  3.8862e+005
            0  1.0000e+000  7.5794e+001

% Frequency Response
Fmax = 200; Wmax = 2*pi*Fmax; [db,mag,pha,w] = freqs_m(b,a,Wmax); pha = unwrap(pha);
%
% Plots
Hf_1 = figure('Units','normalized','position',[0.1,0.1,0.8,0.8],'color',[0,0,0]);
set(Hf_1,'NumberTitle','off','Name','P8.3');
%
subplot(2,1,1);plot(w/(2*pi),mag); axis([0,Fmax,0,1.1]); set(gca,'fontsize',10);
xlabel('Analog frequency in Hz'); ylabel('Magnitude');
title ('Magnitude Response','fontsize',12);
set(gca,'XTickMode','manual','Xtick',[0;Fp;Fs;Fmax]);
magRp = round(10^(-Rp/20)*100)/100;
set(gca,'YTickMode','manual','Ytick',[0;magRp;1]);grid
%
subplot(2,1,2);plot(w/(2*pi),db); axis([0,Fmax,-100,0]); set(gca,'fontsize',10);
xlabel('Analog frequency in Hz'); ylabel('Decibels');
```

```
title ('Log-Magnitude Response','fontsize',12);
set(gca,'XTickMode','manual','Xtick',[0;Fp;Fs;Fmax]);
set(gca,'YTickMode','manual','Ytick',[-100;-As;0]);grid
AS = [' ',num2str(As)];
set(gca,'YTickLabelMode','manual','YTickLabels',['100';AS;'  0']);
```

The system function is given by

$$H_a(s) = 7.7954 \times 10^{22} \left(\frac{1}{s^2 + 142.45s + 51926}\right)\left(\frac{1}{s^2 + 116.12s + 168860}\right)$$
$$\left(\frac{1}{s^2 + 75.794s + 301830}\right)\left(\frac{1}{s^2 + 26.323s + 388620}\right)\left(\frac{1}{s + 75.794}\right)$$

The magnitude-response plots are given in Figure G.3.

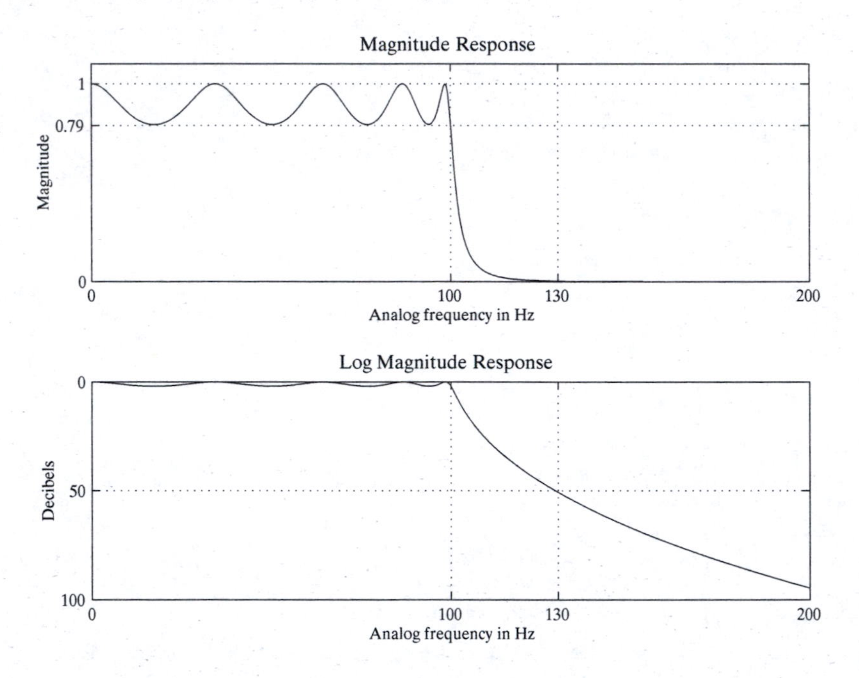

Figure G.3: Analog Chebyshev-I lowpass-filter plots in Problem P8.3

P8.4 Analog Chebyshev-II lowpass-filter design: $\Omega_p = 2\pi$ (250) rad/s, $R_p = 0.5$ dB, $\Omega_s = 2\pi$ (300) rad/s,
$A_s = 45$ dB.

MATLAB script:

```
clear; close all;
% Filter Specifications
Fp = 250; Fs = 300; Rp = 0.5; As = 45;
Wp = 2*pi*Fp; Ws = 2*pi*Fs;
% Filter Design
```

```
[b,a] = afd_chb2(Wp,Ws,Rp,As); format short e;

*** Chebyshev-2 Filter Order = 12
% Rational Function Form
a0 = a(1); b = b/a0, a = a/a0
b =
  Columns 1 through 6
  5.6234e-003            0  1.4386e+006            0  5.9633e+013            0
  Columns 7 through 12
  9.0401e+020            0  6.1946e+027            0  1.9564e+034            0
  Column 13
  2.3171e+040
a =
  Columns 1 through 6
  1.0000e+000  1.5853e+004  1.2566e+008  6.5800e+011  2.5371e+015  7.5982e+018
  Columns 7 through 12
  1.8208e+022  3.5290e+025  5.5639e+028  6.9823e+031  6.9204e+034  4.7962e+037
  Column 13
  2.3171e+040
% Frequency Response
Fmax = 500; Wmax = 2*pi*Fmax; [db,mag,pha,w] = freqs_m(b,a,Wmax); pha = unwrap(pha);
% Impulse Response
%   The impulse response of the designed filter when computed by Matlab is numerically
%   unstable due to large coefficient values.  Hence we will compute the impulse
%   response of the filter with Wp/1000 and Ws/1000 band edges to keep coefficient
%   values small. The actual impulse response is time-scaled and amplitude scaled
%   version of the computed impulse response.
[b,a] = afd_chb2(Wp/1000,Ws/1000,Rp,As); [ha,x,t] = impulse(b,a);

*** Chebyshev-2 Filter Order = 12
t = t/1000; ha = ha/1000;
%
% Plots
Hf_1 = figure('Units','normalized','position',[0.1,0.1,0.8,0.8],'color',[0,0,0]);
set(Hf_1,'NumberTitle','off','Name','P8.4');
%
subplot(2,2,1);plot(w/(2*pi),mag); axis([0,Fmax,0,1]); set(gca,'fontsize',10);
xlabel('Analog frequency in Hz'); ylabel('Magnitude');
title ('Magnitude Response','fontsize',12);
set(gca,'XTickMode','manual','Xtick',[0;Fp;Fs;Fmax]);
magRp = round(10^(-Rp/20)*100)/100;
set(gca,'YTickMode','manual','Ytick',[0;magRp;1]);grid
%
subplot(2,2,2);plot(w/(2*pi),db); axis([0,Fmax,-100,0]); set(gca,'fontsize',10);
xlabel('Analog frequency in Hz'); ylabel('log-Magnitude in dB');
title ('Log-Magnitude Response','fontsize',12);
set(gca,'XTickMode','manual','Xtick',[0;Fp;Fs;Fmax]);
set(gca,'YTickMode','manual','Ytick',[-100;-As;0]);grid
AS = [' ',num2str(As)];
set(gca,'YTickLabelMode','manual','YTickLabels',['100';AS;'  0']);
%
```

```
minpha = floor(min(pha/pi)); maxpha = ceil(max(pha/pi));
subplot(2,2,3);plot(w/(2*pi),pha/pi); axis([0,Fmax,minpha,maxpha]);...
                                                   set(gca,'fontsize',10);
xlabel('Analog frequency in Hz'); ylabel('Phase in pi units');
title ('Phase Response','fontsize',12);
set(gca,'XTickMode','manual','Xtick',[0;Fp;Fs;Fmax]);
phaWp = (round(pha(Wp/Wmax*500+1)/pi*100))/100;
phaWs = (round(pha(Ws/Wmax*500+1)/pi*100))/100;
set(gca,'YTickMode','manual','Ytick',[phaWs;phaWp;0]); grid
%
subplot(2,2,4); plot(t,ha); set(gca,'fontsize',10);
title ('Impulse Response','fontsize',12); xlabel('t (sec)'); ylabel('ha(t)');
```

The filter-design plots are given in Figure G.4.

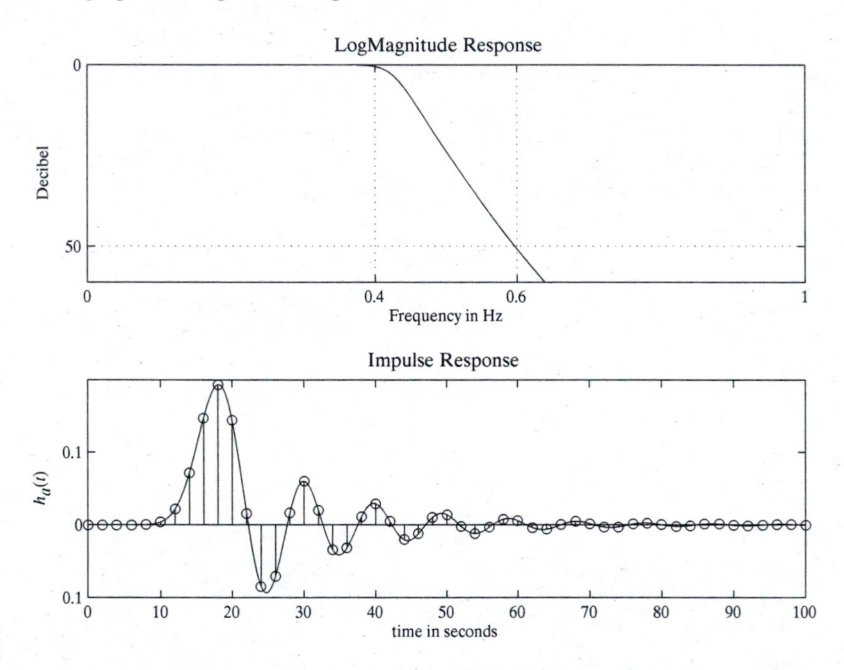

Figure G.4: Analog Chebyshev-II lowpass-filter plots in Problem P8.4

P8.5 MATLAB function `afd.m`:

```
function [b,a] = afd(type,Fp,Fs,Rp,As)
%
% function [b,a] = afd(type,Fp,Fs,Rp,As)
%   Designs analog lowpass filters
% type = 'butter' or 'cheby1' or 'cheby2' or 'ellip'
%   Fp = passband cutoff in Hz
%   Fs = stopband cutoff in Hz
%   Rp = passband ripple in dB
%   As = stopband attenuation in dB
```

```
type = lower([type,' ']); type = type(1:6);
twopi = 2*pi;
if type == 'butter'
    [b,a] = afd_butt(twopi*Fp,twopi*Fs,Rp,As);
elseif type == 'cheby1'
    [b,a] = afd_chb1(twopi*Fp,twopi*Fs,Rp,As);
elseif type == 'cheby2'
    [b,a] = afd_chb2(twopi*Fp,twopi*Fs,Rp,As);
elseif type == 'ellip '
    [b,a] = afd_elip(twopi*Fp,twopi*Fs,Rp,As);
else
    error('Specify the correct type')
end
```

P8.6 Digital Chebyshev-I lowpass-filter design by using impulse invariance. MATLAB script:

```
clear; close all; Twopi = 2*pi;
%% Analog Filter Specifications
Fsam = 8000;                     % Sampling Rate in sam/sec
Fp = 1500;                       % Passband edge in Hz
Rp = 3;                          % Passband Ripple in dB
Fs = 2000;                       % Stopband edge in Hz
As = 40;                         % Stopband attenuation in dB
%
%% Digital Filter Specifications
wp = Twopi*Fp/Fsam;              % Passband edge in rad/sam
Rp = 3;                          % Passband Ripple in dB
ws = Twopi*Fs/Fsam;              % Stopband edge in rad/sam
As = 40;                         % Stopband attenuation in dB
```

(a) $T = 1$. MATLAB script:

```
%% (a) Impulse Invariance Digital Design using T = 1
T = 1;
OmegaP = wp/T;                   % Analog Prototype Passband edge
OmegaS = ws/T;                   % Analog Prototype Stopband edge
[cs,ds] = afd_chb1(OmegaP,OmegaS,Rp,As); % Analog Prototype Design

*** Chebyshev-1 Filter Order =  7
[b,a] = imp_invr(cs,ds,T);       % II Transformation
[C,B,A] = dir2par(b,a),          % Parallel form
C =
     []
B =
   -0.0561    0.0558
    0.1763   -0.1529
   -0.2787    0.2359
    0.1586         0
A =
    1.0000   -0.7767    0.9358
    1.0000   -1.0919    0.8304
```

```
    1.0000   -1.5217    0.7645
    1.0000   -0.8616         0
```

The filter-design plots are shown in Figure G.5.

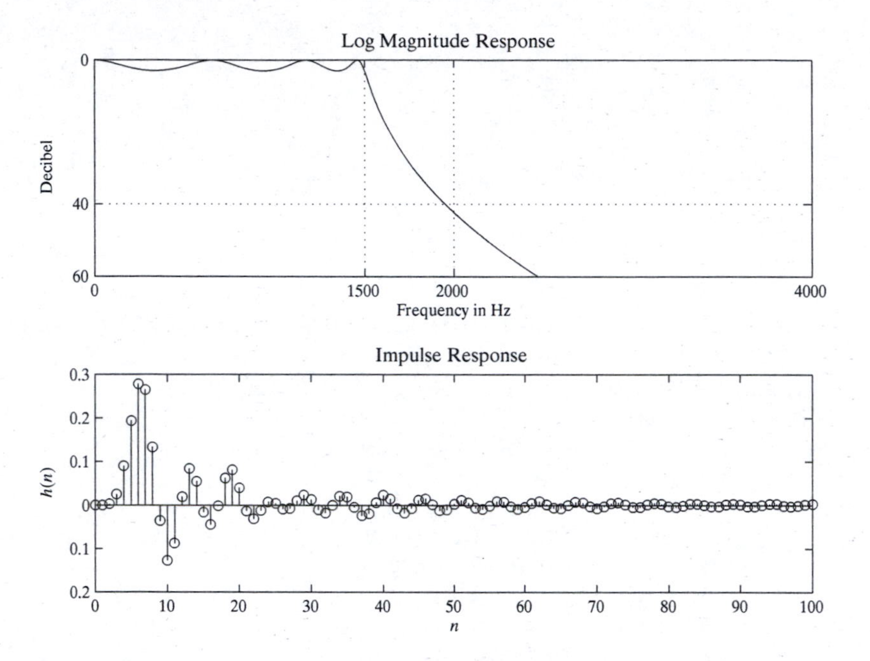

Figure G.5: Impulse-invariance design method with $T = 1$ in Problem P8.6(a)

(b) $T = 1/8000$. MATLAB script:

```
%% (b) Impulse Invariance Digital Design using T = 1/8000
T = 1/8000;
OmegaP = wp/T;                    % Analog Prototype Passband edge
OmegaS = ws/T;                    % Analog Prototype Stopband edge
[cs,ds] = afd_chb1(OmegaP,OmegaS,Rp,As); % Analog Prototype Design

*** Chebyshev-1 Filter Order =  7
[b,a] = imp_invr(cs,ds,T);        % II Transformation
[C,B,A] = dir2par(b,a),           % Parallel form

C =
     []
B =
  1.0e+003 *
   -0.4487    0.4460
    1.4102   -1.2232
   -2.2299    1.8869
    1.2684         0
A =
    1.0000   -0.7767    0.9358
    1.0000   -1.0919    0.8304
```

```
1.0000   -1.5217    0.7645
1.0000   -0.8616         0
```

The filter-design plots are shown in Figure G.6.

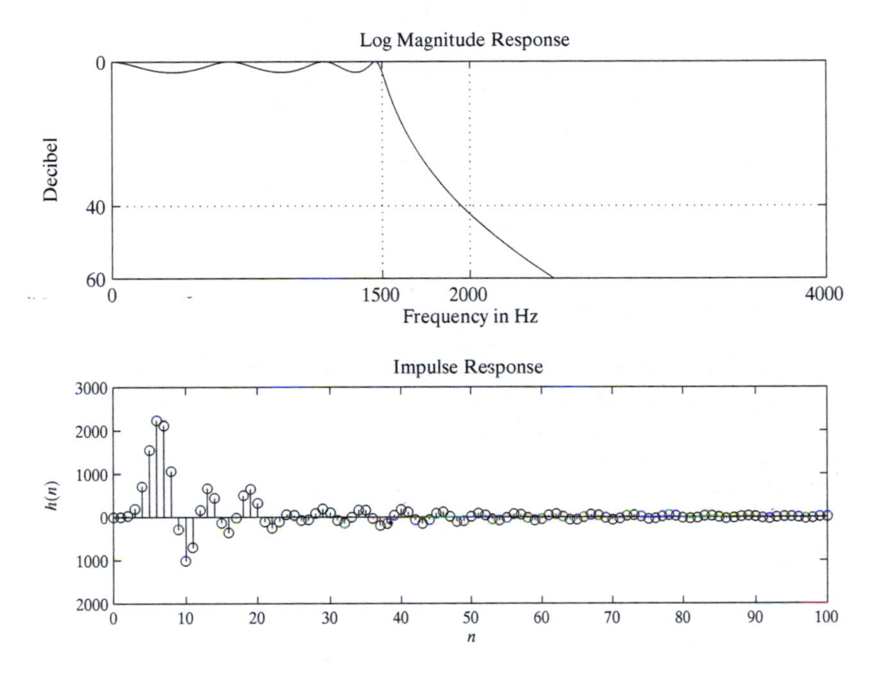

Figure G.6: Impulse-invariance design method with $T = 1/8000$ in Problem P8.6(b)

(c) Comparison: Both the designed system function and the impulse response in part (b) are *similar* to those in part (a) except for an overall gain due to $F_s = 1/T = 8000$. This problem can be avoided if, in the impulse-invariance design method, we set

$$h(n) = T \cdot h_a(nT)$$

P8.7 Digital Butterworth lowpass-filter design by using impulse invariance. MATLAB script:

```
clear; close all;
%% Digital Filter Specifications
wp = 0.4*pi;                    % Passband edge in rad/sam
Rp = 0.5;                       % Passband Ripple in dB
ws = 0.6*pi;                    % Stopband edge in rad/sam
As = 50;                        % Stopband attenuation in dB
%% Impulse Invariance Digital Design using T = 2

T = 2;
OmegaP = wp/T;                  % Analog Prototype Passband edge
OmegaS = ws/T;                  % Analog Prototype Stopband edge
[cs,ds] = afd_butt(OmegaP,OmegaS,Rp,As); % Analog Prototype Design
```

```
*** Butterworth Filter Order = 17
[b,a] = imp_invr(cs,ds,T);         % II Transformation: rational form
% Plots of Log-magnitude Response and Impulse Response
Hf_1 = figure('Units','normalized','position',[0.1,0.1,0.8,0.8],'color',[0,0,0]);
set(Hf_1,'NumberTitle','off','Name','P8.7');
% Frequency response
[db,mag,pha,grd,w] = freqz_m(b,a);
subplot(2,1,1); plot(w/pi,db); axis([0,1,-60,0]); set(gca,'fontsize',10);
set(gca,'XTickMode','manual','Xtick',[0;wp/pi;ws/pi;1]);
set(gca,'YTickMode','manual','Ytick',[-80;-As;0]);grid
AS = [num2str(As)];
set(gca,'YTickLabelMode','manual','YtickLabels',['80';AS;' 0']);
xlabel('Frequency in Hz'); ylabel('Decibel');
title('Log-Magnitude Response','fontsize',12); axis;
% Impulse response of the prototype analog filter
Nmax = 50; t = 0:T/10:T*Nmax; [ha,x,t] = impulse(cs,ds,t);
subplot(2,1,2); plot(t,ha); axis([0,T*Nmax,-0.1,0.2]); set(gca,'fontsize',10);
xlabel('time in seconds','fontsize',10); ylabel('ha(t)','fontsize',10);
title('Impulse Response','fontsize',12); hold on;
% Impulse response of the digital filter
[x,n] = impseq(0,0,Nmax); h = filter(b,a,x);
stem(n*T,h); hold off;
```

The filter-design plots are shown in Figure G.7.

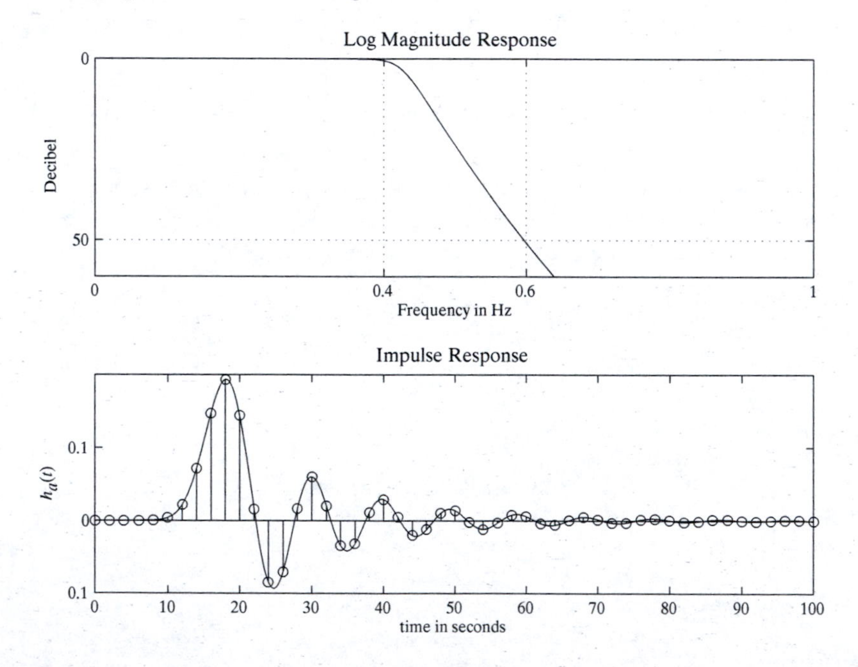

Figure G.7: Impulse-invariance design method with $T = 2$ in Problem P8.7

Comparison: From Figure G.7, we observe that the impulse response $h(n)$ of the digital filter is a sampled version of the impulse response $h_a(t)$ of the analog prototype filter, as expected.

P8.8 MATLAB function dlpfd_ii.m:

```
function [b,a] = dlpfd_ii(type,wp,ws,Rp,As,T)
%
% function [b,a] = dlpfd_ii(type,wp,ws,Rp,As,T)
%   Designs digital lowpass filters using impulse invariance mapping
% type = 'butter' or 'cheby1'
%   wp = passband cutoff in radians
%   ws = stopband cutoff in radians
%   Rp = passband ripple in dB
%   As = stopband attenuation in dB
%    T = sampling interval

if (type == 'cheby2')|(type == 'ellip ')
     error('Specify the correct type as butter or cheby1')
end
Fs = 1/T; twopi = 2*pi; K = Fs/twopi;
% Analog Prototype Specifications: Inverse mapping for frequencies
Fp = wp*K;                          % Prototype Passband freq in Hz
Fs = ws*K;                          % Prototype Stopband freq in Hz
ep = sqrt(10^(Rp/10)-1);            % Passband Ripple parameter
Ripple = sqrt(1/(1+ep*ep));         % Passband Ripple
Attn = 1/(10^(As/20));              % Stopband Attenuation
% Analog Butterworth Prototype Filter Calculation:
[cs,ds] = afd(type,Fp,Fs,Rp,As);
% Impulse Invariance transformation:
[b,a] = imp_invr(cs,ds,T);
[C,B,A] = dir2par(b,a)
```

MATLAB verification, using Problem P8.7:

```
clear; close all; format short e
%% Problem~P8.7 : Butterworth Design
%% Digital Filter Specifications
wp = 0.4*pi;                     % Passband edge in rad/sam
Rp = 0.5;                        % Passband Ripple in dB
ws = 0.6*pi;                     % Stopband edge in rad/sam
As = 50;                         % Stopband attenuation in dB
%
% Impulse Invariance Digital Design using T = 2
T = 2;
[b,a] = dlpfd_ii('butter',wp,ws,Rp,As,T);

*** Butterworth Filter Order = 17
C =
     []
B =
  3.3654e+000 -1.2937e+000
 -8.3434e+000 -6.9148e+000
```

```
  2.9916e-002  2.7095e-001
 -5.2499e+001  2.6839e+001
  1.2550e+002  4.3610e+000
  1.7953e+002 -1.0414e+002
 -5.1360e+002  1.0421e+002
 -1.3402e+002  8.1785e+001
  4.0004e+002            0
A =
  1.0000e+000 -3.9001e-001  4.8110e-001
  1.0000e+000 -4.0277e-001  3.0369e-001
  1.0000e+000 -4.1965e-001  7.8138e-001
  1.0000e+000 -4.3154e-001  1.9964e-001
  1.0000e+000 -4.6254e-001  1.3864e-001
  1.0000e+000 -4.8933e-001  1.0298e-001
  1.0000e+000 -5.0923e-001  8.2651e-002
  1.0000e+000 -5.2131e-001  7.2211e-002
  1.0000e+000 -2.6267e-001            0
%
% Plots of Log-magnitude Response and Impulse Response
Hf_1 = figure('Units','normalized','position',[0.1,0.1,0.8,0.8],'color',[0,0,0]);
set(Hf_1,'NumberTitle','off','Name','P8.8b');
% Log-Magnitude response
[db,mag,pha,grd,w] = freqz_m(b,a);
subplot(2,1,1); plot(w/pi,db); axis([0,1,-60,0]); set(gca,'fontsize',10);
set(gca,'XTickMode','manual','Xtick',[0;wp/pi;ws/pi;1]);
set(gca,'YTickMode','manual','Ytick',[-80;-As;0]);grid
AS = [num2str(As)];
set(gca,'YTickLabelMode','manual','YtickLabels',['80';AS;' 0']);
xlabel('Frequency in pi units'); ylabel('Decibel');
title('Log-Magnitude Response','fontsize',12); axis;
% Magnitude response
subplot(2,1,2); plot(w/pi,mag); axis([0,1,0,0.55]); set(gca,'fontsize',10);
set(gca,'XTickMode','manual','Xtick',[0;wp/pi;ws/pi;1]);
set(gca,'YTickMode','manual','Ytick',[0;0.5]);grid
set(gca,'YTickLabelMode','manual','YtickLabels',['0.0';'0.5']);
xlabel('Frequency in pi units'); ylabel('|H|');
title('Magnitude Response','fontsize',12); axis;
```

The filter-design plots are given in Figure G.8.

P8.9 Digital Butterworth lowpass-filter design using bilinear transformation. MATLAB script:

```
clear; close all;
%% Digital Filter Specifications
wp = 0.4*pi;                    % Passband edge in rad/sam
Rp = 0.5;                       % Passband Ripple in dB
ws = 0.6*pi;                    % Stopband edge in rad/sam
As = 50;                        % Stopband attenuation in dB
```

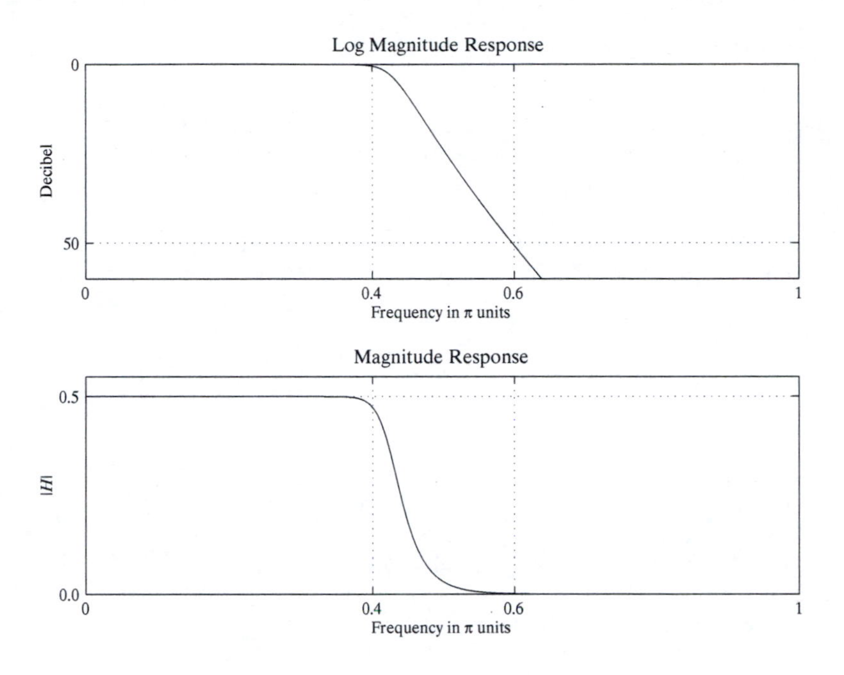

Figure G.8: Digital-filter design plots in Problem P8.8

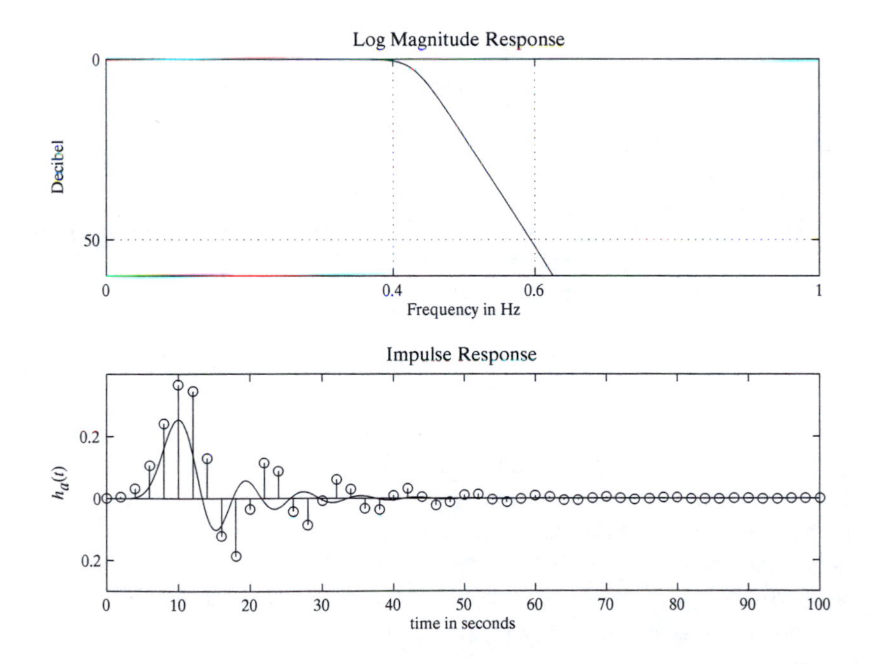

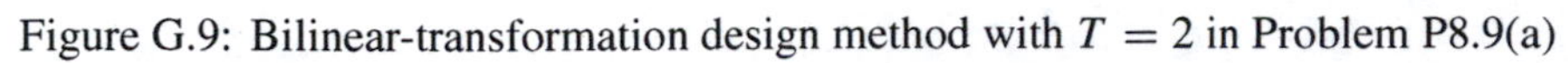

Figure G.9: Bilinear-transformation design method with $T = 2$ in Problem P8.9(a)

(a) $T = 2$. MATLAB script:

```
T = 2;
OmegaP = (2/T)*tan(wp/2);         % Analog Prototype Passband edge
OmegaS = (2/T)*tan(ws/2);         % Analog Prototype Stopband edge
[cs,ds] = afd_butt(OmegaP,OmegaS,Rp,As); % Analog Prototype Design

*** Butterworth Filter Order = 11
[b,a] = bilinear(cs,ds,1/T)       % Bilinear Transformation
b =
  Columns 1 through 7
    0.0004    0.0048    0.0238    0.0715    0.1429    0.2001    0.2001
  Columns 8 through 12
    0.1429    0.0715    0.0238    0.0048    0.0004
a =
  Columns 1 through 7
    1.0000   -1.5495    2.5107   -2.1798    1.7043   -0.8997    0.4005
  Columns 8 through 12
   -0.1258    0.0309   -0.0050    0.0005    0.0000
%
% Plots of Log-magnitude Response and Impulse Response
Hf_1 = figure('Units','normalized','position',[0.1,0.1,0.8,0.8],'color',[0,0,0]);
set(Hf_1,'NumberTitle','off','Name','P8.9a');
% Frequency response
[db,mag,pha,grd,w] = freqz_m(b,a);
subplot(2,1,1); plot(w/pi,db); axis([0,1,-60,0]); set(gca,'fontsize',10);
set(gca,'XTickMode','manual','Xtick',[0;wp/pi;ws/pi;1]);
set(gca,'YTickMode','manual','Ytick',[-80;-As;0]);grid
AS = [num2str(As)];
set(gca,'YTickLabelMode','manual','YtickLabels',['80';AS;' 0']);
xlabel('Frequency in Hz'); ylabel('Decibel');
title('Log-Magnitude Response','fontsize',12); axis;
% Impulse response of the prototype analog filter
Nmax = 50; t = 0:T/10:T*Nmax; [ha,x,t] = impulse(cs,ds,t);
subplot(2,1,2); plot(t,ha); axis([0,T*Nmax,-0.3,0.4]); set(gca,'fontsize',10);
xlabel('time in seconds','fontsize',10); ylabel('ha(t)','fontsize',10);
title('Impulse Response','fontsize',12); hold on;
% Impulse response of the digital filter
[x,n] = impseq(0,0,Nmax); h = filter(b,a,x);
stem(n*T,h); hold off;
```

The filter-design plots are shown in Figure G.9.

Comparison: If we compare filter orders from the two methods, then the bilinear transformation gives a lower order than the impulse-invariance method. This implies that the bilinear-transformation design method is a better one in all aspects. If we compare the impulse responses, then we observe from Figure G.9 that the digital impulse response is *not* a sampled version of the analog impulse response, as was the case in Figure G.7.

(b) Use of the `butter` function. MATLAB script:

```
[N,wn] = buttord(wp/pi,ws/pi,Rp,As);
[b,a] = butter(N,wn)
```

```
b =
  Columns 1 through 7
    0.0005    0.0054    0.0270    0.0810    0.1619    0.2267    0.2267
  Columns 8 through 12
    0.1619    0.0810    0.0270    0.0054    0.0005
a =
  Columns 1 through 7
    1.0000   -1.4131    2.3371   -1.9279    1.5223   -0.7770    0.3477
  Columns 8 through 12
   -0.1066    0.0262   -0.0042    0.0004    0.0000
%
% Plots of Log-magnitude Response and Impulse Response
Hf_2 = figure('Units','normalized','position',[0.1,0.1,0.8,0.8],'color',[0,0,0]);
set(Hf_2,'NumberTitle','off','Name','P8.9b');
% Frequency response
[db,mag,pha,grd,w] = freqz_m(b,a);
subplot(2,1,1); plot(w/pi,db); axis([0,1,-60,0]); set(gca,'fontsize',10);
set(gca,'XTickMode','manual','Xtick',[0;wp/pi;ws/pi;1]);
set(gca,'YTickMode','manual','Ytick',[-80;-As;0]);grid
AS = [num2str(As)];
set(gca,'YTickLabelMode','manual','YtickLabels',['80';AS;' 0']);
xlabel('Frequency in Hz'); ylabel('Decibel');
title('Log-Magnitude Response','fontsize',12); axis;
% Impulse response of the digital filter
Nmax = 50; [x,n] = impseq(0,0,Nmax); h = filter(b,a,x);
subplot(2,1,2); stem(n,h); axis([0,Nmax,-0.3,0.4]); set(gca,'fontsize',10);
xlabel('n','fontsize',10); ylabel('h(n)','fontsize',10);
title('Impulse Response','fontsize',12);
```

The filter-design plots are shown in Figure G.10.

Comparison: If we compare the plots of filter responses in part (0a) with those in part (0b), then we observe that the `butter` function satisfies stopband specifications *exactly* at ω_s. Otherwise the two designs are essentially *similar*.

P8.10 Digital Chebyshev-I lowpass-filter design by using bilinear transformation. MATLAB script:

```
clear; close all; Twopi = 2*pi;
%% Analog Filter Specifications
Fsam = 8000;                    % Sampling Rate in sam/sec
Fp = 1500;                      % Passband edge in Hz
Rp = 3;                         % Passband Ripple in dB
Fs = 2000;                      % Stopband edge in Hz
As = 40;                        % Stopband attenuation in dB
%
%% Digital Filter Specifications
wp = Twopi*Fp/Fsam;             % Passband edge in rad/sam
Rp = 3;                         % Passband Ripple in dB
ws = Twopi*Fs/Fsam;             % Stopband edge in rad/sam
As = 40;                        % Stopband attenuation in dB
```

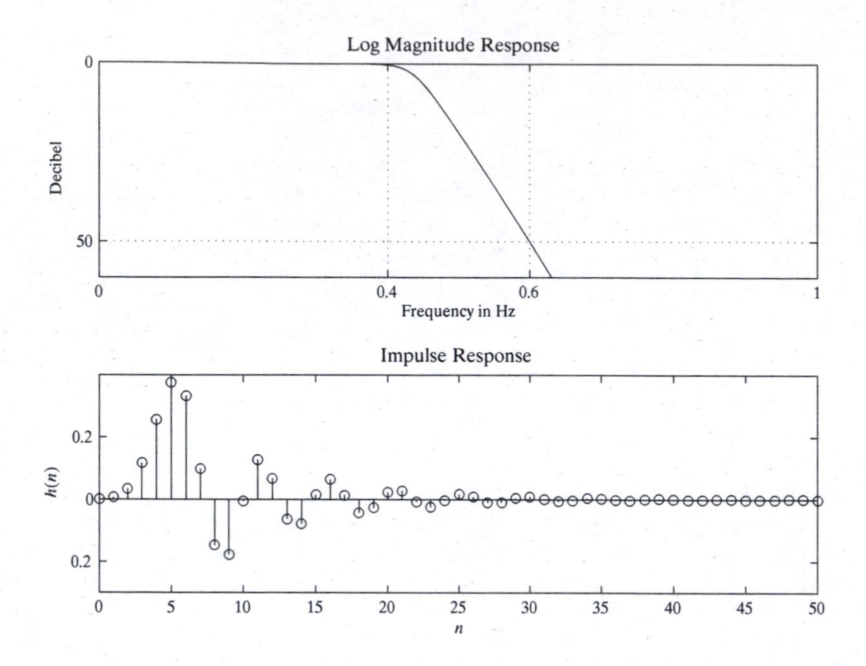

Figure G.10: Butterworth-filter design using the `butter` function in Problem P8.9(b)

(a) $T = 1$. MATLAB script:

```
%% (a) Bilinear Transformation Digital Design using T = 1
T = 1;
OmegaP = (2/T)*tan(wp/2);          % Analog Prototype Passband edge
OmegaS = (2/T)*tan(ws/2);          % Analog Prototype Stopband edge
[cs,ds] = afd_chb1(OmegaP,OmegaS,Rp,As); % Analog Prototype Design

*** Chebyshev-1 Filter Order =  6
[b,a] = bilinear(cs,ds,1/T);       % Bilinear Transformation
[C,B,A] = dir2cas(b,a),            % Cascade form
C =
    0.0011
B =
    1.0000    2.0126    1.0127
    1.0000    1.9874    0.9875
    1.0000    2.0001    0.9999
A =
    1.0000   -0.7766    0.9308
    1.0000   -1.1177    0.7966
    1.0000   -1.5612    0.6901
```

The filter-design plots are shown in Figure G.11.

(b) $T = 1/8000$. MATLAB script:

```
%% (b) Impulse Invariance Digital Design using T = 1/8000
T = 1/8000;
```

```
OmegaP = (2/T)*tan(wp/2);          % Analog Prototype Passband edge
OmegaS = (2/T)*tan(ws/2);          % Analog Prototype Stopband edge
[cs,ds] = afd_chb1(OmegaP,OmegaS,Rp,As); % Analog Prototype Design

*** Chebyshev-1 Filter Order =  6
[b,a] = bilinear(cs,ds,1/T);       % II Transformation
[C,B,A] = dir2cas(b,a),            % Cascade form
C =
    0.0011
B =
    1.0000    2.0265    1.0267
    1.0000    1.9998    1.0001
    1.0000    1.9737    0.9739
A =
    1.0000   -0.7766    0.9308
    1.0000   -1.1177    0.7966
    1.0000   -1.5612    0.6901
```

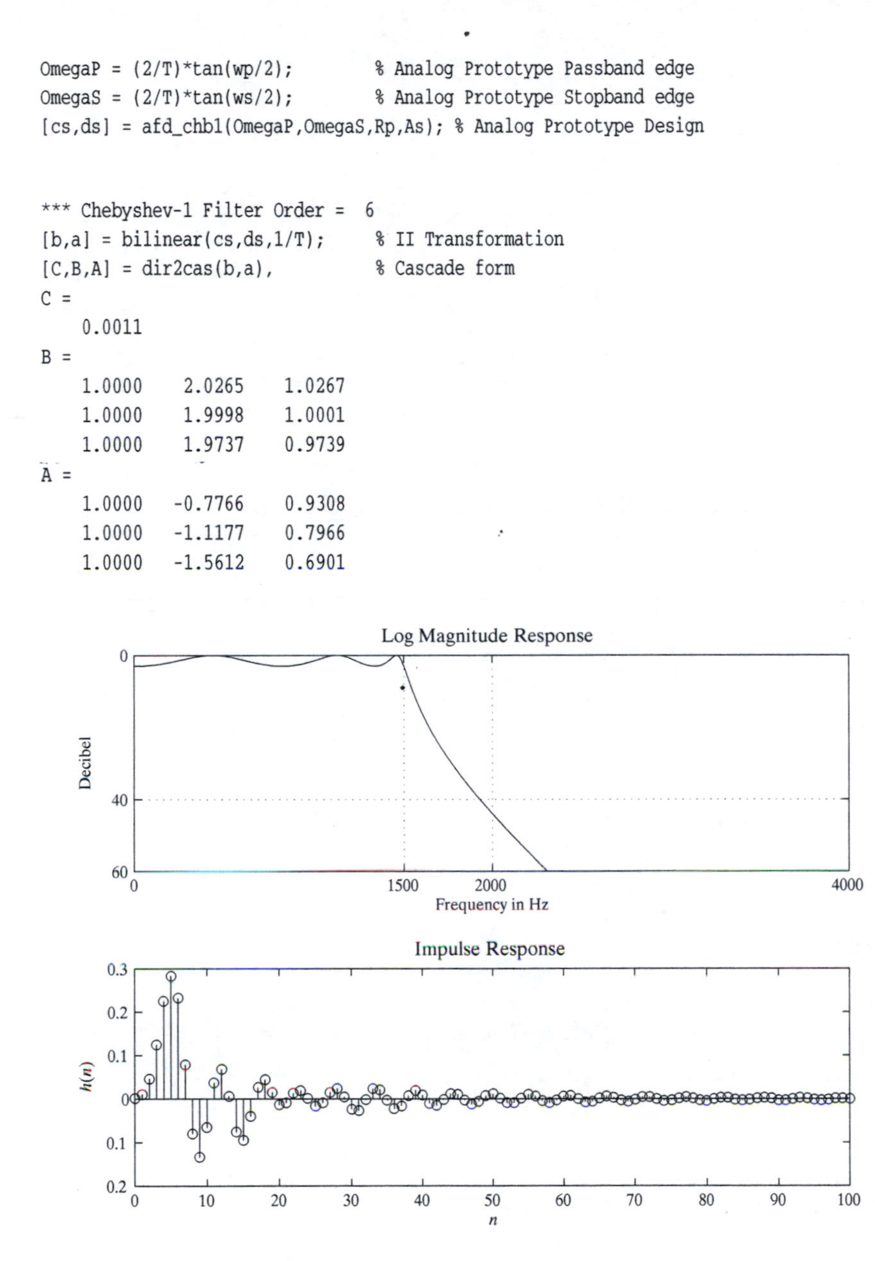

Figure G.11: Bilinear-transformation design method with $T = 1$ in Problem P8.10(a)

The filter-design plots are shown in Figure G.12.

(c) Comparison: If we compare the designed system function and the plots of system responses in part (0a) with those in part (0b), then we observe that these are *exactly the same*. If we compare the impulse-invariance design in Problem 8.6 with this one, then we note that the order of the impulse-invariance-designed filter is one higher. This implies that the bilinear-transformation design method is a better one in all aspects.

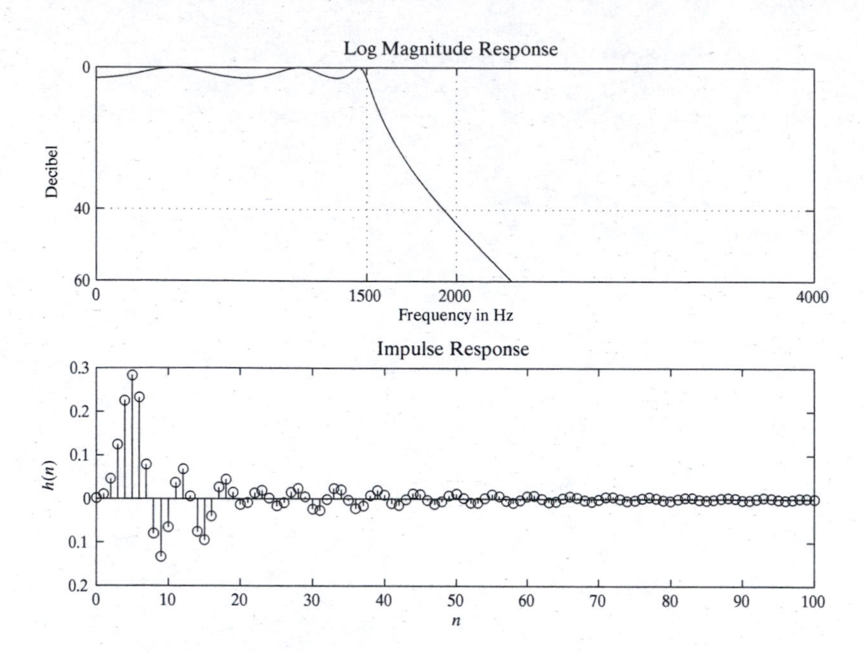

Figure G.12: Bilinear-transformation design method with $T = 1/8000$ in Problem P8.10(b)

P8.11 Digital lowpass-filter design by using elliptic prototype.

MATLAB script using the bilinear function:

```
clear; close all;
%% Digital Filter Specifications
wp = 0.4*pi;                    % Passband edge in rad/sam
Rp = 1;                         % Passband Ripple in dB
ws = 0.6*pi;                    % Stopband edge in rad/sam
As = 60;                        % Stopband attenuation in dB
%
%% (a) Bilinear Transformation Digital Design using T = 2
T = 2;
OmegaP = (2/T)*tan(wp/2);       % Analog Prototype Passband edge
OmegaS = (2/T)*tan(ws/2);       % Analog Prototype Stopband edge
[cs,ds] = afd_elip(OmegaP,OmegaS,Rp,As); % Analog Prototype Design

*** Elliptic Filter Order =  5
[b,a] = bilinear(cs,ds,1/T);    % Bilinear Transformation
%
% Plots of Log-magnitude Response and Impulse Response
Hf_1 = figure('Units','normalized','position',[0.1,0.1,0.8,0.8],'color',[0,0,0]);
set(Hf_1,'NumberTitle','off','Name','P8.11a');
% Frequency response
[db,mag,pha,grd,w] = freqz_m(b,a);
subplot(2,1,1); plot(w/pi,db); axis([0,1,-80,0]); set(gca,'fontsize',10);
set(gca,'XTickMode','manual','Xtick',[0;wp/pi;ws/pi;1]);
```

```
set(gca,'YTickMode','manual','Ytick',[-80;-As;0]);grid
AS = [num2str(As)];
set(gca,'YTickLabelMode','manual','YtickLabels',['80';AS;' 0']);
xlabel('Frequency in Hz'); ylabel('Decibel');
title('Log-Magnitude Response','fontsize',12); axis;
% Impulse response of the prototype analog filter
Nmax = 50; t = 0:T/10:T*Nmax; [ha,x,t] = impulse(cs,ds,t);
subplot(2,1,2); plot(t,ha); axis([0,T*Nmax,-0.3,0.4]); set(gca,'fontsize',10);
xlabel('time in seconds','fontsize',10); ylabel('ha(t)','fontsize',10);
title('Impulse Response','fontsize',12); hold on;
% Impulse response of the digital filter
[x,n] = impseq(0,0,Nmax); h = filter(b,a,x);
stem(n*T,h); hold off;
```

The filter-design plots are shown in Figure G.13.

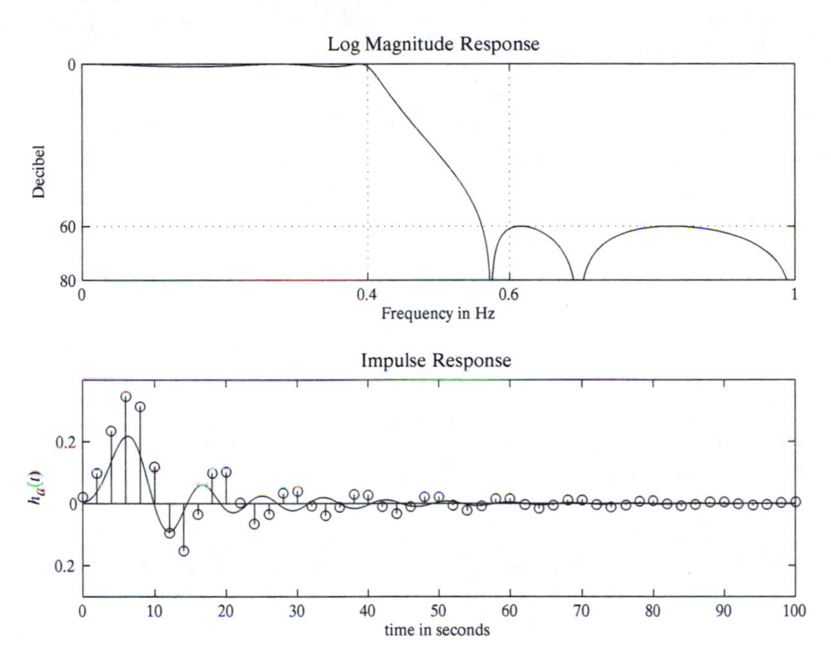

Figure G.13: Digital elliptic lowpass-filter design using the bilinear function in Problem P8.11(a)

MATLAB script, using the `ellip` function:

```
%% (b) Use of the 'Ellip' function
[N,wn] = ellipord(wp/pi,ws/pi,Rp,As);
[b,a] = ellip(N,Rp,As,wn);
%
% Plots of Log-magnitude Response and Impulse Response
Hf_2 = figure('Units','normalized','position',[0.1,0.1,0.8,0.8],'color',[0,0,0]);
set(Hf_2,'NumberTitle','off','Name','P8.11b');
% Frequency response
```

```
[db,mag,pha,grd,w] = freqz_m(b,a);
subplot(2,1,1); plot(w/pi,db); axis([0,1,-80,0]); set(gca,'fontsize',10);
set(gca,'XTickMode','manual','Xtick',[0;wp/pi;ws/pi;1]);
set(gca,'YTickMode','manual','Ytick',[-80;-As;0]);grid
AS = [num2str(As)];
set(gca,'YTickLabelMode','manual','YtickLabels',['80';AS;' 0']);
xlabel('Frequency in Hz'); ylabel('Decibel');
title('Log-Magnitude Response','fontsize',12); axis;
% Impulse response of the digital filter
Nmax = 50; [x,n] = impseq(0,0,Nmax); h = filter(b,a,x);
subplot(2,1,2); stem(n,h); axis([0,Nmax,-0.3,0.4]); set(gca,'fontsize',10);
xlabel('n','fontsize',10); ylabel('h(n)','fontsize',10);
title('Impulse Response','fontsize',12);
```

The filter-design plots are shown in Figure G.14. From these two Figures, we observe that both functions give the same design, one in which the digital-filter impulse response is not a sampled version of the corresponding analog-filter impulse response.

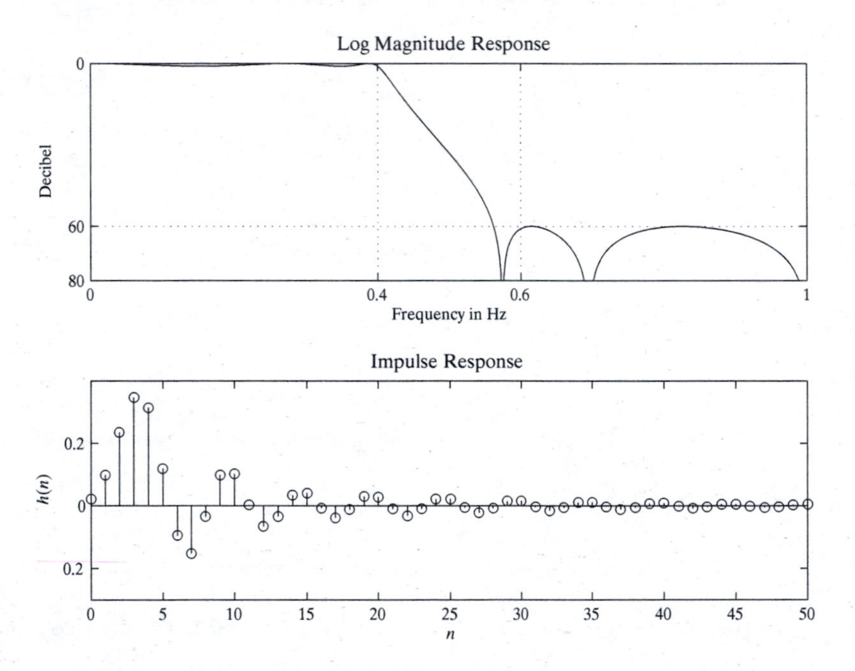

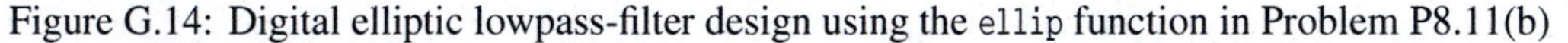

Figure G.14: Digital elliptic lowpass-filter design using the ellip function in Problem P8.11(b)

P8.12 Digital elliptic highpass filter design using bilinear mapping.

MATLAB function dhpfd_bl.m:

```
function [b,a] = dhpfd_bl(type,wp,ws,Rp,As)
%
% function [b,a] = dhpfd_bl(type,wp,ws,Rp,As,T)
%   Designs digital highpass filters using bilinear
```

```
%     b = Numerator polynomial of the highpass filter
%     a = Denominator polynomial of the highpass filter
% type = 'butter' or 'cheby1' or 'cheby2' or 'ellip'
%    wp = passband cutoff in radians
%    ws = stopband cutoff in radians (ws < wp)
%    Rp = passband ripple in dB
%    As = stopband attenuation in dB

% Determine the digital lowpass cutoff frequencies:
wplp = 0.2*pi;
alpha = -(cos((wplp+wp)/2))/(cos((wplp-wp)/2));
wslp = angle(-(exp(-j*ws)+alpha)/(1+alpha*exp(-j*ws)));
%
% Compute Analog lowpass Prototype Specifications:
T = 1; Fs = 1/T;
OmegaP = (2/T)*tan(wplp/2);
OmegaS = (2/T)*tan(wslp/2);

% Design Analog Chebyshev Prototype Lowpass Filter:
type = lower([type,' ']); type = type(1:6);
if type == 'butter'
     [cs,ds] = afd_butt(OmegaP,OmegaS,Rp,As);
elseif type == 'cheby1'
     [cs,ds] = afd_chb1(OmegaP,OmegaS,Rp,As);
elseif type == 'cheby2'
     [cs,ds] = afd_chb2(OmegaP,OmegaS,Rp,As);
elseif type == 'ellip '
     [cs,ds] = afd_elip(OmegaP,OmegaS,Rp,As);
else
     error('Specify the correct type')
end

% Perform Bilinear transformation to obtain digital lowpass
[blp,alp] = bilinear(cs,ds,Fs);

% Transform digital lowpass into highpass filter
Nz = -[alpha,1]; Dz = [1,alpha];
[b,a] = zmapping(blp,alp,Nz,Dz);
```

(a) Design using the dhpfd_bl function:

```
clear; close all;
Hf_1 = figure('Units','normalized','position',[0.1,0.1,0.8,0.8],'color',[0,0,0]);
set(Hf_1,'NumberTitle','off','Name','P8.12')

%% Digital Filter Specifications
type = 'ellip';                      % Elliptic design
ws = 0.4*pi;                         % Stopband edge in rad/sam
As = 60;                             % Stopband attenuation in dB
wp = 0.6*pi;                         % Passband edge in rad/sam
Rp = 1;                              % Passband Ripple in dB
```

```
%
%% (a) Use of the dhpfd_bl function
[b,a] = dhpfd_bl(type,wp,ws,Rp,As)

*** Elliptic Filter Order =  5
b =
    0.0208   -0.0543    0.0862   -0.0862    0.0543   -0.0208
a =
    1.0000    2.1266    2.9241    2.3756    1.2130    0.3123
% Plot of Log-magnitude Response
% Frequency response
[db,mag,pha,grd,w] = freqz_m(b,a);
subplot(2,1,1); plot(w/pi,db); axis([0,1,-80,0]); set(gca,'fontsize',10);
set(gca,'XTickMode','manual','Xtick',[0;ws/pi;wp/pi;1]);
set(gca,'YTickMode','manual','Ytick',[-80;-As;0]);grid
AS = [num2str(As)];
set(gca,'YTickLabelMode','manual','YtickLabels',['80';AS;' 0']);
xlabel('Frequency in Hz'); ylabel('Decibel');
title('Design using the dhpfd_bl function','fontsize',12); axis;
```

The filter frequency-response plot is shown in the top row of Figure G.15.

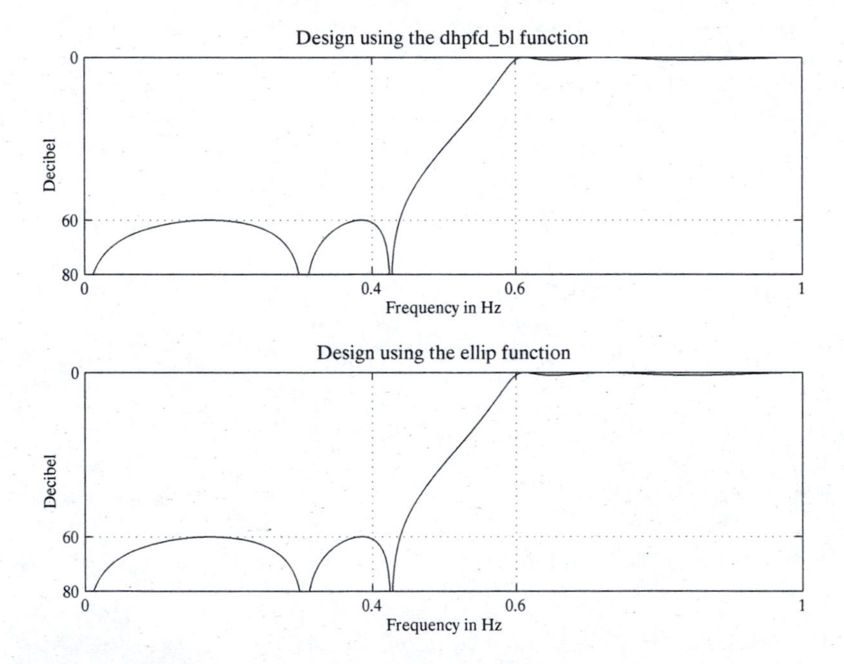

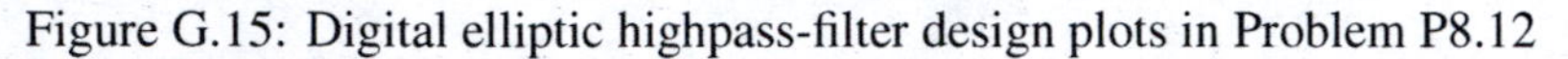
Figure G.15: Digital elliptic highpass-filter design plots in Problem P8.12

(b) Design using the `ellip` function:

```
%% (b) Use of the 'Ellip' function
[N,wn] = ellipord(wp/pi,ws/pi,Rp,As);
[b,a] = ellip(N,Rp,As,wn,'high')
b =
```

```
    0.0208   -0.0543    0.0862   -0.0862    0.0543   -0.0208
a =
    1.0000    2.1266    2.9241    2.3756    1.2130    0.3123
%
% Plot of Log-magnitude Response
% Frequency response
[db,mag,pha,grd,w] = freqz_m(b,a);
subplot(2,1,2); plot(w/pi,db); axis([0,1,-80,0]); set(gca,'fontsize',10);
set(gca,'XTickMode','manual','Xtick',[0;ws/pi;wp/pi;1]);
set(gca,'YTickMode','manual','Ytick',[-80;-As;0]);grid
AS = [num2str(As)];
set(gca,'YTickLabelMode','manual','YtickLabels',['80';AS;' 0']);
xlabel('Frequency in Hz'); ylabel('Decibel');
title('Design using the ellip function','fontsize',12); axis;
```

The filter frequency-response plot is shown in the bottom row of Figure G.15. The MATLAB scripts and Figure G.15 all indicate that we designed essentially the same filter.

P8.13 Digital Chebyshev-II bandpass-filter design using bilinear transformation. MATLAB script:

```
clear; close all;
Hf_1 = figure('Units','normalized','position',[0.1,0.1,0.8,0.8],'color',[0,0,0]);
set(Hf_1,'NumberTitle','off','Name','P8.13')

%% Digital Filter Specifications
ws1 = 0.3*pi;                   % Lower Stopband edge in rad/sam
ws2 = 0.6*pi;                   % Upper Stopband edge in rad/sam
 As = 50;                       % Stopband attenuation in dB
wp1 = 0.4*pi;                   % Lower passband edge in rad/sam
wp2 = 0.5 *pi;                  % Upper passband edge in rad/sam
 Rp = 0.5;                      % Passband Ripple in dB
%
%% Use of the 'cheby2' function
ws = [ws1,ws2]/pi; wp = [wp1,wp2]/pi;
[N,wn] = cheb2ord(wp,ws,Rp,As)
N =
     5
wn =
    0.3390    0.5661
[b,a] = cheby2(N,As,wn)
b =
  Columns 1 through 7
    0.0050   -0.0050    0.0087   -0.0061    0.0060    0.0000   -0.0060
  Columns 8 through 11
    0.0061   -0.0087    0.0050   -0.0050
a =
  Columns 1 through 7
    1.0000   -1.3820    4.4930   -4.3737    7.4582   -5.1221    5.7817
  Columns 8 through 11
   -2.6221    2.0882   -0.4936    0.2764
%
```

```
% Plot of the filter Responses
% Impulse response
Nmax = 50; [x,n] = impseq(0,0,Nmax); h = filter(b,a,x);
subplot(2,1,1); stem(n,h); axis([0,Nmax,-0.15,0.15]); set(gca,'fontsize',10);
xlabel('n','fontsize',10); ylabel('h(n)','fontsize',10);
title('Impulse Response','fontsize',12);
% Frequency response
[db,mag,pha,grd,w] = freqz_m(b,a);
subplot(2,1,2); plot(w/pi,db); axis([0,1,-70,0]); set(gca,'fontsize',10);
set(gca,'XTickMode','manual','Xtick',[0,ws,wp,1]);
set(gca,'YTickMode','manual','Ytick',[-70;-As;0]);grid
AS = [num2str(As)];
set(gca,'YTickLabelMode','manual','YtickLabels',['70';AS;' 0']);
xlabel('Frequency in pi units'); ylabel('Decibel');
title('Log-Magnitude Response','fontsize',12); axis;
```

The filter impulse and log-magnitude response plots are shown in Figure G.16.

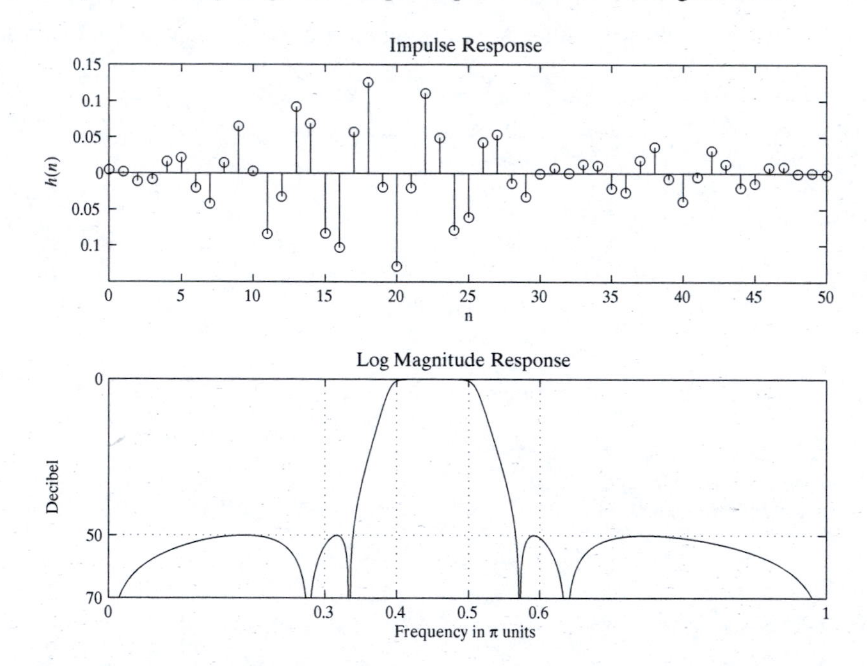

Figure G.16: Digital Chebyshev-II bandpass-filter design plots in Problem P8.13

Index